International Conference of the Carpathian Euro-Region Specialists in Industrial Systems 2019

IOP Conference Series: Materials Science and Engineering
Volume 749

Baia Mare, Romania
11 - 12 April 2019

Editors:

Nicolae Ungureanu
Radu Cotetiu
Mihai Banica
Cristian Barz

ISBN: 978-1-7138-2658-3
ISSN: 1757-8981

Printed from e-media with permission by:

Curran Associates, Inc.
57 Morehouse Lane
Red Hook, NY 12571

Some format issues inherent in the e-media version may also appear in this print version.

Institute of Physics
Dirac House, Temple Back
Bristol BS1 6BE UK

Phone: 44 1 17 929 7481
Fax: 44 1 17 920 0979

techtracking@iop.org

Additional copies of this publication are available from:

Curran Associates, Inc.
57 Morehouse Lane
Red Hook, NY 12571 USA
Phone: 845-758-0400
Fax: 845-758-2633
Email: curran@proceedings.com
Web: www.proceedings.com

TABLE OF CONTENTS

MECHANICAL ENGINEERING

INDUSTRIAL ENGINEERING

ELECTRICAL ENGINEERING

MATERIAL ENGINEERING

Preface

The International Conference of the Carpathian Euro-Region Specialists in Industrial Systems (CEurSIS), 12th edition, is a conference held by the Faculty of Engineering, North University Center of Baia Mare, Technical University of Cluj Napoca, Romania. The objective is to ensure the conditions for high quality research and professional interactions in the benefit of science, technology and collaboration.

The first edition of The International Meeting of the Carpathian Region Specialists, focused on Gears, was organized in 1996 under the coordination of Professor Eugen PAY Ph.D., Dr. H.C.. The following editions had a continuously grown multidisciplinary approach of the research and technological development activities, so the 2019 meeting comprises a wide spectrum of engineering. The scientific activities was completed with social and cultural activities.

Until now more than 450 specialists from Romania, Poland, Hungary, Germany, Czech Republic, Austria, Serbia, Ukraine, Slovakia and Italy, with more than 570 papers, attended the CEurSIS Conference.

We would like to thank everyone involved in organizing the conference and reviewing the papers presented..

The Editors

Professor Nicolae UNGUREANU, Ph.D., Dr. H.C.

Professor Radu COTETIU, Ph.D.

Assoc. Prof. Mihai Banica, Ph.D.

Senior Lecturer Cristian BARZ, Ph.D.

IOP Conf. Series: Materials Science and Engineering **749** (2020) 011002 doi:10.1088/1757-899X/749/1/011002

Peer review statement

All papers published in this volume of *IOP Conference Series: Materials Science and Engineering* have been peer reviewed through processes administered by the proceedings Editors. Reviews were conducted by expert referees to the professional and scientific standards expected of a proceedings journal published by IOP Publishing.

A texture based descriptors used for real time tool condition monitoring

A Antic[1,a], M Milosevic[1], R Cep[2] and N Ungureanu[3]

[1] University of Novi Sad, Faculty of Technical Sciences, Novi Sad, Serbia
[2] VŠB-Technical University of Ostrava, Faculty of Mechanical Engineering, Ostrava, Czech Republic
[3] Technical University of Cluj-Napoca, Faculty of Engineering, Department of Engineering and Technology Management, Baia Mare, Romania

E-mail: [a] antica@uns.ac.rs

Abstract. Tool Condition Monitoring Systems (TCMs), in the task of tool wear recognition, are highly dependent on the choice of descriptors. It is very important to extract the appropriate marking that contains information on the tool wear condition from the sensor signal. We consider the portions of the Short-Term Discrete Furrier Transform (STDFT) spectrum obtained by speaking from a vibration signal sensor as a 2D textured image. This is achieved by setting the STDFT timeline as the first dimension and the frequency scales as the second dimension of the resulting textured image. We divide the resulting textured image into special particular patches of texture, focusing on a portion of the frequency range of interest. Applying the set filter bank, 2D textons are allocated for each predefined frequency band. From 2D textons, for each frequency band, we extract probability density function (PDF) data in the form of lower order statistical moments. The applied method gives robust descriptors of the Tool wear state.

1. Introduction

Preventing possible damages to the workpiece and machine tool due to the unexpected breakage of the tool in the machining process requires the introduction of constant monitoring of the condition of tool wear, which provide a quick response to the unexpected failure of the tool error. Many academic and practical studies have been conducted to monitor the state of the real-time tool monitoring (TCM) in the turning process, which is a very important contribution to increasing the safety and stability of the production process [1]. The state of the machining process and its control play an important role in predicting tool wear. Different signal processing techniques are applied to obtain crucial information. The development of information technology has made it possible to use multiple sensors to effectively monitor of tool wear condition [2]. Tool fracture prediction can be implemented by effectively monitoring changes occurring on the tool itself. Cho et al., in their work, divided the tool changes, and tool wear states, into the following categories: tool breakage, tool crunching, and tool wear [3]. The main task in using a modern machining system is to develop and deploy reliable and robust TCM systems.

In research studies and practical solutions, different methods are applied to collect tool failure information using the appropriate sensors and thus the corresponding signals used in TCMs. The basic division is in two groups: direct methods (consisting of laser, optical and ultrasonic sensors that allow direct measurement, for example in [4] and indirect, based on sensors that bring to the state of processing by monitoring cutting forces, vibrations, temperature, instantaneous consumption etc. The application of the direct method is very expensive and difficult to apply in the workspace of the machine, indirect methods much more economical and more frequent in practice. Among indirect methods, vibration signal monitoring (i.e. accelerometers) is highly correlated with surface roughness in turning operations [5], as well as wear of cutting tools [6], therefore, these signals and their processing are highly applicable in TCM tasks.

Modern TCM systems based on artificial intelligence and their precision in determining the state of wear and tear depends primarily on the choice of descriptors, ie. a feature extracted from different sensor signals. The fact is that unless the descriptors describe the signal appropriately, other techniques, such as feature extraction or feature selection, as well as recognition methodology, are also ineffective. Bahr in [7], and Tsai et. all in [8] were the first to apply descriptors derived from vibrational signals in TCM tasks. In fact, they used RMS and / or the mean value of the vibration sensor signal to detect an increase in the vibration magnitude, which corresponds to an increase in the vibration signal energy generated by the flank wear. Also, in [9], the mean and peak values of vibration sensor signals are used in the TCM task. In [6], Dimila analysed the correlation of vibration signal features to cutting-tool wear, in both the time and the frequency domains, during turning operations. Time domain features were deemed to be more sensitive to the cutting condition than tool wear, whereas certain peak values in the frequency domain correlated well with the measured wear values. In [10], the influence of power spectral density (PSD) for prediction of surface robustness is analysed. In [11], the average, variance, and single harmonic from the vibration sensor signal are used as descriptors in order to efficiently model the tool condition. In [12], authors used wavelet analysis and lower order moments, such as average value, standard deviation, power value, kurtosis value, harmonics frequency, skew value, etc., in order to express work-piece and spindle vibrations in the X, Y and Z directions. In [13], authors analysed five harmonic values of vibration signals in the frequency domain and the signal average in the time domain.

This study proposes tool wear monitoring strategy in which new descriptors based on image texture and feature extraction have been implemented to be applied in TCMs that use vibration sensor signals. In fact, the textons approach to constructing descriptors is completely new in the field of TCM to the authors knowledge. The proposed descriptors obtained from the vibrational sensor signals are based on the spectral domain. They rely on what the module of the Short-Term Discrete Furrier Transform (STDFT) spectrum obtained from a particular sensor signal will view as a 2D textured "image". We identified time and its scale as the first dimension, and frequency and frequency scale as the second dimension of the STDFT sensor signal module of interest. The 2D textured image obtained in this way is then divided into separate discrete narrow 2D texture patches, which cover a portion of the frequency band of interest. Then, by applying the appropriate filter bank for each predefined frequency band, we extract 2D textons (see [14]), i.e., small dimensional vectors in the filter response space. Our goal is to use the aforementioned filter response for the corresponding tectons, to encode fine differences in the texture structure of the aforementioned texture patches that correspond to the degree of tool wear.

This approach is based on the assumption, which was later confirmed in the experiments, that the basic physical process of cutting and forming a chip is generated by different states of tool wear directly related to the structure of the aforementioned textons. However, the main problem we face when trying to use a texture-based approach is that most features, i.e. Descriptors used in texture recognition are inappropriate in this particular application. The reason for this is that the discriminability of texture patches belonging to different classes of tool wear states is contained in vertical lines distributed over the entire frequency range, and many of the techniques used in texture analysis could not account for this. Therefore, it is crucial in our approach to select the appropriate filter bank that will be able to efficiently extract the presence information of these lines, for each frequency band. In this paper, we apply a particular filter bank, proposed in [14], which is used in texture recognition problems, which has the aforementioned ability. We apply it to 2D patches of texture, obtained from the STDFT vibration sensor signal, to efficiently extract information on said lines, in the form of a texton, i.e. characteristic in filter response space.

We additionally use the modal probability distribution for each frequency range, which describe each particular utterance in the filter response space, and use them as low-dimensional, robust descriptors. The marks are obtained by using the first four statistical moments for the final marks to be obtained, which is a special saying that will be used in the training or recognition process. In this way, information on the texton probability distributions corresponding to each frequency band is extracted and placed in small features vector to be used effectively in the training and recognition in TCM tasks.

2. Apparatus used in experiments

The experiment setup and processing were performed on a CNC GU 600 lathe manufactured by INDEX installed in the laboratory of the Faculty of Technical Sciences in Novi Sad. Investigation of the tool wear process involved monitoring the mechanism of dominant wear through the following parameters: wear belt, crater wear, and tool life. During the turning process, a vibrating signal was detected on the tool shank. Generated signals from the machining process were recorded for each tool pass. The placement of the sensors on the tool as well as the workpiece dimensions used in this experiment are shown in Figure 1. On the experiment we used two cutting speeds, 180 to 250m/min, correlated with 0.15 and 0.3mm/rev feed rates. In the experiment was used 20x20mm cross section of the tool shank. The machining process was performed with P25 tool inserts marked TNMM 110408. Used sensors set is Accelerometers Kistler 8002 was fixed onto the tool holder, where the measured acceleration of vibrations signal. All signals was sampled at 625 kHz, using A/D converter NI 625 USB, National Instruments. Workpiece material was 42CrMo4, 310 HB hardness and 950MPa.

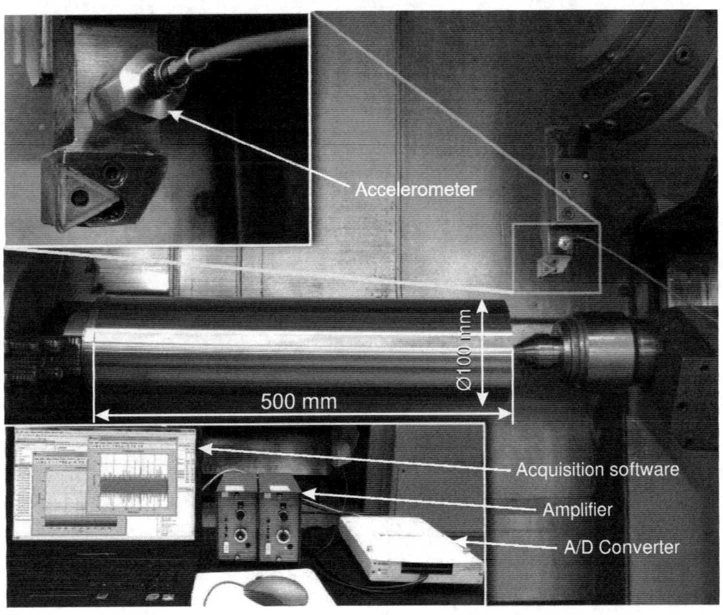

Figure 1. Experimental setup and acquisition system

The features extraction and selection of the descriptors used in the training as well as in the recognition phase of the classifier is performed from a signal of different lengths obtained from the vibration sensor. The focus in studies is on the problem of recognizing 2D texture from the perspective of tool wear recognition and classification. The most challenging problems present in the problem of recognizing the actual texture with 3D variations come from the variability of those textures, ie. the fact that classical texture is primarily a function of the following variables: texture surface, its albedo (i.e., the reflection of the corresponding surface factor), lighting, cameras, and its viewing position.

We have noticed that the texture corresponding to the STDFT spectrogram of the vibration sensor signal that we consider to be seen as a 2D texture does not possess this kind of variability [15]. By identifying the presence of vertical lines in the performed resulting STDFT (viewed as texture that is of interest), our goal is to effectively isolate them, and / or to obtain reliable information about their presence. Therefore, imposes the need to extract texture features used by certain specific filter banks able to identify them.

To extract the relevant characteristics to be used in the actual task of identifying tool wear states in a TCM system, we first extract the relevant descriptors in the form of textons characteristics obtained in the appropriate filter output space, and then model their probability distribution for each band frequency using the first four moments of those outputs, as final, compressed very robust descriptors.

3. Texture based features – forming the texton

The task is to identify the discrete time frame k of the STDFT spectrogram of the uterance of interest, as the discrete $X - axis$ of the textured image and the discrete frequency ω, as the discrete $Y - axis$ of the corresponding textured images. The hypothesis we confirm in our experiments is that the shape of the STDFT spectrogram shown in 2D space discriminativly for different tool wear states is highly correlated with the characteristics (number, position, structure of the vertical lines) that appear in the corresponding textured images (see Figure 2). In fact, the condition of tool wear is tightly related to the characteristics of those lines. The lines actually represent sudden jumps in the time frequency range of the signal received from certain sensors, at different time and frequency scales, induced by the cutting process and the tool wear basically. Obtaining information with respect to these lines is directly related to the selection of the filter bank to be used for that particular task.

The novel texture based descriptors in the task of TCM and named them the Texture Based Tool Condition Descriptors (TBTCD). The TBTCD descriptors are obtained by considering the STDFT spectrogram of the windowed time frames delivered from the previously mentioned utterance, as the 2D textured image.

The feature extraction procedure is as follows: A standard technique is applied for most texture recognition approaches it is divide the STDFT textured image into texture patches spread across in different filter bands. Then, using the appropriate filter bank, we extract texton-based descriptors as the feature vectors in the low dimensional filter response space. As already mentioned, it is crucial to select the appropriate filter bank that is able to efficiently extract the presence of the mentioned lines for each frequency band. The best results are achieved in [16] for the mentioned tasks are those that use the reactions of spatially invariant filter banks, i.e. filter banks based on a set of maximum response filters, such as (BFS, MR8, MR4 and MRS4), which have been reported and discussed in the classification assignment material [14, 17], as well as the Leung and Malik filter bank. A rotationally invariant, multi-scale MR8 filter has been shown to produce better results than any of the previously mentioned filters in texture recognition applications. In this application for the TCM task, we decided to use the MR8 filter, shown in Figure 3, without excluding the last two isotropic components, as they do not extract the corresponding as texton features.

Let $F(x, y)$ be the textured image that corresponds, i.e., is identified with the STDFT spectrogram of the vibration sensor signal of the particular utterance s. The STDFT spectrogram is defined as the square module of the spectral density, i.e., $|S(k,\omega)|^2$, where

$$S(k,\omega) = \sum_{n=-\frac{K}{2}}^{\frac{K}{2}} s(n)w(k-n)e^{-ikn} \tag{1}$$

is the STDFT of signal s. In (1), by k we denote the discrete time frame, by ω we denote discrete time frequency, i.e., discrete frequency beam, while w is the windowing sequence that we use (we use Hamming experiments), with the fixed length K. We identified x axes of the textured image with the Discreet time frame k of the mentioned STDFT, for $k = 0, \cdots, k_{max}$, so that it holds $x_{max} = k_{max}$. Actually, to every discreet time frame k in STDFT, there is a corresponding discreet index x of the texture patch in the textured image F. Also ω, the discreet frequencies $\omega = 0, \cdots, \omega_{max}$ are identified with discreet Y axes of the textured image F so that it holds $y_{max} = \omega_{max}$. We now have the following interpretation of STDFT in the form of textured image:

$$F(x,y) = |S(k,\omega)|^2 \tag{2}$$

$x = 0,\cdots,x_{max}$, $y = 0,\cdots,y_{max}$
$k = 0,\cdots,k_{max}$, $\omega = 0,\cdots,\omega_{max}$

We further do all our analysis on textured image $F(x,y)$ obtained as it is explained previously, where for the easier modeling, without loss of generalization, we consider the continuous case $x \in [0,x_{max}]$ and $y \in [0,y_{max}]$.

We briefly describe the filter that we use: Let

$$G(\sigma_x,\sigma_y,\theta,x,y) = \frac{1}{2\pi\sigma_x\sigma_y} e^{(A(\theta)^T[x \quad y])\begin{bmatrix}\sigma_x & 0\\ 0 & \sigma_y\end{bmatrix}\left(A(\theta)\begin{bmatrix}x\\ y\end{bmatrix}\right)} \tag{3}$$

be the directed anisotropic Gaussian kernel. The fixed terms $\sigma_1 > 0$ and $\sigma_2 > 0$ denote scales in $x \in [0,x_{max}]$, $y \in [0,y_{max}]$direction, respectively, while θ and $A(\theta) = \begin{bmatrix}cos\theta & -sin\theta\\ sin\theta & cos\theta\end{bmatrix}$, for $\theta \in [0,2\pi]$, denote the orientation of the kernel (3) and its 2D rotational matrix, respectively. The MR8 filter bank that we use in our application consist of 38Gaussian filters of the form (3), at6 different orientations $\theta_j \in \left\{{}^{j\pi}/_6 \mid j = 0,\cdots,5\right\}$, and 3 different scales $(\sigma_x,\sigma_y) \in \{(1,3),(2,6),(4,12)\}$, and also 6 Laplacians of Gaussians, defined as $\Delta G(\sigma_x,\sigma_y,\theta,x,y)$, where $\Delta = \partial_x^2 + \partial_y^2$is the Laplace operator of differentiation, evaluated at the same orientations and scales (for more details see [13, 18]). The filter outputs are collapsed, by taking only the maximum values, across all orientations, obtaining 6 different filter responses. Additional two responses are isotropic, thus unable to extract characteristic lines in the textured image F, so we do not use them in application.

Further we proceed with the procedure of texton extraction. We divide the interval $[0,y_{max}]$ where texture image F have its values into M sub-intervals $\{[y_{q-1,}y_q] \mid q = 1,\cdots,M\}$,$y_0 = 0$, $y_M = y_{max}$ where for every index $x = 0,\cdots,x_{max}$, the sub-patch of $F(x,y)$, is defined as matrix of pixels bounded by $[y_{q-1,}y_q]$ on Y access and by $[x,x+T]$ on x access. Hire, T is predefined width of the patch. We denote such described textured patch as $P_{x,q}$, $q = 1,\cdots,M$. The texton in the form of filter response are then formed for every interval $[y_{q-1,}y_q]$ and discrete coordinate x. We note, that the interval $[y_{q-1,}y_q]$ of the textured image F, corresponds to the frequency band$[\omega_{q-1},\omega_q]$ of the STDFT spectrogram of particular utterance.

Actually, for each X and each different band q, texton $v_{x,q}$ is composed of $Q = 6$ components of the MR8 filter bank.

Thus, we have:

$$v_{x,q}^l = \left[v_{x,q}^{l,(1)} \middle| v_{x,q}^{l,(2)} \middle| l \in L\right] \tag{4}$$

where

$$v_{x,q}^{l,(1)} = max_\theta\, G(\sigma_x,\sigma_y,\theta,x,y) * P_{x,q}$$

$$v_{x,q}^{l,(2)} = max_\theta\, \Delta G(\sigma_x,\sigma_y,\theta,x,y) * P_{x,q}$$

$$\theta \in \left\{{}^{j\pi}/_6 \mid j = 0,\cdots,5\right\}$$

and

$$L = \{(\sigma_x,\sigma_y)|(1,3),(2,6),(4,12)\}.$$

In (4), for simplicity we enumerate: $l = 1,2,3$.

3.1. Using moments for describing the probability distribution in the filter response space

We model the probability distribution of textons described in the previous section, for every band q separately. As a final result, we obtain compressed information about the mentioned distributions in the form of robust features.

We proceed as follows: For the fixed utterance u, for fixed band$q = 1, \cdots, M$, we have the set of observation feature, i.e., texton descriptor vectors
$\{v_{x,q} | x = 0, \cdots, x^u_{max}\}$

$$v_{x,q} = [v^1_{x,q}, v^2_{x,q}, v^3_{x,q}], \tag{5}$$

where $v^l_{x,q}$ given by (4). It is obtained for all patches $P_{x,q}$. It holds that, where $x^u_{max} = k^u_{max}$ (see (2)) is the number of time samples in the utterance u which, of course, can vary, dependent of utterance. We consider $v_{x,q}$ to be the observations, i.e., realizations of the random variable V^u_q with the probability distribution function (pdf) $p_{V^u_q}$. As, for fixed band q, there is a unique correspondence between the probability distribution $p_{V^u_q}$of V^u_q and its characteristic function given by

$$k_{V^u_q}(t) = E[e^{iV^u_q t}] = \sum_{j=0}^{\infty} \frac{(it)^j}{j!} E\left((V^u_q)^j\right) \tag{6}$$

where one could observe that (5) is given as an expansion which depend only on moments $m^{u,j}_q = E\left((V^u_q)^j\right)$, where the averaging is conducted regarding x. Thus, by taking only first few moments into account we manage to (roughly) approximate pdf $p_{V^u_q}$, and yet obtain robust features i.e., descriptors which can be used in the recognition part of TCM task

The final feature vector assigned to each utterance u is obtained either as P moments per 3 scale, or P moments per scale case. For the first case, for each band $q = 1, \cdots, M$, one vector containing Pmoment components $m^{u,j}_q j = 1, \cdots, P$ is obtained by averaging vectors $v_{x,q}$ given by (4), corresponding to patch $P_{x,q}$where the averaging is conducted regarding x. Then, all the mentioned component $m^{u,j}_q$ s are concatenated into the single vector containing MPcomponents:

$$v^u = [m^{u,1}_1 \cdots m^{u,P}_1 | \cdots | m^{u,1}_M \cdots m^{u,P}_M] \tag{7}$$

uniquely representing the utterance u. For the second case, a vector containing $3MP$ moment components $m^{u,j,l}_q j = 1, \cdots, P$; a P for each scale $l = 1,2,3$, is obtained also by averaging in x. Actually, for the second case, we obtain:

$$v^u_l = [m^{u,1,l}_1 \cdots m^{u,P,l}_1 | \cdots | m^{u,1,l}_M \cdots m^{u,P,l}_M],$$

$$v^u = [v^u_1 v^u_2 v^u_3]. \tag{8}$$

4. Experimental setup

The classification of the wear state is made by an indirect method by extracting features from the vibration signal generated on the tool shank during the cutting process. Experimental tests were conducted on more than 150 recorded vibration signals. The signals are corresponding with 3 tool wear states (50 signals for each state), based on the wear band (new plate, i.e. no wear, little wear, i.e. up to 0.25 mm, and high wear, i.e. above 0.5 mm). The tool wear process is a continuous process because the wear condition of the tool is constantly changing and the boundaries between classes cannot be uniquely defined. The analysis of the results was carried out using fivefold cross-validation, i. a sample of 120 vibration signals used for training the classifier and another 30 vibration signals were used for the test (while all three categories were presented equally in the test and in the control set). An average value of 5 results were its analysed to obtain a final estimate.

Forming the training set started by computing the vibration signal spectrum, using a Short-Term Discrete Furrier Transform for the frequency range of M = 256 samples of equal length of Hamming windows, with M / 4 overlap between segments (see Figure 2). The signals are 125000 samples in length, with a sampling frequency range over the 10 kHz. As a result of the applied transformations of signal, is obtained by matrix containing 129 frequencies (M / 2 + 1 for real signals and an equal

IOP Conf. Series: Materials Science and Engineering **749** (2020) 012001 doi:10.1088/1757-899X/749/1/012001

number of frequency points, although a limited frequency range for subsequent transformations) and 650 time bins (number of segments) was used. Each segment represents an estimate of the short-term time-localized frequency content of the vibration input signals. Each spectrograph was normalized to exclude the DC component. The convolution is performed between the spectrum and the corresponding filter from the MR8 filter bank described in 3.1 and it is shown in Figure 3. We have selected P = 3 moments: variance, skewness and kurtosis, while the first moment, i.e. Mean, excluded from consideration because we only considered normalized spectrograms, i.e. correspondingly normalized textured images, representing a total of 9 marks per signal.

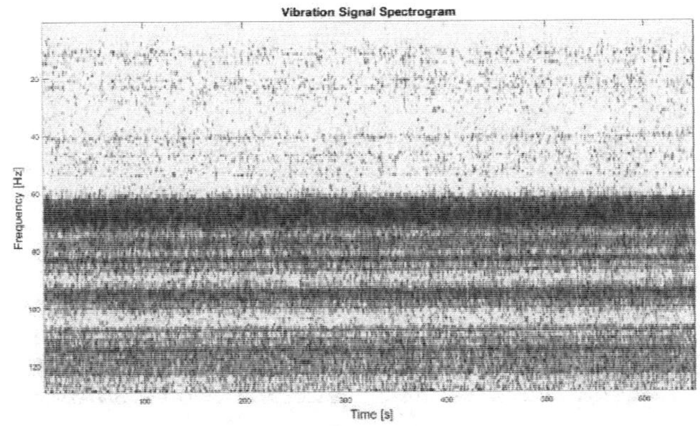

Figure 2. Vibration signal spectrogram

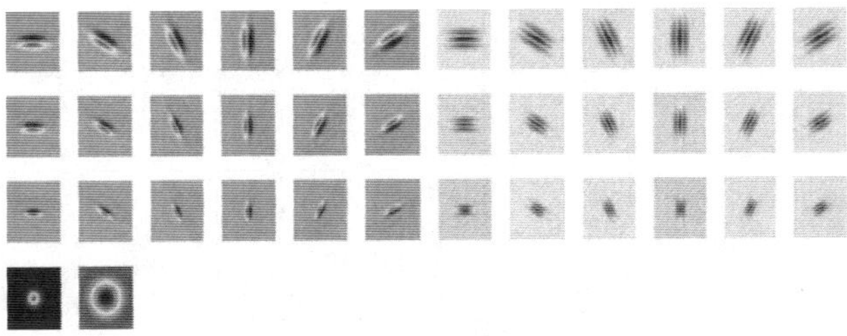

Figure 3. MR8 filter bank

Classification is done using the Fuzzy classifier with the selected feature set. The classifier generates the structure of the Mamdani-type Fuzzy Inference System (FIS) system from input coefficients, where we used the Fuzzy C-Means (FCM) clustering algorithm extracting a set of rules that model the behaviour of the data. Each input signal is represented by a vector of 3x3 (by scale) or 9 coefficients and is classified into one of three output states labelled with numerical values 1, 2 and 3, depending on the tool wear condition (band wear). Instead of being bound to only one cluster, each vector has a degree of affiliation across multiple clusters, depending on the distance from the center of the cluster. In FCM clustering, the mean cluster is the mean of all vectors, weighted by the degree of their affiliation to a particular cluster, i.e. inverse to the center of the cluster. During the initial initialization of the classifier, each vector is randomly linked to the clusters. Attachment levels change after a certain number of iterations, i.e. until the optimum is found. Center of clusters are calculated multiple times until they begin to converge.

5. Experimental results

In Table 1, the average classification accuracy for 5 folds (F1 to F5) is presented for fuzzy classification conducted per each scale separately (3 moments per scale). The output values are averaged for all 3 scales in order to obtain the final result. These values are classified in the following

IOP Conf. Series: Materials Science and Engineering **749** (2020) 012001 doi:10.1088/1757-899X/749/1/012001

way: less than 1.5 - class 1, 1.5 to 2.5 - class 2, more than 2.5 - class 3. The final results for 5 folds, including the target result, are presented in Fig. 4. The results are presented for 250 clusters (this was around the optimal value).

Table 1 The average fuzzy classification accuracy (separate scales)

Tool Wear	new plate (class 1)	0.25 mm (class 1)	0.5 mm or more (class 3)
scale 1	96 %	100 %	98 %
scale 2	98 %	100 %	100 %
scale 3	96 %	98 %	100 %
final classification	98 %	100 %	100 %

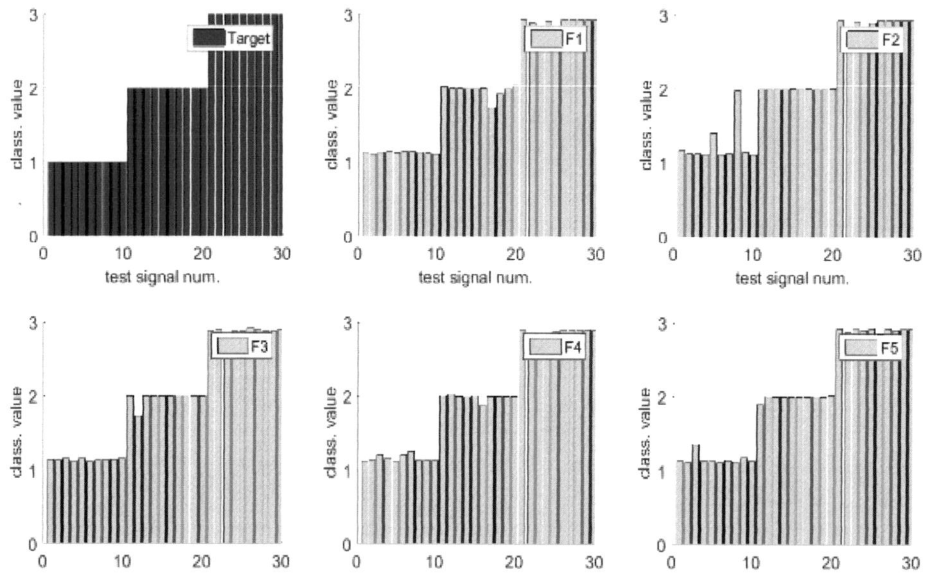

Figure 4. The final results for fuzzy classification (separate scales) for 5 fold

The classification accuracy obtained by using 9 coefficients (3 moments per 3 scales) for 5 folds is given in Table 2. The optimal number of clusters was smaller in this case, having in mind the fact that we had the same amount of data (number of signals) in higher dimensional space. Better results were obtained in the case of sum of the output values for separate scales pondered by 1/3, than in the case of classification for all 3 scales together.

Table 2 The average fuzzy classification accuracy (3 scales together)

Tool Wear	new plate (class 1)	0.25 mm (class 1)	0.5 mm or more (class 3)
3 scales	98 %	100 %	90 %

Acknowledgments
This paper presents a segment of the research on the project "Contemporary approaches in the development of special solutions bearing in mechanical engineering and medical prosthetics" project number TR 35025, financed by the Ministry of Education and Science of the Republic of Serbia.

References
[1] Siddhpura A and Paurobally R 2013 International Journal of Advanced Manufacturing Technology, vol. 65, p.371–393,
[2] Abellan-Nebot J V and Subirón F R 2010 International Journal of Advanced Manufacturing Technology, vol. 4,7 p.237–257,
[3] Cho S, Binsaeid S and Asfour S 2010 International Journal of Advanced Manufacturing Technology vol. 46, p.681-694.
[4] Wong Y S, Nee A Y C, Li X Q and Reisdorf C 1997 Journal of Materiel Processing Technology, vol. 63, p.205–210
[5] Jang D Y, Choi Y G, Kim H G and Hsiao A 1996 International Journal Machine Tools and Manufacturing, vol. 36, p.453–464.
[6] Dimla D E 2002 International Journal of Advanced Manufacturing Technology, vol. 19, p.705–713.
[7] Bahr B, Motavalli S and Arfi T 1997 International Journal of Computer Integrated Manufacturing, vol. 10,.314–323
[8] Tsai Y H, Chen J C and Lou S J 1999 International Journal Machine Tools and Manufacturing, vol. 39, p.583–605
[9] Haber R E, Jimenez J E, Peres CR and Alique J R 2004 Sensors and Actuators A: Physical, vol. 116, ISSN: 0924-4247 p.539–545.
[10] Abouelatta O B and Madl J 2001 Journal of Materials Processing Technology, vol. 118, p.269–277
[11] Chen J C and Chen W L 1999 Journal of Intelligent Manufacturing, vol. 10, p.187–197
[12] Al-Habaibeh A and Gindy N 2000 Journal of Materials Processing Technology, vol. 107, p.243–251
[13] Kuo R J and Cohen P H 1999 Neural Networks, vol. 12, p.355–370
[14] Varma M and Zisserman A 2009 IEEE Transactions on Pattern Analysis and Machine Intelligence, 31, 11:2032 - 2047.
[15] Varma M and Zisserman A 2005 International Journal of Computer Vision, vol. 62, (1-2), p.61-81
[16] Lazebnik S, Schmid C and Ponce J 2005 IEEE Trans. Pattern Analysis and Machine Intelligence, 27, 8:1265-1278.
[17] Varma M 2004 Statistical Approaches to Texture Classification, PhD thesis, Univ. of Oxford.
[18] Varma M and Zisserman A 2004 Image and Vision Computing, vol. 22, (14), ISSN: 0262-8856 p.1175-1183

IOP Conf. Series: Materials Science and Engineering **749** (2020) 012002 doi:10.1088/1757-899X/749/1/012002

Vibration analysis of Storz type firehose couplings

C Babut and V Diciuc[a]

Technical University of Cluj-Napoca, Faculty of Engineering, Department of Engineering and Technology Management, Baia Mare, Romania

E-mail: [a] vlad.diciuc@cunbm.utcluj.ro

Abstract. This paper presents the results of an investigation on the effects of the vibrations in the Storz- type firehose couplings. The data processed using a Fast Fourier Transform analyser and a Constant Percentage Bandwidth filter of the vibrations recorded in several couplings are shown. The frequency spectrum analysis of the signals acquired using a "Machine Diagnostics Toolbox", vibration measurement system, "Pulse", type 3560, Brüel&Kjær Sound&Vibration. Measurement strongly indicates a connection with the vibration-induced, complex fretting process, that occurs in the stress-concentration areas, leads to the nucleation and growth of micro-cracks and ultimately to the fracturing of the couplings' lugs.

1. Introduction

1.1. General presentation and background

This research is part of an investigation on the breaking of the Storz type firehose couplings, used by the Romanian firefighters. During firefighting operations, fire hoses are connected to the fire truck or to the hydrants with couplings. When needed, hose lines can be extended with couplings which make these components very important parts of fire control operations.

The Storz type couplings ensure the connection of two parts by interlocking lugs and flanges then rotating these with a quarter of a turn.

A Storz coupling in operating position is shown in figure 1, and a detail of a coupling between two hose sections is shown in figure 2.

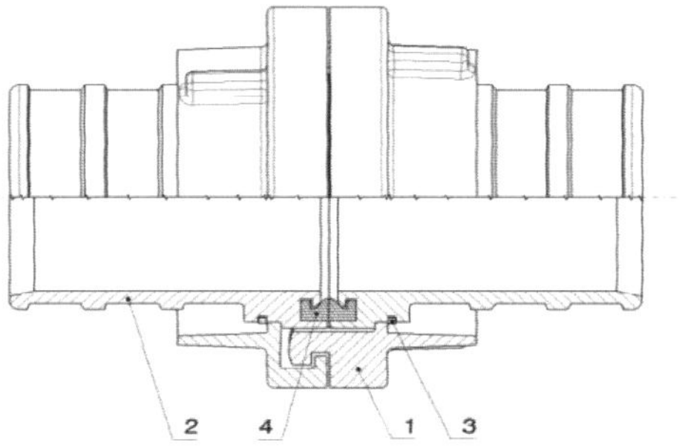

1- body with hooking lugs
2- pipe/tail piece
3- safety ring
4- sealing ring/gasket

Figure 1. Storz coupling in operating position

Figure 2. Two hose sections with Storz coupling

The damaging of the couplings leads to serious delays in the fire extinguishing actions and may have important effects on the outcome of these operations.

Table 1. Romanian dimensional standards for Storz couplings [1]

Type	A	B	C	D
Diameter [mm]	96	62	43	18
Mass [kg]	2,1	1,67	0,86	0,24

The Storz couplings used by Romanian firefighters are fabricated from die cast aluminium, more specifically- AlSi5Cu1Mg alloy. The alloy's chemical composition, according to [2] and [3], is given in table 2.

Table 2. Chemical composition of AlSi5Cu1 alloy [2], [3]

Chemical composition [%]					Impurities [%]			
Cu	Si	Mg	Mn	Al	Fe	Zn	Pb	Ni
1-1.5	5-6	0.3-0.6	0.2-0.5	rest	0.8	0.5	0.2	0.3

One of the most common and potentially dangerous damages sustained by these couplings during firefighting operations is breaking of the hooking lugs. One example is presented in figure 3.

Figure 3. Storz coupling with a broken lug

The authors' hypothesis is that this type of damage appears mainly due to cracking as a result of microcracks coalescence. The microcracking is thought to appear as an effect of vibration induced fretting wear, occurring in the stress concentration areas of the couplings, during fire extinguishing operations.

The objectives of this research were to acquire data on the vibrations appearing in the couplings, to analyse the spectrum of the vibrations in the couplings, using a Fast Fourier Transform (FFT) analyser and a Constant Percentage Bandwidth (CPB) filter, to identify the frequencies of maximum magnitude, in terms of Power Spectral Density, and investigate the correlation with vibration induced fretting wear and fatigue.

1.2. Research literature review

The vibration frequency effects on fretting wear was initially investigated for different steel types [4]. The research showed an increase of debris at low frequencies, less than 1Hz, for a given number of fretting cycles and displacement amplitudes.

Researches aimed at accelerating fretting tests for steel specimens showed an increase in fretting wear in the frequency range of 102Hz-2,104Hz [5].

Investigations on steel behaviour at partial slip and low amplitude displacement fretting conditions indicated that maximum wear occurred at frequencies in the range of 100Hz-300Hz [6].

Similar work has been done for several pure metals, including iron, copper and aluminium, with specimens subjected to constant amplitude loading scheme and variable frequencies, of 10, 60 and 80 Hz in variable humidity conditions [7].

For aluminium, maximum wear appeared at 60Hz, for normal loading, in the range of 105 fretting cycles and 2% relative humidity in the contact [7].

For other alloys, such as Al7075-T6, European standard EN AW-7075, it was found that the maximum fretting wear for cyclic normal load occurred at a frequency of 1Hz [8]. Research on more advanced alloys, such as Ti-6A1-4V, showed a decrease of fretting fatigue life for combined high-cycle fatigue/low-cycle fatigue testing at 200Hz [9].

Another study, on the 2024-T4 aluminium alloy, European standard EN AW-2024, crack nucleation and failure in specimens at frequencies in the range of 100-350Hz [10]. The effects of frequency variations, of phase difference between normal and tangential vibrations on fretting wear and fatigue has not been extensively studied [11].

Though earlier research results seemed to indicate a decrease in fretting wear rate at higher frequencies, the literature review conclusion is that significant wear, crack nucleation and failure of specimens were observed at frequencies in the range of 1-350Hz. This is due to the specific designs of the experiments, that were conducted on pure metals and alloys, with a wide range of different parameters such frequency, humidity, temperature, number of cycles, displacement amplitude and loading.

Therefore, we can infer with a high degree of confidence, that frequencies between 1-350Hz are an important factor in increasing the fretting wear.

2. Methodologies

2.1. Experimental setup presentation

The experimental setup schematic, presented in figure 4, consists of a fire engine, two hose lines, first of which is connected to the fire engines 'pump with a Storz coupling and the second hose that extends the first one using a coupling of the same type. The second hose line is connected through another similar coupling to a nozzle.

Data was acquired with a Vibration Analysis System- VAS. The system used was a "Machine Diagnostics Toolbox", Brüel&Kjaer PulseTM 9727, based on the "Pulse" 3560 multi analyser, with the 7910 software bundle installed.

The system used allows multichannel real time simultaneous Fast Fourier Transform (FFT) and 1/n octave, e.g. Constant Percentage Bandwidth (CPB) spectrum analysis, for vibration diagnostics. The CPB filter used was at 1/12 octave. The data were acquired through four channels, from four sensors, placed consecutively on each of the couplings. The schematic of sensor placing is shown in figure 5.

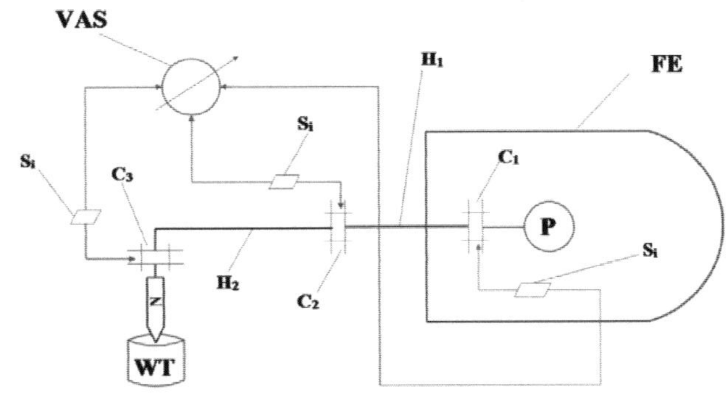

VAS- vibration analysis system;
FE- fire engine;
$H_{1,2}$- hose lines
P- pump;
$C_{1...3}$- couplings;
WT- water tank;
S_i- sensors; i= 1...4;
N- nozzle

Figure 4. Schematic of the experimental setup

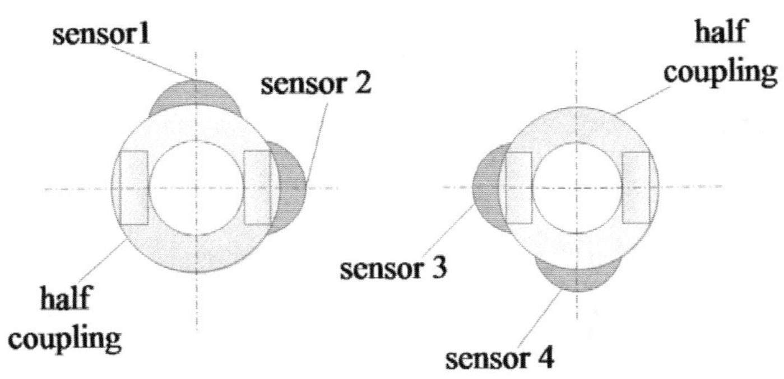

Figure 5. Schematic of the sensor placement

The sensors used were one ISOTRON accelerometer, IEPE TEDS ENDEVCO type, 752A12 model, sensitivity of 97.74mV/g at 100Hz or 9.967mV/m/s^2 at 100 Hz and transverse sensitivity of 5%, one ISOTRON accelerometer, IEPE TEDS ENDEVCO type, 752A12 model, sensitivity of 100,6mV/g at 100Hz or 10.26 mV/m/s2 at 100 deHz and transverse sensitivity of 0,5%, one I-TEDS ENDEVCO type accelerometer, 752A12 model, sensitivity of 107.6mV/g at 100Hz or 10.98mV/m/s^2 at 100Hz and transverse sensitivity of 1% and one I-TEDS ENDEVCO type accelerometer, 752A12 model, sensitivity of 107.4mV/g at 100Hz or 10.95mV/m/s2 at 100Hz and transverse sensitivity of 0.8%.

The fire engine- FE used was APCAR 12215 type. Though outdated, it is still widely used, mainly for its 9000 litre water tank.

The fire engine is fitted with the PSI 50/8 centrifugal pump, that allows flows up to 5000 l/min, and working pressures up to 23 bar. PSI 50/8 pump characteristic curves are shown in figure 6.

The hoses were regular 20 m lines, fitted with "C" type Storz couplings. The first line was connected to the fire engine. The second line was connected with a variable flow nozzle- N, as shown in figure 4, TURA-P SF 225 type. To discharge the water, a 1000 litre tank- WT, as shown in figure 4, was used.

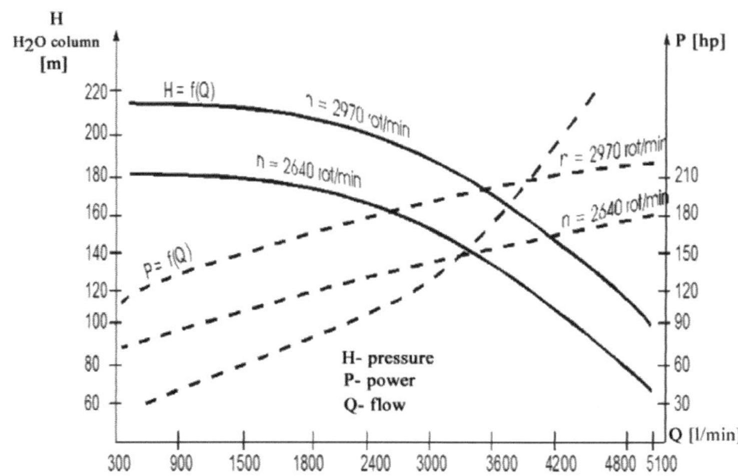

Figure 6. PSI 50/8 pump characteristic curves

2.2. Experimental process and data acquisition details

The data acquisition and processing flow was designed around the "Machine Diagnostics Toolbox", Brüel&Kjaer PulseTM 9727 vibration analysis system's capabilities.

To acquire the data, three series of in-situ measurements, one for each of the couplings, were designed. To ensure that measurements are taken in conditions as close to real working parameters as possible, each series consisted of four runs of the fire engine's pump, at working pressures of 2.5 bar, 5 bar, 7.5 bar and 10 bar.

The duration of the runs was chosen at 20 seconds, for practical reasons, i.e. to allow the system to acquire the data needed for the signal processing and avoid the overload of the analyser's hard drive. The data processing flowchart is shown in figure 7.

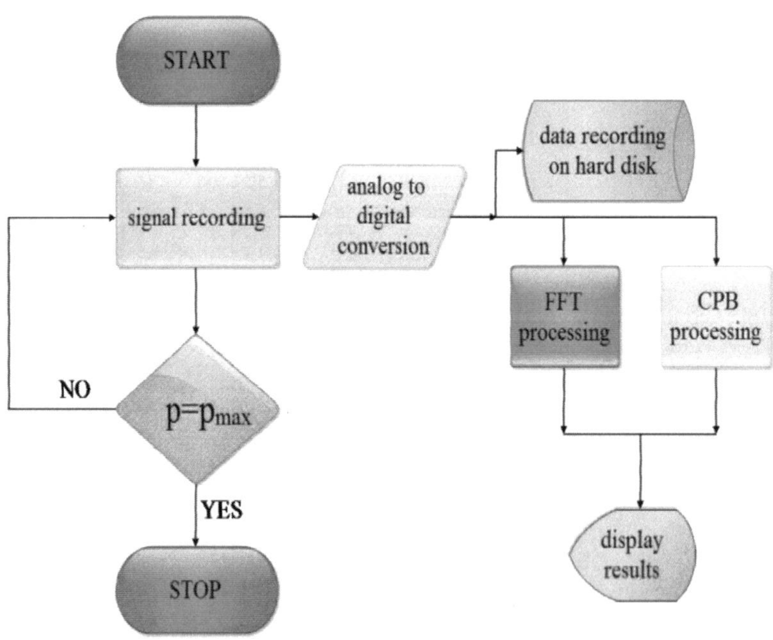

Figure 7. Data processing flowchart

IOP Conf. Series: Materials Science and Engineering **749** (2020) 012002 doi:10.1088/1757-899X/749/1/012002

The functional block diagram of the experiment is shown in figure 8.

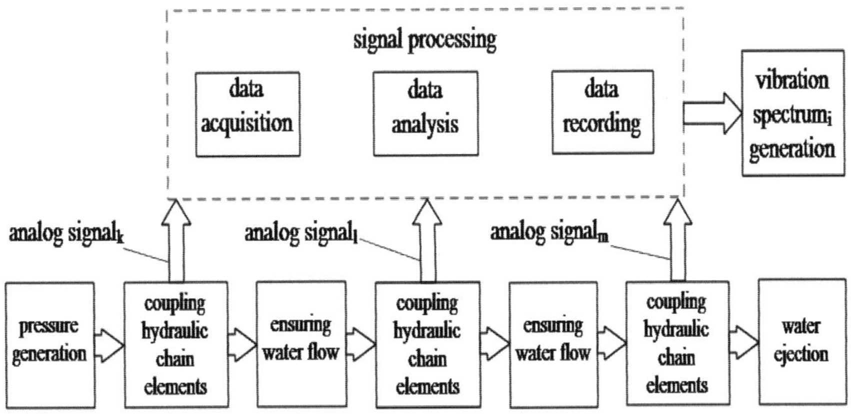

k, l, m- ordinal number of analog signals recorded in the couplings
k=l=m= 1...4;
i= ordinal number of vibration spectrum
i=1...12

Figure 8. Functional diagram of the experiment

3. Results

Due to the extent of the data, only the spectrums obtained at 2.5bar for the first coupling will be shown in graphic format. The rest of the data will be shown in numerical format, and discussed in correlation with the aforementioned results.

The FFT processing results for the signals acquired in the first coupling, the one that connects the first hose line with the fire engine's pump, are shown in figure 8, figure 9, figure 10 and figure 11.

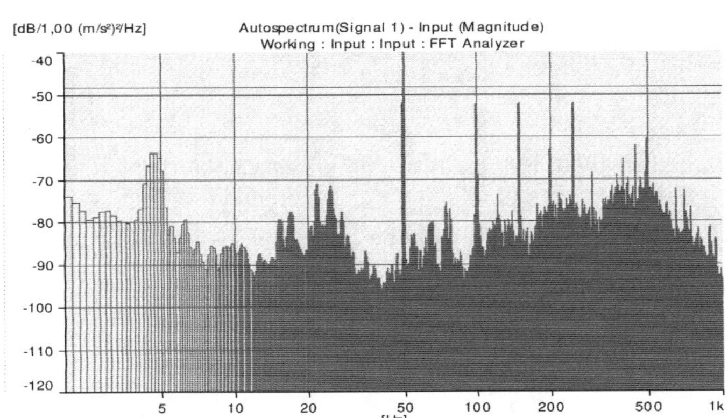

Figure 8. FFT spectrum for sensor 1, coupling 1, at 2.5bar

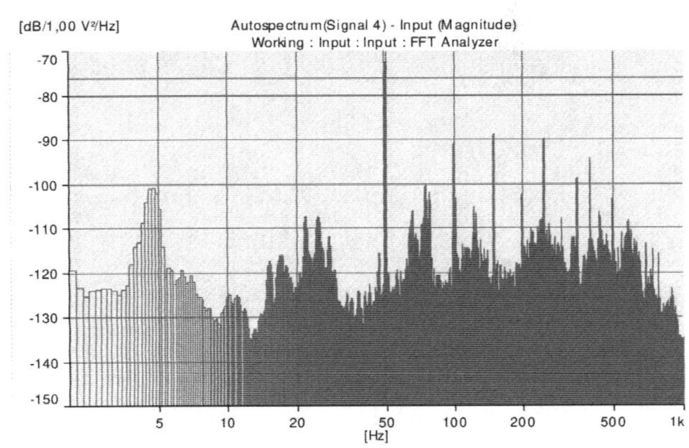

Figure 9. FFT spectrum for sensor 4, coupling 1, at 2.5bar

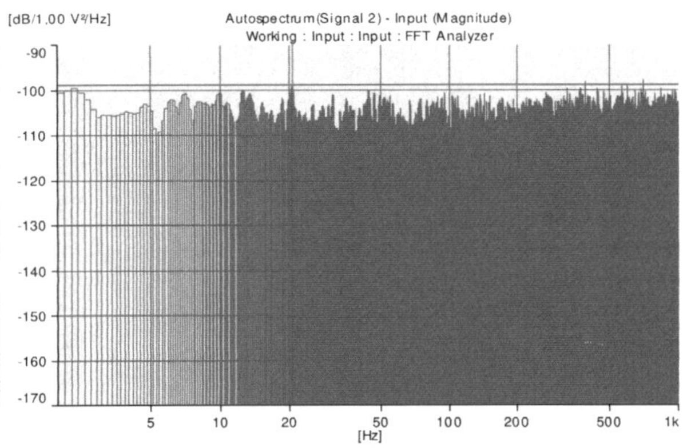

Figure 10. FFT spectrum for sensor 2, coupling 1, at 2.5bar

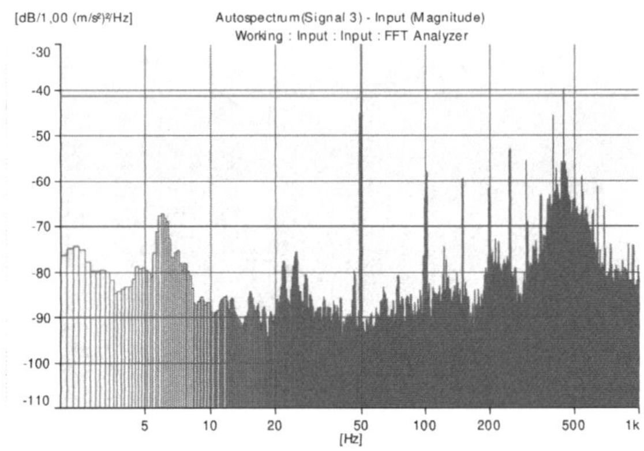

Figure 11. FFT spectrum for sensor 3, coupling 1, at 2.5bar

4. Conclusions

The FFT analyzer identified the frequencies for the peak PSD values of the acquired signals. Most of them were found to be in the range of 1Hz to 350Hz, described in the research literature review as a domain of vibrations that increase the fretting wear. A number of these frequencies were in the 400Hz-450Hz range, with one "spike" at 721,6Hz, for the coupling connecting the second hose and the nozzle, measured at a pressure of 10 bar. This can be attributed to a resonance effect, occurring as a result of the operating conditions during in-situ measurements.

The CPB filter processing of the vibrations recorded in the three couplings also indicated a predominance of the frequencies in the range of 1Hz to 350Hz, with a number of them in the range of 400Hz-500Hz. A very high "peak" was found at 917.3Hz. This can be also attributed to a resonance resulting from the operating conditions

The results showed that the frequencies that carry the most energy, found at maximum values of the PSD identified by the FFT and CPB analysis are in the vibration range of 1Hz to 350Hz, thus indicating that the investigated couplings are prone to damage through fretting wear and fatigue.
This strongly indicates that fretting is one of the main factors in the damaging of the Storz type couplings, described by the authors in previous works.

References

[1] *** 1992 *Manual pentru cunoasterea accesoriilor,utilajelor şi autospecialelor de stingere a incendiilor* Bucuresti Editura Ministerului de Interne

[2] *** 2010 *Romanian Standards Association Aluminum and aluminum alloys. Castings, Chemical composition sand mechanical properties SR EN 1706* Bucureşti

[3] Niţu M I 2010 *Cercetări privind îmbunătăţirea unor proprietăţi fizico-mecanice la aliajele de aluminiu turnabile prin intermediul tratamentelor termice*, Teză de doctorat Braşov Universitatea "Transilvania"

[4] Feng I M and Uhlig H M 1954 *Fretting Corrosion of Mild Steel in Air and in Nitrogen* J. App. Mec. 21 395-400

[5] Söderberg S, Bryggman U and McCullough T 1986 *Frequency Effects in Fretting Wear* J. Wear 110 I 19-34

[6] Bryggman U and Söderberg S 1986 *Contact conditions in fretting* J. Wear 110 I pp.1-17

[7] Goto H and Buckley D H 1984 *Effect of Humidity on Fretting Wear of Several Pure Metals* NASA Tech. P. 2403 pp.4-10

[8] Majzoobi G and Abbasi F 2018 *An Investigation into the Effect of Normal Load Frequency on Fretting Fatigue Behavior of Al7075-T6* J. Trib. Tr. 61 III pp.547-559

[9] Cortez R, Mall S and Calcaterra J. R 2000 *Interaction of High-Cycle and Low-Cycle Fatigue on*

Fretting Behavior of Ti-6-4 Fretting Fatigue: Current Technology and Practices ed D W Hoeppner et al ASTM STP 1367 pp.183-198

[10] Matik J F and Farris T N *High 2003 Frequency Fretting Fatigue Experiments Fretting Fatigue: Advances in Basic Understanding and Applications* ed Mutoh Y et al ASTM International STP 1425 pp.252-272

[11] Waterhouse R B 1992 *Fretting wear ASM Handbook Friction, Lubrication, and Wear Technology* vol 18 (ASM International) pp.159-161

IOP Conf. Series: Materials Science and Engineering **749** (2020) 012003 doi:10.1088/1757-899X/749/1/012003

Optimising Aluminium Structures Using the FEA Method

H Cioban

Technical University of Cluj-Napoca, Faculty of Engineering, Department of Engineering and Technology Management, Baia Mare, Romania

E-mail: horia.cioban@cunbm.utcluj.ro

Abstract. As a request of a partner and subject of a research contract, this paper intend to present different stages of studies regarding the behaviour of some Aluminium structures under different types of loads. Starting from existing examples, a company from aerospace industry intended to produce similar products but on a larger scale. On a virtual product created in a CAD environment and using a FEA method is possible to optimize aluminium structures and to apply the conclusions and results given by FEA reports to the real products. Tests made on manufactured structures validate the findings of these studies.

1. Introduction

In Romania, the government statistics indicate a 70% increase in the last 3 years revenue from the aerospace industry. The main products manufactured are aircrafts, helicopters, gliders, engines, landing gears, parts and subassemblies [1]. Studies made in 2019 identify the main trends in R&D focused on materials (aluminium and composites), electric and hybrid engines, mechanical solutions, electronics and software [2].

Using CAD-CAM-CAE software, aircrafts manufactures are beginning to redesign outdated models or create new models. The aluminium structure of the airplane is subject to some of these changes in design activity.

2. Requirements of the studies

Until few years ago, the maximum length of the beams in the aluminium structure of the aircrafts was up to 6m. The progress of concepts and the new technologies allow the length to increase up to 9m.

The aim of the project is to design and to optimize two products: trolley for parts (figure 1) and container for three trolleys (figure 2).

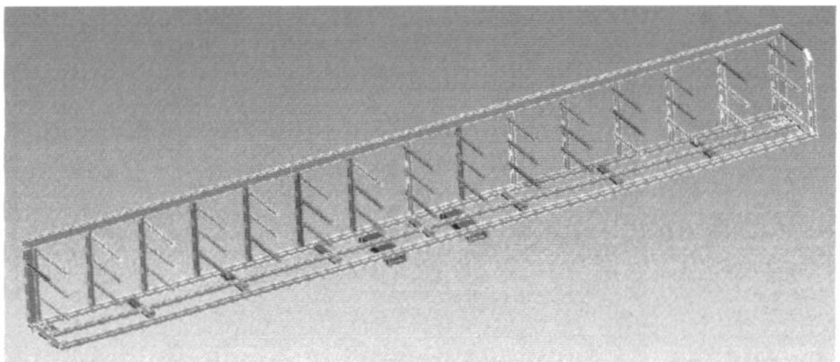

Figure 1. Trolley for long beams transportation

Figure 2. Container for three trolleys

These extruded aluminium beams are parts of the airplane structure, and transport from the manufacturer's sites to the assembly workshop is a problem that needs to be solved.

In design studies, it started from similar existing products used for 6m beams. The main purpose is to get minimal deformations of these welded aluminium structures.

2.1. Additional constrains

Additional requirements in these studies were related to operating and transport conditions. Thus, it was required to know how the winch is behaving in forklift lifting, and the container at lifting with forklift and bridge crane. Other simulated situations where: deformations of the container under the conditions of 2 overlapped containers, which are the deformations if the container remains suspended in only 3 points, or if the conditions required by the road transport rules are met, where three-ways accelerations can occur, as in figure 3.

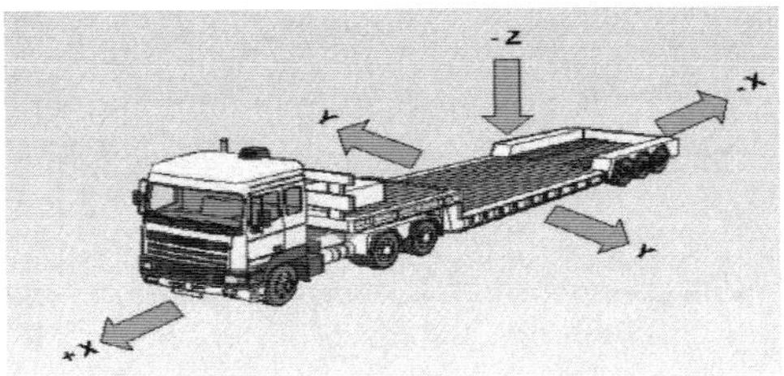

Figure 3. Simulation conditions imposed by road transport regulations

Each of these simulations showed the critical areas in the aluminium structures of the trolley and the container and required changes of the product geometry. After each modification in the project, the stages required by FEA simulation were resumed.

2.2. Type of analysis

The analysis used to study the behaviour of these two welded assemblies is known as *static structural*, that means the loads and displacements are in the elastic area of the Hooke's law diagram. FEA analysis concepts accept some simplifications of the model's geometry:

- a welded assembly having bonded contacts between parts can be simplified to a single part-one body;
- the welding seam can be ignored;
- avoid rounded/chamfered edges; all profiles must have sharp edges;
- avoid small gaps between parts; the model has to be continuous.

3. Pre-processing stage

The pre-processing phase is also known as the model preparation. It is the most labor-intensive step because it implies that only the part of the project that is relevant to the study is chosen and the parts that do not influence the results are excluded from the analysis. Thus, the wheels were removed from the analysis made on the trolley and from the container only the mainframe were retained, and it was ignored the exterior walls and doors, as in figures 4 and 5.

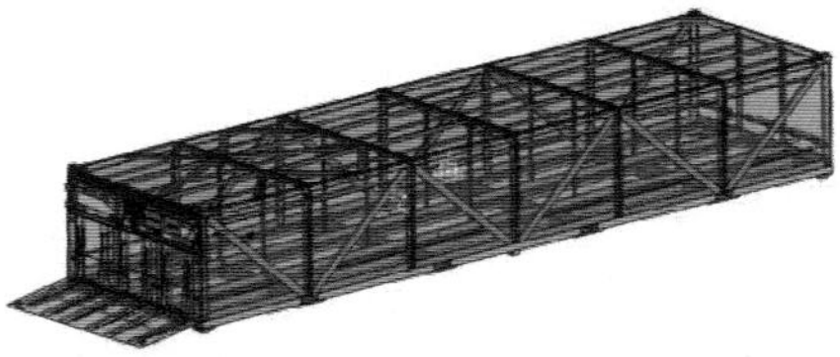

Figure 4. The complete container design

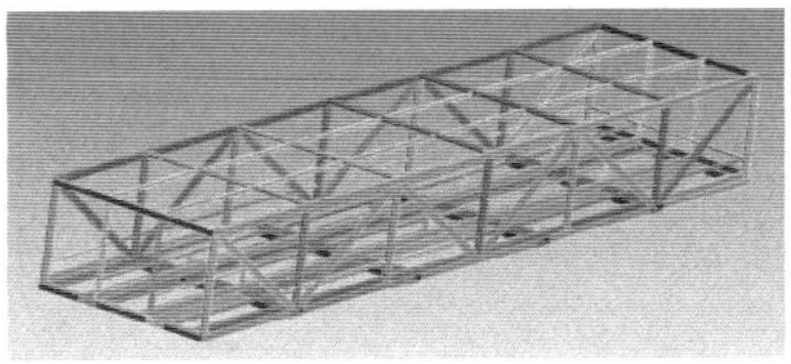

Figure 5. Geometry subjected to FEA analysis

3.1. Material properties

The choice of the materials used was imposed by the initial conditions of the design theme. Aluminium alloys have been required especially because of the lower weight than steel alloys, although aluminium, due to high elasticity, has deformations greater than steel under the same stress conditions [3].

Most FEA software has an internal library of materials or provides functions to create or import data to define the properties of the materials used in analyses. In figures 6 and 7 the properties of a steel alloy and of an aluminium alloy are compared.

Property	Value	Unit
Density	7850	kg m^-3
Isotropic Secant Coefficient of Thermal Expansion		
Coefficient of Thermal Expansion	1.2E-05	C^-1
Reference Temperature	22	C
Isotropic Elasticity		
Derive from	Young's Modulu...	
Young's Modulus	2E+11	Pa
Poisson's Ratio	0.3	
Bulk Modulus	1.6667E+11	Pa

Figure 6. Properties of a steel alloy

Property	Value	Unit
Density	2770	kg m^-3
Isotropic Secant Coefficient of Thermal Expansion		
Coefficient of Thermal Expansion	2.3E-05	C^-1
Reference Temperature	22	C
Isotropic Elasticity		
Derive from	Young's Modulus...	
Young's Modulus	7.1E+10	Pa
Poisson's Ratio	0.33	
Bulk Modulus	6.9608E+10	Pa

Figure 7. Properties of an aluminium alloy

Significant differences can be seen in density (3 times higher in steel than aluminium alloy and giving a larger mass at the same volume), in Young's modulus that gives the elasticity property (higher at aluminum) and in Bulk modulus.

3.2. Meshing stage

This stage involves dividing the 3D model studied into geometric elements of tetrahedron or parallelogram shape, called finite elements. The results of the analysis will be more accurate if the finite elements will have more regular shape. Also, a deeper division of the 3D model leads to results closer to reality.

The meshing stage involves a strong hardware and time-consuming resources. There are often situations when errors occur, and the analysis cannot continue until they are eliminated. For this reason, it is not recommended to directly import the CAD model into the FEA application, but to reconstruct the geometry from scratch, under the simplifying conditions presented in paragraph 2.2. For example, chamfers and fillets are removed from a beam section. The 3D model must be continuous and gaps, even small, are not accepted. Figures 8 and 9 display the result of the meshing stage of the trolley and the container, the reports showing that the trolley 3D model was divided into 37069 finite elements and the container in 36862 finite elements.

Object Name	*Mesh*
State	Solved
Bodies	186
Active Bodies	177
Nodes	129423
Elements	37069

Figure 8. Meshing the trolley

Object Name	*Mesh*
State	Solved
Nodes	64874
Elements	35862

Figure 9. Meshing the container

At the first running of the meshing phase, the default settings of the software parameters are used. After all errors in geometry are eliminated and the meshing is successful finished is recommended to refine the mesh and to reduce the size of elements, mostly at the critical areas. Finally, the number of the finite elements can reach hundreds of thousands and even millions.

3.3. Boundary conditions

This is the last step of the model preparation and contains applying of constrains like fixed supports, defining loads (forces, pressure or moments) or applying external forces (earth gravity, accelerations

IOP Conf. Series: Materials Science and Engineering **749** (2020) 012003 doi:10.1088/1757-899X/749/1/012003

etc.) [4]. Boundary conditions are defined by the real environment in which the trolley and the container will be used.

In the trolley case, the position of the fixed supports depends on the handling conditions: it is supported on wheels or is raised on forklift. In the container case, the position of the fixed supports is different when it rests on the feet or is lifted by fork-lift or bridge crane. Each new boundary condition generates a new FEA study.

Earth gravity and 2700N forces (the weight of carried parts) were applied on the trolley, as in figure 10. Accelerations generated by the vehicle's movement were added in the case of road transport.

At the container, 27000 N forces were applied given by the weight of the 3 loaded trolleys applied to the runway surfaces, or 24000 N, the weight of an empty container placed on the top of another container. In each case, gravity was taken into account.

Figure 10. Applying of boundary condition on trolley

Figure 11. Applying of boundary condition on container

When the all model preparations are finished, the finite element analysis can go further to the next step: processing using a solver.

4. Processing stage

The boundary conditions are included in equations and distributed on each finite. Depending of number of finite elements of the model, a system of millions of equations (the total stiffness matrix) has to be solved and this is the role of the solver. Sometimes the processing stages takes too long or even the solver cracks. For these reasons a strong hardware is required.

The results consist of displacements of the single nodes, distortions, stress and nodal forces.

5. Post processing stage

After the processing is finished FEA applications display the results in a very intuitive graphical mode. On the mesh of the 3D model, each finite element is coloured using a graduated scale, from blue to red. Interpretation of results is thus very easy, and critical areas can easily be highlighted.

5.1. Equivalent (von-Mises) Stress

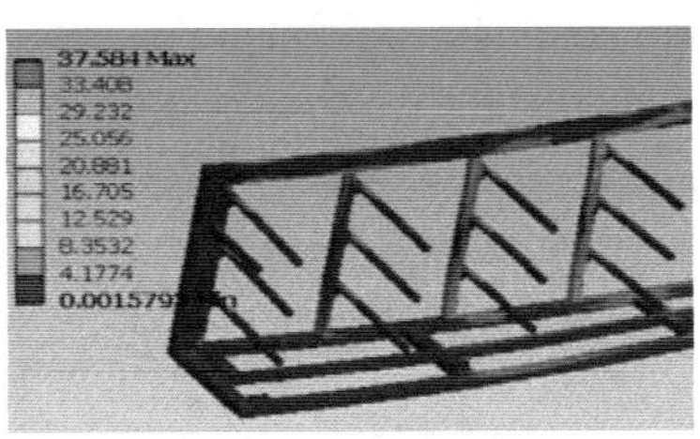

Figure 12. Equivalent (von-Mises) Stress at the 1st version of the trolley

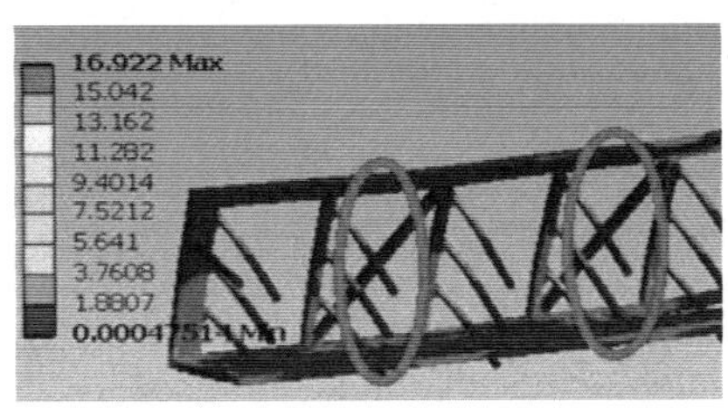

Figure 13. Equivalent (von-Mises) Stress at the optimized version of the trolley

One of the solver results is the equivalent tensile stress. The maximum value calculated for each finite element has to be in the elastic area of the Hooke's law diagram. The resulting values for both the trolley and the container show that under the conditions imposed by the initial data of the study topic the loads are within acceptable limits.

In the case of the winch, figures 12 and 13 show the results of the analysis for the initial model, respectively after the optimization of the geometry by the addition of hardening beams.

For the container, the results are shown in figures 14 and 15 for lifting with side-loaders and bridge cranes at the 1st version and optimized version.

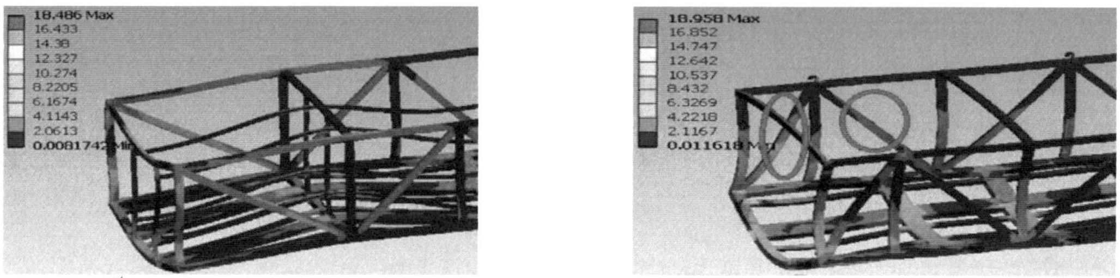

a. 1st version of the container b. Optimized version of the container

Figure 14. Equivalent (von-Mises) Stress (side loaders lifting)

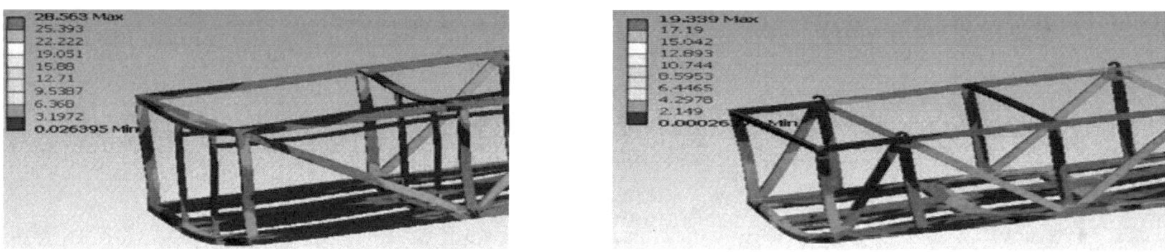

a. 1st version of the container b. Optimized version of the container

Figure 15. Equivalent (von-Mises) Stress (bridge cranes lifting)

The values calculated in each case are compared to the admissible value of the aluminum tensile yield strength (280 MPa). In all cases, the maximum values are lower than the admissible values, so the loads to which the winch and the container will be subjected will not affect the mainframe. Changes to geometry show a better behaviour on constrains and loads.

5.2. Total deformation

In the optimization of the geometry of the two aluminium structures, the calculation of the displacements of the finite elements gives a total deformation. Figures 16 and 17 show the deformations of the trolley and container structures, comparing the original with the optimized version.

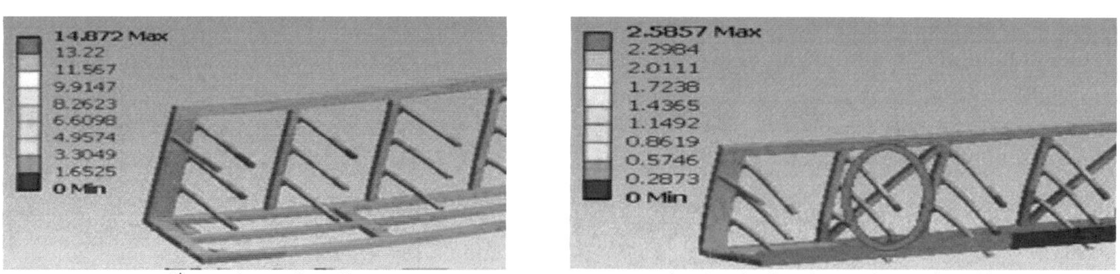

a. 1st version of trolley b. Optimized version of trolley

Figure 16. Total deformation of trolley (side loaders lifting)

IOP Conf. Series: Materials Science and Engineering **749** (2020) 012003 doi:10.1088/1757-899X/749/1/012003

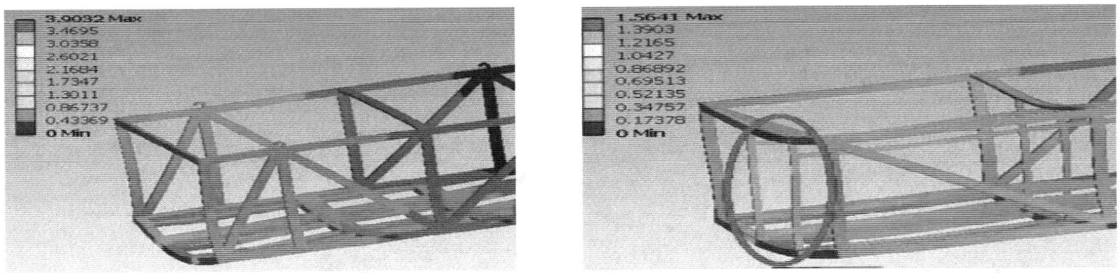

a. 1st version of container b. Optimized version of container

Figure 17. Total deformation of container (side loaders lifting)

Changes to the geometry of the welded structures have led to decreasing of deformations from 14.8mm to 2.5mm to the trolley and from 3.9mm to 1.5mm to the container.

6. Conclusions

Studies using the finite element method have proven to be extremely useful in the cases presented. One of the possible solutions is that of using another material with better physical properties, but was excluded from the initial data of the theme: aluminium alloy was imposed. Geometry optimization was the only possible option.

Also starting from the initial conditions 3 situations has resulted for the trolley and 4 for the container: lifting with side loader, lifting with bridge crane (with 3 and 4 supporting decks) and transport behaviour. From the first version of the geometry to the final version, there were at least 6 intermediate stages, resulting in over 50 FEAs performed. The results of the analyses showed a behaviour of the welded aluminium structures within the permissible limits, with enough safety margin. Measurements made on prototypes showed deformations close to those calculated on virtual products. Currently, both the trolley and the container are in operation.

References

[1] Costescu M 2016 *Aerospace in Romania,* http://investromania.gov.ro/web/wp-content/uploads/2016/07/Aerospace_EN.pdf

[2] PwC Advisory 2018 *Turning the Aerospace and Defence industry into a source of added value for the economy*, Atlantic Black Sea Security Forum, Bucharest June 5, http://aspeninstitute.ro/wp-content/uploads/2018/05/PwC-Presentation_Atlantic-Black-Sea-Security-Forum_June5_2018.pdf

[3] Moldovan Ş A, Banica M, Năsui V 2017 *Defining the machining process for complex parts made from aluminium alloy extruded profiles by correcting raw material deviations*, MATEC Web of Conferences Vol. 112 pp.01011

[4] Butnar L A, Basarman A P, Tranisan D I, Groza I 2017 *FEA Deformations Analysing of the Polyurethane Armchair Seat* Materiale Plastice Vol.54 Issue 3 pp.593-595

IOP Conf. Series: Materials Science and Engineering **749** (2020) 012004 doi:10.1088/1757-899X/749/1/012004

On jerk in the kinematic study of the rigid body

I Craciun[a] and M Ungureanu

Technical University of Cluj-Napoca, Faculty of Engineering, Department of Engineering and Technology Management, Baia Mare, Romania

E-mail: [a] ioana.craciun@cunbm.utcluj.ro

Abstract. Jerk, as the derivative of acceleration, is used more and more in the kinematic study of the systems characterized by short travels and requiring a high precision of positioning, including elevators, CNC machine tools, and industrial robots. The acceleration modelling using jerk variation serve as a basis for generating the functioning tahogrames of the mechanical systems. Since peak values of jerk and acceleration are closely connected to safety and also to comfort in the mechanical systems where human presence is involved, a study concerning the positions of the systems where these values are reached will be performed in this paper. The position vector of the points where peak values of jerk are reached will be determined. Spatial motion as well as the particular motions of the rigid body (translation movement, rotation around a fixed axis, helical movement, the motion of planar mechanisms, spherical motion) will be considered. Since positioning is an important aim for the industrial robots, jerk variation in the various points of the robot's parts may provide important information about peak values of acceleration, as well as about inertial forces involved in positioning. Also, jerk is currently used in speed modelling of elevators and robots, a study considering the positions corresponding to peak values of acceleration has not yet been performed.

1. Introduction

Speed modelling using the variation of acceleration is frequently applied in certain mechanical systems, where the functioning cycle supposes to move from one point to another, to stay in that point, and move further. Such functioning cycles have a phase of acceleration, a phase of even motion, and a phase of deceleration. Another issue associated to safety and comfort during movement is linked to the presence of jerks during acceleration and deceleration phases. Jerk, the acceleration's derivative with respect to time, is usually considered as dynamic parameter in the study of these systems, such as robots or elevators. Peak values of jerk are associated with high inertial forces, with impact on the positioning as well as on the reliability of the system. If jerk limitation is necessary, considering a rigid body that executes a spatial motion, this paper aims to find the positions where peak values of jerk are reached, in order to generate trajectories that will avoid those positions. [1]

2. Theoretical considerations

The motion of a rigid body is studied by respect to two Cartesian frames, one fixed, and the other one firmly attached to the rigid body, and moving together with it. [2] The frames are shown in figure 1.

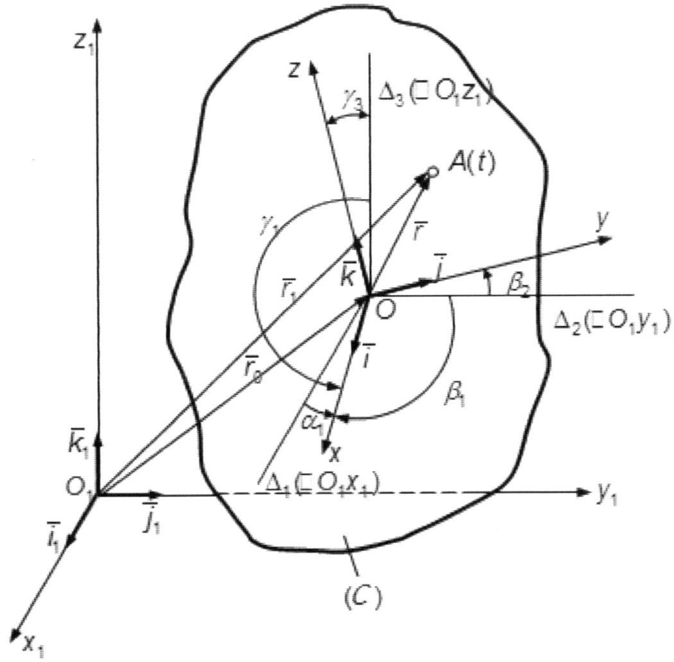

Figure 1. Choosing Cartesian frames for studying the movement

In order to calculate the acceleration of a point belonging to the rigid body Euler's formula for acceleration is written:

$$\bar{a} = \bar{a}_o + \bar{\varepsilon} \times \bar{r} + \bar{\omega} \times \left(\bar{\omega} \times \bar{r}\right) \tag{1}$$

in which is the position vector of the considered point with respect to a frame connected to the rigid body and moving with it. The derivative of acceleration by respect to time will be calculated:

$$\frac{d\bar{a}}{dt} = \frac{d\bar{a}_o}{dt} + \frac{d\bar{\varepsilon}}{dt} \times \bar{r} + \bar{\varepsilon} \times \frac{d\bar{r}}{dt} + \frac{d\bar{\omega}}{dt} \times \left(\bar{\omega} \times \bar{r}\right) + \bar{\omega} \times \left(\frac{d\bar{\omega}}{dt} \times \bar{r}\right) + \bar{\omega} \times \left(\bar{\omega} \times \frac{d\bar{r}}{dt}\right) \tag{2}$$

The following notations will be used:

$\bar{j} = \dfrac{d\bar{a}}{dt}$ - the linear jerk,

$\bar{j}_o = \dfrac{d\bar{a}_o}{dt}$ - the linear jerk of point O, origin of the frame,

$\bar{\varphi} = \dfrac{d\bar{\varepsilon}}{dt}$ - the angular jerk.

The expression of jerk is:

$$\bar{j} = \bar{j}_o + \bar{\varphi} \times \bar{r} + 2\bar{\varepsilon} \times \left(\bar{\omega} \times \bar{r}\right) + \bar{\omega} \times \left(\bar{\varepsilon} \times \bar{r}\right) + \bar{\omega} \times \left(\bar{\omega} \times \left(\bar{\omega} \times \bar{r}\right)\right) \tag{3}$$

3. Expression of the position vector of the point having a given value of jerk

Given values of speed and acceleration, angular speed and angular acceleration, as well as speed and acceleration of the frame origin O are known, the position vector of the point having a given value of jerk can be determined.

Relation (3) will be transformed as:

$$\bar{j}-\bar{j}_o=(\bar{\varphi}-\omega^2\bar{\omega})\times\bar{r}+2(\bar{\varepsilon}\cdot\bar{r})\bar{\omega}+(\bar{\omega}\cdot\bar{r})\bar{\varepsilon}-3(\bar{\omega}\cdot\bar{\varepsilon})\bar{r} \tag{4}$$

Relation (4) is considered as an equation in $\bar{r}$.
In order to find vector, we calculate the cross product to the left with $\left(\bar{\varphi}-\omega^2\bar{\omega}\right)$.

$$\begin{aligned}&\left(\bar{\varphi}-\omega^2\bar{\omega}\right)\times\left(\bar{j}-\bar{j}_o\right)=\left[\left(\bar{\varphi}-\omega^2\bar{\omega}\right)\cdot\bar{r}\right]\cdot\left(\bar{\varphi}-\omega^2\bar{\omega}\right)-\left(\bar{\varphi}-\omega^2\bar{\omega}\right)^2\cdot\bar{r}+2\left(\bar{\varepsilon}\cdot\bar{r}\right)\left[\left(\bar{\varphi}-\omega^2\bar{\omega}\right)\times\bar{\omega}\right]+\\&+\left(\bar{\omega}\cdot\bar{r}\right)\left[\left(\bar{\varphi}-\omega^2\bar{\omega}\right)\times\bar{\varepsilon}\right]-3\left(\bar{\omega}\cdot\bar{\varepsilon}\right)\left[\left(\bar{\varphi}-\omega^2\bar{\omega}\right)\times\bar{r}\right]\end{aligned} \tag{5}$$

From relation (4) we have:

$$\left(\bar{\varphi}-\omega^2\bar{\omega}\right)\times\bar{r}=-2\left(\bar{\varepsilon}\cdot\bar{r}\right)\bar{\omega}-\left(\bar{\omega}\cdot\bar{r}\right)\bar{\varepsilon}+3\left(\bar{\omega}\cdot\bar{\varepsilon}\right)\bar{r}+\bar{j}-\bar{j}_o \tag{6}$$

After substituting (6) in (5):

$$\begin{aligned}&\left[-\left(\bar{\varphi}-\omega^2\bar{\omega}\right)-9\left(\bar{\omega}\cdot\bar{\varepsilon}\right)^2\right]\cdot\bar{r}+\left[\left(\bar{\varphi}-\omega^2\bar{\omega}\right)\cdot\bar{r}\right]\cdot\left(\bar{\varphi}-\omega^2\bar{\omega}\right)+\left(\bar{\varepsilon}\cdot\bar{r}\right)\left\{2\left[\left(\bar{\varphi}-\omega^2\bar{\omega}\right)\times\bar{\omega}\right]+6\left(\bar{\omega}\cdot\bar{\varepsilon}\right)\cdot\bar{\omega}\right\}+\\&+\left(\bar{\omega}\cdot\bar{r}\right)\left\{\left[\left(\bar{\varphi}-\omega^2\bar{\omega}\right)\times\bar{\varepsilon}\right]+3\left(\bar{\omega}\cdot\bar{\varepsilon}\right)\cdot\bar{\varepsilon}\right\}=3\left(\bar{\omega}\cdot\bar{\varepsilon}\right)\left(\bar{j}-\bar{j}_o\right)+\left(\bar{\varphi}-\omega^2\bar{\omega}\right)\times\left(\bar{j}-\bar{j}_o\right)\end{aligned} \tag{7}$$

Introducing notations:

$$\bar{A}_1=\bar{\varphi}-\omega^2\bar{\omega} \tag{8}$$

$$\bar{B}_1=2\cdot\left[\left(\bar{\varphi}-\omega^2\bar{\omega}\right)\times\bar{\omega}\right]+6\left(\bar{\omega}\cdot\bar{\varepsilon}\right)\cdot\bar{\omega} \tag{9}$$

$$\bar{C}_1=\left(\bar{\varphi}-\omega^2\bar{\omega}\right)\times\bar{\varepsilon}+3\left(\bar{\omega}\cdot\bar{\varepsilon}\right)\cdot\bar{\varepsilon} \tag{10}$$

$$D_1=-\left(\bar{\varphi}-\omega^2\bar{\omega}\right)^2-9\left(\bar{\omega}\cdot\bar{\varepsilon}\right)^2 \tag{11}$$

$$\bar{E}_1=\left(\bar{\varphi}-\omega^2\bar{\omega}\right)\times\left(\bar{j}-\bar{j}_o\right)+3\left(\bar{\omega}\cdot\bar{\varepsilon}\right)\cdot\left(\bar{j}-\bar{j}_o\right) \tag{12}$$

Equation becomes:

$$\bar{A}_1\left(\bar{A}_1\cdot\bar{r}\right)+\bar{B}_1\left(\bar{\varepsilon}\cdot\bar{r}\right)+\bar{C}_1\left(\bar{\omega}\cdot\bar{r}\right)+D_1\cdot\bar{r}=\bar{E}_1 \tag{13}$$

The dot product of relation (13) with $\bar{\omega}$ is:

$$\left(\bar{\omega}\cdot\bar{A}_1\right)\left(\bar{A}_1\cdot\bar{r}\right)+\left(\bar{\omega}\cdot\bar{B}_1\right)\left(\bar{\varepsilon}\cdot\bar{r}\right)+\left(\bar{\omega}\cdot\bar{C}_1+D_1\right)\left(\bar{\omega}\cdot\bar{r}\right)=\left(\bar{\omega}\cdot\bar{E}_1\right) \tag{14}$$

Notating:

$$F_1=\left(\bar{\omega}\cdot\bar{C}_1\right)+D_1 \tag{15}$$

Generally $F_1\neq 0$, so we get:

$$\left(\bar{\omega}\cdot\bar{r}\right)=\frac{\left(\bar{\omega}\cdot\bar{E}_1\right)-\left(\bar{\omega}\cdot\bar{A}_1\right)\left(\bar{A}_1\cdot\bar{r}\right)-\left(\bar{\omega}\cdot\bar{B}_1\right)\left(\bar{\varepsilon}\cdot\bar{r}\right)}{F_1} \tag{16}$$

IOP Conf. Series: Materials Science and Engineering **749** (2020) 012004 doi:10.1088/1757-899X/749/1/012004

With relation (16), relation (13) becomes:

$$\overline{A_2}\left(\overline{A_1}\cdot\bar{r}\right)+\overline{B_2}\left(\bar{\varepsilon}\cdot\bar{r}\right)+D_2\cdot\bar{r}=\overline{E_2} \tag{17}$$

Where:

$$\overline{A_2}=\overline{A}_1\cdot F_1-\overline{C}_1\cdot\left(\overline{\omega}\cdot\overline{A_1}\right) \tag{18}$$

$$\overline{B_2}=\overline{B}_1\cdot F_1-\overline{C}_1\cdot\left(\overline{\omega}\cdot\overline{B_1}\right) \tag{19}$$

$$D_2=D_1\cdot F_1 \tag{20}$$

$$\overline{E_2}=\overline{E}_1\cdot F_1-\overline{C}_1\cdot\left(\overline{\omega}\cdot\overline{E_1}\right) \tag{21}$$

The dot product of relation (21) with $\bar{\varepsilon}$ is:

$$\left(\bar{\varepsilon}\cdot\overline{A_2}\right)\left(\overline{A_1}\cdot\bar{r}\right)+\left(\bar{\varepsilon}\cdot\overline{B_2}\right)\left(\bar{\varepsilon}\cdot\bar{r}\right)+D_1\cdot\left(\bar{\varepsilon}\cdot\bar{r}\right)=\left(\bar{\varepsilon}\cdot\overline{E_2}\right) \tag{22}$$

We notate:

$$F_2=\left(\bar{\varepsilon}\cdot\overline{B_2}\right)+D_2 \tag{23}$$

Which generally is not null, therefor:

$$\left(\bar{\varepsilon}\cdot\bar{r}\right)=\frac{\left(\bar{\varepsilon}\cdot\overline{E_2}\right)-\left(\bar{\varepsilon}\cdot\overline{A_2}\right)\left(\overline{A_1}\cdot\bar{r}\right)}{F_2} \tag{24}$$

Equation (17) becomes:

$$\overline{A_3}\left(\overline{A_1}\cdot\bar{r}\right)+D_3\cdot\bar{r}=\overline{E_3} \tag{25}$$

Where:

$$\overline{A_3}=\overline{A}_2\cdot F_2-\overline{B}_2\cdot\left(\bar{\varepsilon}\cdot\overline{A_2}\right) \tag{26}$$

$$D_3=D_2\cdot F_2 \tag{27}$$

$$\overline{E_3}=\overline{E}_2\cdot F_2-\overline{B}_2\cdot\left(\bar{\varepsilon}\cdot\overline{E_2}\right) \tag{28}$$

With

$$F_3=\left(\overline{A_1}\cdot\overline{A_3}\right)+D_3 \tag{29}$$

Which generally is not null, we get:

$$\left(\overline{A_1}\cdot\bar{r}\right)=\frac{\left(\overline{A_1}\cdot\overline{E_3}\right)}{F_3}=\frac{\left(\overline{A_1}\cdot\overline{E_3}\right)}{\left(\overline{A_1}\cdot\overline{A_3}\right)+D_3} \tag{30}$$

Using relation (30), we get:

$$\bar{r}=\frac{1}{D_3}\left[\overline{E_3}-\overline{A_3}\frac{\left(\overline{A_1}\cdot\overline{E_3}\right)}{\left(\overline{A_1}\cdot\overline{A_3}\right)+D_3}\right] \tag{31}$$

4. Conclusions

Although the relation above has been calculated considering a movement of the rigid body on which no restriction is imposed, it can be easily extended on the particular movements of the rigid body (translation movement, rotation around an axis or a fixed point, roto–translation). An appropriate choice of the reference frames will lead to the values of the terms A1 to E3. Given the speed and acceleration values, the geometric location of the points that have the maximum jerk during the movement can be found [3].

References

[1] Harris C M, Crede Ch E 1968, 1969 *Şocuri si vibratii* vol. I II III Editura Tehnica Bucuresti
[2] Craciun I 2015 *Mecanica: Cinematica* Editura RISOPRINT Cluj-Napoca ISBN 978-973-53-1700-3
[3] Ungureanu M, Craciun I, Cosma M, Damian F 2011 *Study of the jerk at the start and stop of an electrical powered elevator* SCIENTIFIC BULLETIN Serie C Volume XXIII Fascicle: Mechanics Tribology Machine Manufacturing Technology ISSN 1224-3264 pg.101-106

Consideration about determination of vibro-acustic behaviour of the cycloid reducer

A Dascalescu[a] and F Suciu

Technical University of Cluj-Napoca, Faculty of Engineering, Department of Engineering and Technology Management, Baia Mare, Romania

E-mail: [a] anamaria.dascalescu@cunbm.utcluj.ro

Abstract. In order to achieve the favourable operating characteristics of the cycloidal reducer, it was conducted the necessary determination of the parameters which influence its vibro-acoustic behaviour and measurements of the vibrations of the cycloidal gears in initial stages of exploitation and after hours of exploitation. The conducted measurements presented in the paper allows to determine how the various deviations from the reference profile of the surfaces in contact cause different mechanical and acoustic vibrations, making possible assessments of the influence of deviations of the cycloid profile of the satellite wheels on the dynamic behaviour of the cycloidal reducers and the comparison of the results with the requirements of the standards in the field. In order to appreciate the effect of vibration on the mechanical system and the necessary measures for the attenuation of the phenomenon, the study is focus on the vibration content in frequency, which is done by the frequency analysis method.
Keywords: cycloid reducer, frequency of vibration, stand, measurement equipment

1. Introduction

In terms of constructive solutions, cycloid reducer ensures smooth and silent operation with extremely low vibrations, even under the most powerful loads, having aligned shafts and all the component machine parts are perfectly symmetrical, the centrifugal forces being compensated by the use of two satellite wheels disposed at 1800 or three satellite wheels disposed at 1200 [3].

To obtain these extremely particularly favourable operating characteristics of the cycloid reducer, it is necessary to obtain its machine parts with high precision involving more expensive manufacturing technology and cutting tools, especially for the profiled surface of cycloid satellite wheels.

The difficulty of dynamic behaviour modelling of cycloid gears imposes the need of vibrations measurements during the initial stages of research, design and exploitation.

Along with the development of vibration and noise measurement techniques, signal processing, the continuous test and control systems of the gears have also been improved. Research on gears demonstrates that the general maintenance of vibration characteristics, general level, peaks in the frequency domain and in noise measurements [2].

2. Parameters influencing the vibro-acustic response of cycloid gears

The vibration behaviour of gears determines to a large extent the acoustic behaviour, between the vibration level and the sound pressure level is a qualitative and quantitative correspondence. Theoretical and experimental research on dynamics mechanism of the vibration and acoustic noise

IOP Conf. Series: Materials Science and Engineering **749** (2020) 012005 doi:10.1088/1757-899X/749/1/012005

generation have led to determining the main factors that decisively influence the vibro-acoustic behaviour of gears [4].

By analogy with the involute gear, in figure 1 we present the grouping of the parameters influencing the vibro-acoustic response of the cycloid gears [1].

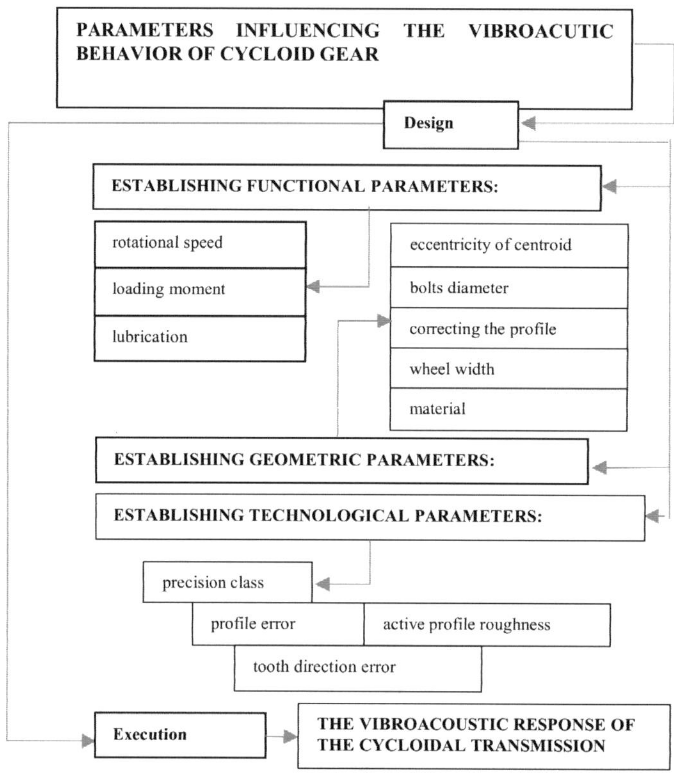

Figure 1. Parameters influencing the vibro-acustic response of cycloid gears

3. Vibro-acoustic determinations at cycloid reducer

It is proposed to study the vibro-acoustic behaviour of the cycloid reducer with bolts depending on the deviations of the cycloid profile, recognizing that they decisively influence the vibration and noise characteristics of the cycloid gears, raising the most important manufacturing technology issues.

Two series of acoustic pressure and mechanical vibration measurements were performed on a cycloid reducer with a transmission ratio i = 17, mounted on a test stand, over a large operating hours, period during which changes of deviations of the active cycloid profile occurred, as well as changes of the vibro-acoustic behaviour of the reducer.

We compare the levels of noise peaks and vibration peaks corresponding to the two measurement moments and correlate the obtained dates with the active profile changes that occurred during the reducer functioning between the two moments. Conclusions were drawn about the influence of the increases deviations of the profile of the planetary wheels on the vibro-acoustic behaviour of the cycloid reducer and comparison of the results with the requirements of industry standards has been made.

In order to appreciate the effect of vibration on the mechanical system and the measures that are necessary for the attenuation of the phenomenon, it is especially necessary to specify the content in frequency, which is done by the frequency analysis method. The frequency spectrums obtained are compared with the reference spectrum corresponding to the normal functioning of the reducer. In the literature it is stated that the upper limit of the vibration level at which defects are to be searched is 2-3 times (6-10 dB) the acceptable vibration level.

3.1. Stand for vibro-acoustic measurements of cycloid gear

In figure 2 we present the stand used in vibration and noise measurements of the cycloid reducer. The essential components of the stand are: (1) the drive electric motor, (2) the cycloid reducer whose drive shaft is also the motor shaft, (3) the elastic coupling, (4) the EPM type magnetic brake.

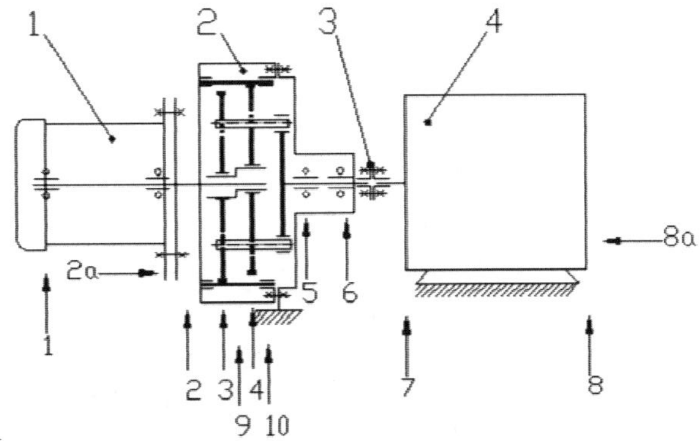

Figure 2. The stand for vibro-acoustic measurements of cycloid gear

The measuring points: 1, 2, 2a, 3, 4, 5, 6, 7, 8, 8a, 9, 10 coincide with the mounting position of the vibration transducer and are located as close as possible to the moving parts: on the gear housing, on the bearing and on the fixed wheel body in its gearing planes with profiled wheels.

Points of measurement in two directions are used: radial measurements in horizontal direction (x) and vertical direction (y) and axial measurements at the front planes of the bearing in the vicinity of the planetary wheel. Mechanical vibration is characterized by its characteristics: displacement, velocity, acceleration, evolution over time, frequency spectrum and amplitude spectrum. Measurements made in the frequency spectrum are: (a) in the displacement range, with the amplitude expressed in [μm], focuses on low vibration components; (b) vibration measurements in the velocity range, with amplitude expressed in [mm/s], advantageous because machine vibrations have a spectrum of frequencies with relatively constant speed components over a wide range of frequencies up to 1kHz; (c) vibration measurements in the acceleration range, with the amplitude expressed in [μV], focuses on high frequency components [5].

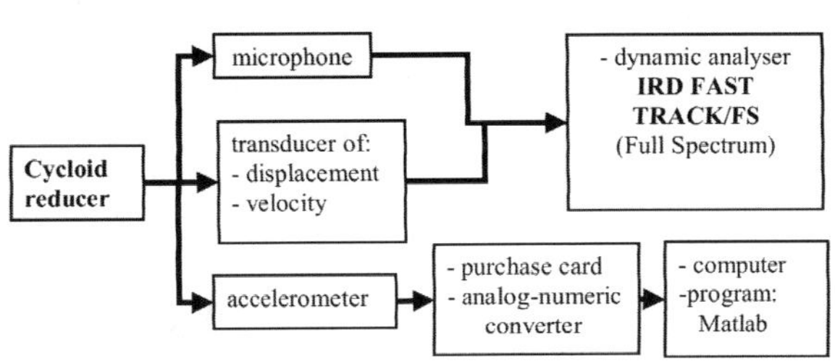

Figure 3. Measuring equipment used to determine the vibro-acoustic behaviour of the cycloid reducer

Usage of specialized software allows performing some analysis functions: digital analysis in frequency or amplitude, spectral comparison, analysis and time mediation. The necessary measuring

IOP Conf. Series: Materials Science and Engineering **749** (2020) 012005 doi:10.1088/1757-899X/749/1/012005

equipment used is presented in figure 3. The frequency content of the signal received from the transducer gives indications of the machine's operating condition.

3.2. Experimental methodology

Frequency spectrum analysis for research, technological or monitoring purposes involves, as the first step, the determination of the frequencies through calculations. No all calculated frequencies find significant correlation in the spectrum, some overlap, others have insignificant importance.

The coincidences between the measured frequencies and the obvious amplitude peaks and the calculated frequencies provide diagnostic data.

By reporting of all the values at the fundamental frequency, f_1, $f_1=n_1/60$, n_1- rotation of the driving shaft (rot/min), it is identified the order of the different spectral components.

3.2.1 Vibration displacement and vibration velocity measurements. In order to obtain quantitative and qualitative information, were carried out vibration displacement and vibration velocity measurements with the analyser IRD FAST TRACK/FS of IRD Mechanalysis, combining the data collection, data transfer and automated computer analysis software PMpower. The analyser IRD (full Spectrum) is a full spectrum data collector with the frequency range between 10Hz and 25kHz with 35 variable frequency ranges. It is a real-time analyser capable of performing a full FFT (Fast Fourier Transform) spectral analysis with the possibility of instantly providing a graphical representation of the spectrum.

In figures 4, 5, 6, 7 we present the representative spectrograms corresponding to the two process of measurements in point 9 of the cycloid reducer. On registered diagrams, figure 4, 5, the signification of notations are: RC - cycloid reducer at first set of measurements; RC1 - cycloid reducer at second set of measurements; n-the point of measurements (from 1 to 10 as figure 2); H,V ore A represent the direction of measurements (horizontal, vertical ore axial); mm/s the vibration characteristics measured (speed); μm the vibration characteristic measured (displacement).

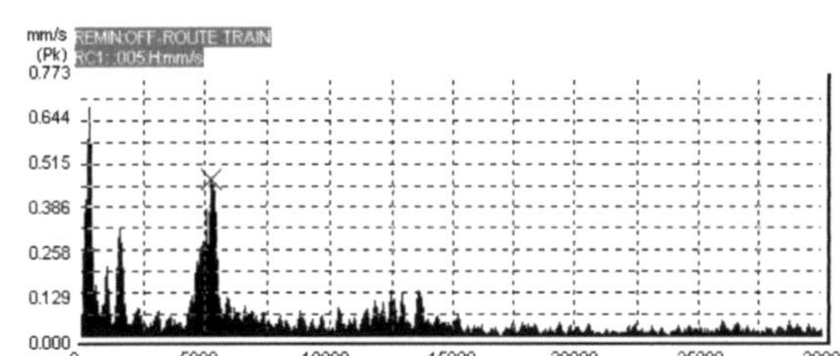

Figure 4. The amplitude of the vibrations in the displacement range in radial point 9 - the case of RC

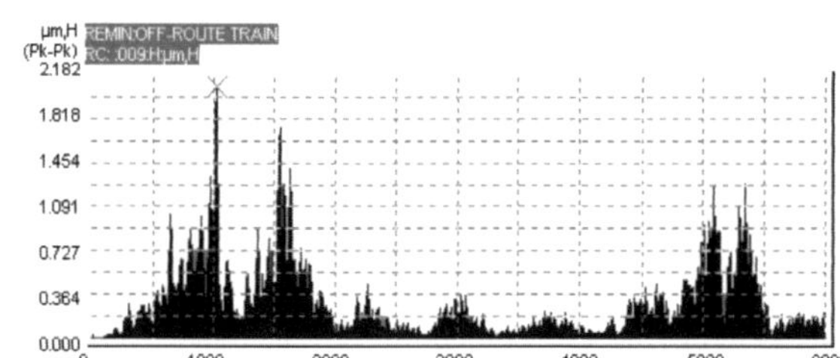

Figure 5. The amplitude of the vibrations in the displacement range in radial point 9 - the case of RC1

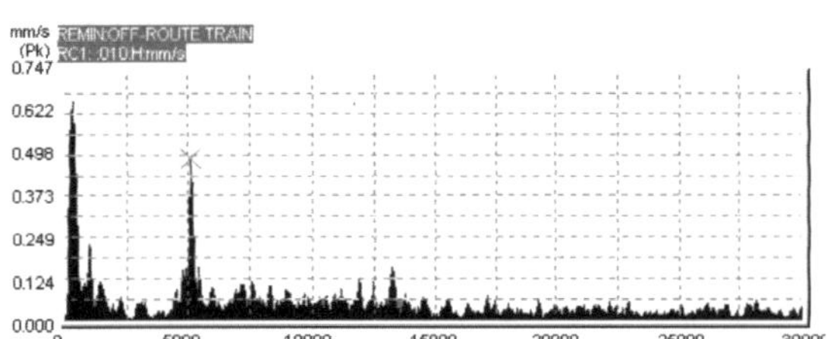

Figure 6. The amplitude of the vibrations in the speed range in radial point 9 in the case of RC

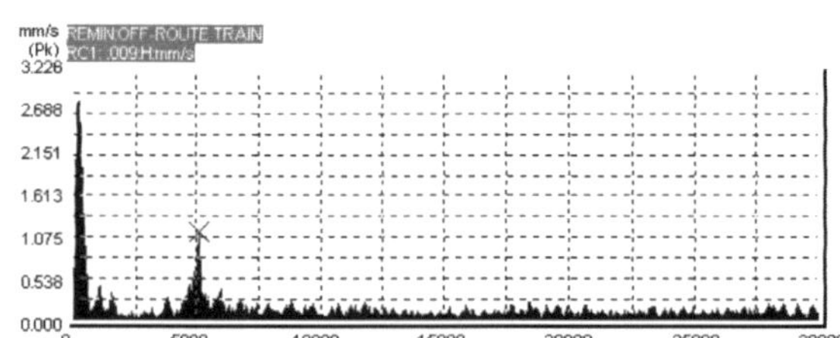

Figure 7. The amplitude of the vibrations in the speed range in radial point 9 in the case of RC1

3.2.2 Vibration acceleration measurements. To obtain accurate recordings in the field of vibration acceleration, the function of the transducer is provided by an accelerometer (piroelectric captor) fixed by gluing to the cycloid reducer measuring points.

The electrical signals are transmitted to an acquisition board that performs analogue-to-digital conversion including the sampling and coding of the analogue signals provided by the vibration captor.

The requirement for sampling is not to lose or distort the information contained in the analogous form of the signal. Therefore, in practice, the sampling theorem and the conditions regarding the choice of the number of static independent samples is respected. The recordings were made with a Sampling Frequency of 180284 Hz and with an acquisition time of 5.54 μs. From the Acquisition Plate, the values are taken in binary system by the computer and stored in the .txt file. It is used the computer and MatLab's specialized software to perform analysis functions: digital analysis in frequency or amplitude, spectral comparison, analysis and mediation in time. It is obtained a spectral vibration analyse using FFT algorithm, figure 8 and 9.

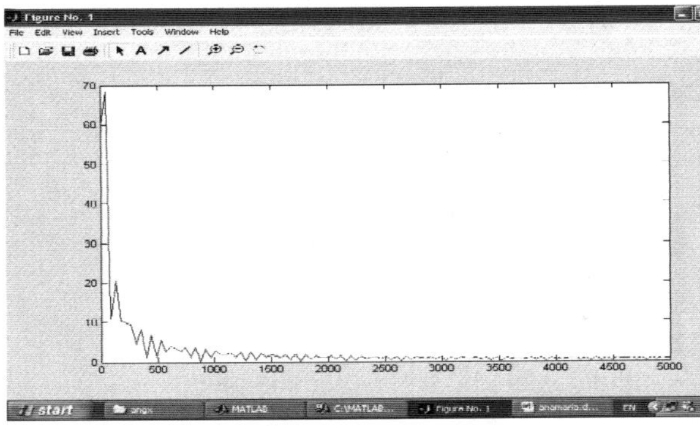

Figure 8. Initial RC recording in the range 0÷5KHz

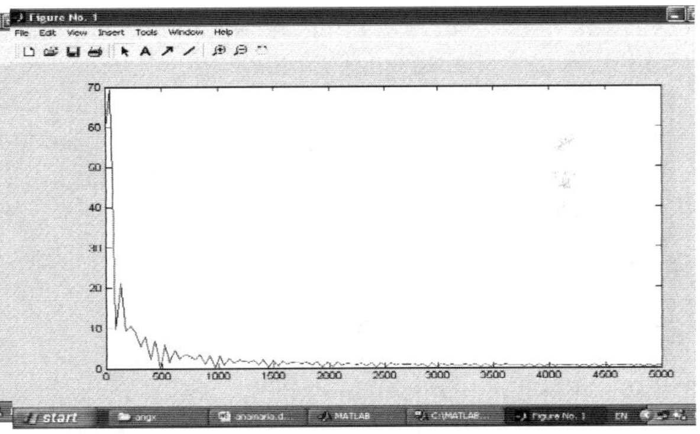

Figure 9. Final RC1 recording in the range 0÷5KHz

To increase the resolution in spectral analysis, digital processing allowed the use of a window-type function, the use of spectrums zooming effect around the significant frequency, the division of spectra obtained in areas of interest in relation to the source, and the comparison of spectra, figure 10.

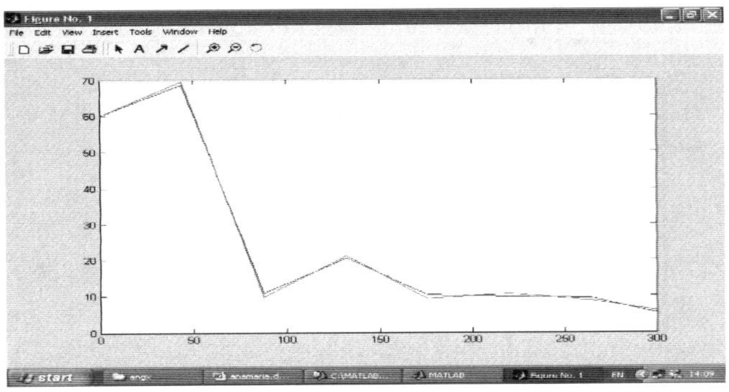

Figure 10. Comparison by overlap of RC and RC1 and "Zoom" in the range 0÷300Hz

Appreciation and highlighting of changes and trends are favoured by comparing frequency spectra. The readings highlighted the coincidence of the frequency peaks with the harmonics of the fundamental frequency and with the harmonics of the gearing frequency, the maximum amplitude being located in the low frequencies and corresponding to the frequency $4f_1$ = 44Hz, being below

IOP Conf. Series: Materials Science and Engineering **749** (2020) 012005 doi:10.1088/1757-899X/749/1/012005

70dB. The next peak was identified at $12f_1 = 132Hz$ but is much lower, just over 20dB. When reading spectrogram, it can be concluded that the rest of the frequencies have amplitude below 10dB.

The results of the theoretical analysis are in good agreement with the experimental results, allowing the understanding the physical phenomenon and the adoption of qualitative measures to reduce the noise and vibrations.

4. Conclusions

Difficulties in modelling the structure properties or estimating the dynamic loads of the cycloid gears during normal operation impose the need to measure the noise and vibrations of the gears in the initial stages of research and design as well as in operation.

The different deviations from the reference profile of the contact surfaces determine different levels of mechanical and acoustic vibrations, thus making assessments of the influence of deviations of the cycloid profile of the satellite wheels on the dynamic behaviour of the cycloid reducers and comparing the results with the requirements of the industry standards.

Measurements in the deviation and vibration speed range recorded significant peaks at frequencies in the vicinity of the 5000 Hz area corresponding to the bearings operating frequencies. They also pointed to low frequency bands.

Acceleration measurements and computerized data processing have shown that the peaks correspond to the harmonics of the fundamental frequency with the maximum values in the low frequency range characteristic of the deviation in alignment, shaft inclination.

An uneven deviation of the profile of the cycloid satellite wheel causes the axle-box to lose its co-axiality with the output shaft. The effect in the vibro-acoustic behaviour is the presence of the peak frequencies in the low frequency band.

References

[1] Banica M 2006 *Optimizarea Dinamicii Angrenajelor (Optimization of Gear Dynamics)* Publisher Risoprint Cluj-Napoca ISBN 978-973-751-385-4

[2] Darabont A, Iorga I and Ciodaru M 1983 *Masurarea zgomotului si a vibratiilor in tehnica* Bucuresti Ed. Tehnica

[3] Dascalescu A 2005 *Contribution to cinematic and dynamic study of planetary gearing with cycloid teeth and bolts* PhD.Thesis

[4] Gafitanu M, Cretu S and Dragan B 1989 *Diagnosticarea vibroacustica a masinilor si utilaje* Bucuresti Ed. Tehnica

[5] Gafitanu M and Focsa V 1980 *Vibratii si zgomote* Ed. Junimea Iasi

IOP Conf. Series: Materials Science and Engineering **749** (2020) 012006 doi:10.1088/1757-899X/749/1/012006

Applying Taguchi method to control the thermal expansion effect on machining process of aluminium alloy extruded profile

S A Moldovan and M Cosma[a]

Technical University of Cluj-Napoca, Faculty of Engineering, Department of Engineering and Technology Management, Baia Mare, Romania

E-mail: [a] marius.cosma@cunbm.utcluj.ro

Abstract. In this paper we present a technological problem encountered in the machining accuracy of the parts for aerospace, made of aluminium alloy extruded profile with 7 meters length. Those parts have very tight tolerances and on milling process appear deviations due to thermal expansion effect, that influence the repeatability of machining processes. Through several tests and recording all values during the milling process, was defined through Taguchi method the main factors which have impact on the machined parts.

1. Introduction

The application methodology for this research is to analyse and control the thermal expansion influence in the machining process of the parts for aerospace industry made of aluminium alloy extruded profiles with 7000 mm length (figure 1).

Figure 1. Aerospace part

Machining process for those parts are define thorough next steps:

- Drill the middle top holes with the diameter 19.7 mm;
- Drill the sides' top holes with the diameter 7.3 mm;
- Milling the pocket's sides from front view.

2. Defining the parameters which influence the thermal expansion

The environment temperature has a major role to define the parameters which have impact on the thermal expansion.

The first parameter which influence the thermal expansion on the extrusion is the machine frame temperature. The temperature was track in different days and different hours of the day to can have a bigger range of results, the record values are: 21°C, 22°C and 23°C.

The second parameter is the extrusion temperature which the same is tracked in different days and different hours of the day, the record values are: 20°C, 22°C and 25°C.

For this experimental research we used the Taguchi method, selecting an orthogonal matrix, on Table 1. The total number of experiments made on this research are equal with 9.

Table 1. Taguchi orthogonal matrix

Experiment number	Frame temperature (°C)	Extrusion temperature (°C)
1	21	20
2	21	22
3	21	25
4	22	20
5	22	22
6	22	25
7	23	20
8	23	22
9	23	25

3. Experimental procedure

The experimental configuration consists through using [5]:

- CNC machine Handtmann PBZ NT 1000;
- extruded profile from aluminium alloys;
- digital control thermometer;
- sensors installed on machine frame to can record the frame temperature;
- CMM (coordinate measuring machine).

3.1. Handtmann PBZ NT 1000 machine

This 5 axis machine is used in special for the profile materials up to 10000 mm length, having a flexible clamping system allowing to machine different profile cross sections (figure 2) [3].

Figure 2. Handtmann PBZ NT 1000 clamping system

3.2. Temperature measurement

The measurement equipment used to define and verify the temperature fluctuation are:

- Digital thermometer used for reading, monitoring and recording the raw material fluctuations (figure 3);
- Sensors used for frame temperature reading, mounted inside of the CNC machine.

Figure 3. Digital thermometer

3.3. Machined parts control

The machining experimental results are made by CMM (coordinate measuring machine), placed in a room with temperature control environment.

The CMM have orthogonal axes to each other in a typical three-dimensional coordinate system. Each axis has a scale system that indicates the location of that axis. The machine will read the input from the touch probe, as directed by the operator or programmer. The machine then uses the X, Y, Z coordinates of each of these points to determine size and position with micrometre precision typically (figure 4) [4].

Figure 4. CMM machine

IOP Conf. Series: Materials Science and Engineering **749** (2020) 012006 doi:10.1088/1757-899X/749/1/012006

4. Taguchi analysed

4.1. Experiment preparation
The Taguchi method is efficient because allows quickly to achieve 70-90% from desired optimizations with minimum costs. The rest of 10-30% could be achieve using one or two complementary experiments limited on 2-4 parameters, considered to be most important [1].

The principles with need to be respected to choose the factors are:

- Identifying the factors which influence assure reproducibility of results.
- Factors quality needs to be checked.
- Factors which are truly independents.
- The values of the factors level which needs to be checked.

4.2. Experimental planning
The aim of this experimental research is to realize stability for this technological process.
From big amount of parameters which influence have been chosen:

- Frame temperature Tb - measured in [°C];
- Extrusion temperature Tp - measured in [°C].

Based on the customer requirements the geometrical deviations is ±0,2 mm. In Table 2 are presented the factors and the level on which is working.

Table 2. Factors and workings levels

Factors	Minimum	Maximum
Frame temperature (°C)	21	23
Raw material temperature (°C)	20	25

The research is to create a plan which all possible combinations (table 3).

Table 3. Taguchi orthogonal matrix

Experiment number	Frame temperature (°C)	Extrusion temperature (°C)	Maximum deviations obtain (mm)
1	21	20	-0.095
2	21	22	0.264
3	21	25	0.715
4	22	20	-0.147
5	22	22	0.158
6	22	25	0.622
7	23	20	-0.258
8	23	22	0.090
9	23	25	0.563

The average deviations based on the defined parameters are shown in figure 5.

Using the incoming dates was created also the residual diagram for the distribution probability, histogram, fits numbers and measurement order, 4 in 1 diagram (figure 6).

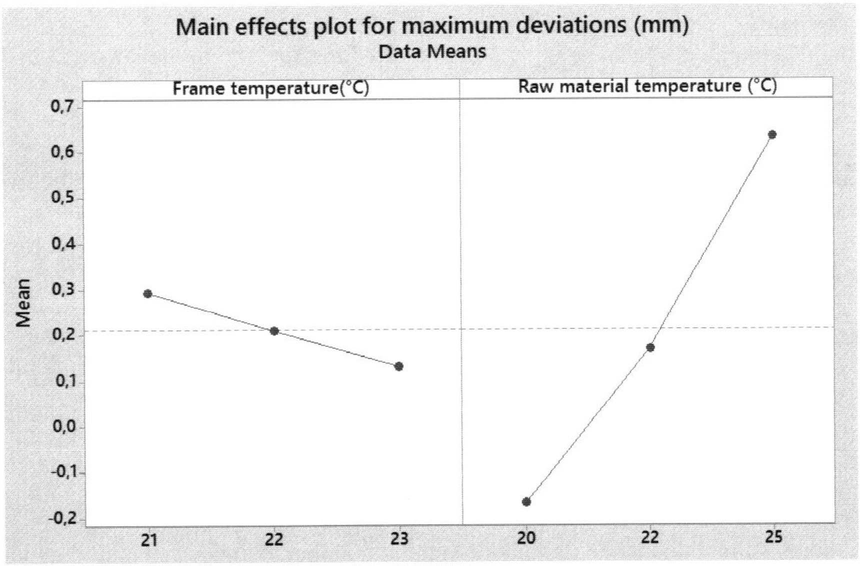

Figure 5. Average deviations based on the defined parameters

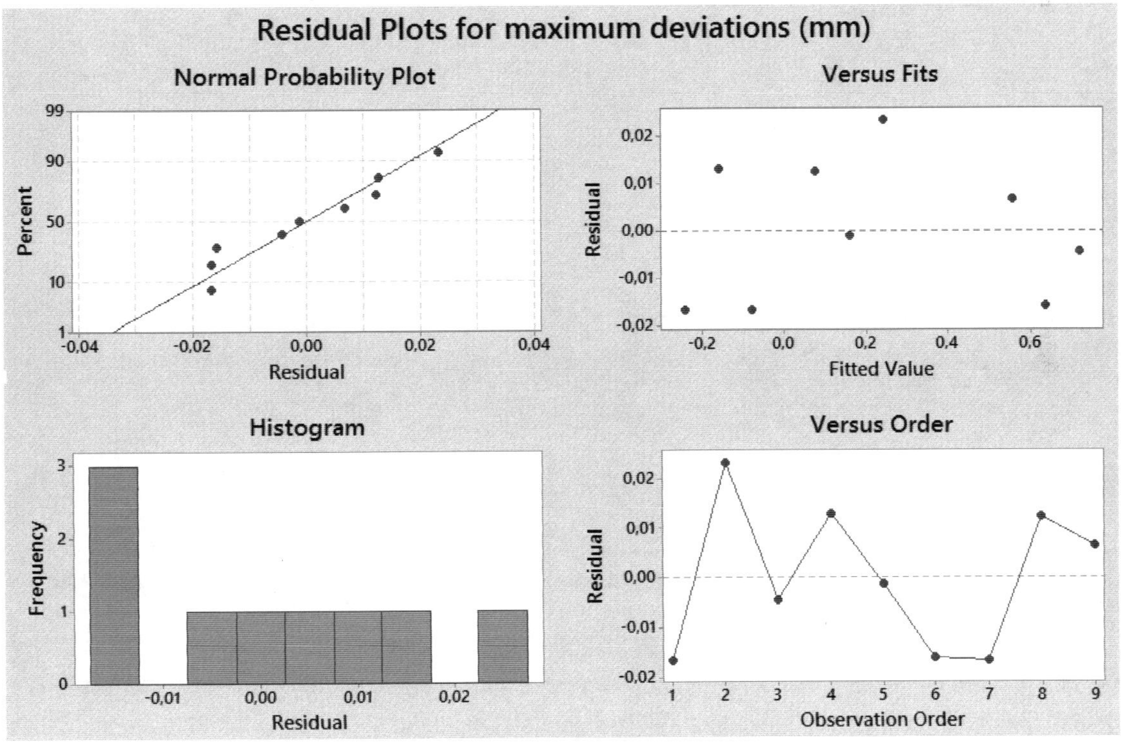

Figure 6. Residual diagram

5. Conclusions

It is visible that the extrusion temperature has the biggest impact on the holes positioning deviations. How long the extrusion temperature is close to the normal temperature (20°C) the deviation is smaller.

Experimental research brings a solution to can track and correct the deviations due to the thermal effect. This research methodology improves the machining process control [2].

IOP Conf. Series: Materials Science and Engineering **749** (2020) 012006 doi:10.1088/1757-899X/749/1/012006

6. Acknowledgments

The experimental research conducted with the support of the machining department of SC Universal Alloy Corporation Europe SRL, Dumbrăvița (Research contract Nr.3657/09.02.2016) [6].

References

[1] Taguchi G 1991 *Taguchi Methods: Research and Development* Quality Engineering Series vol. 1 ASI Press Dearborn MI

[2] Tîțu M, Oprean C and Boroiu A 2011 *Cercetarea experimentală aplicată în creşterea calităţii produselor şi serviciilor* Editura AGIR Bucureşti

[3] *** *Training Guide PBZ NT Ver.* 2010

[4] *** *Metris Bridge CMM User Guide* 2016

[5] ***https://www.handtmann.de/en/machining-centres/products/pbz-profile-machining-centres/pbz-nt/ 2016

[6] *** http://www.universalalloy.com/Manufacturing-Capabilities.html 2016

IOP Conf. Series: Materials Science and Engineering **749** (2020) 012007 doi:10.1088/1757-899X/749/1/012007

Theoretical investigation of the tapered thread joint surface contact pressure in the dependence on the profile and the geometric parameters of the threading turning tool

O R Onysko[a], V B Kopey and V G Panchuk

Ivano-Frankivsk National Technical University of Oil and Gas, Department of Computerized Machine-Building, Ivano-Frankivsk, Ukraine

E-mail: [a] onysko.oleg@gmail.com

Abstract. The tapered screw joints are widely used in the formation of the drill string. They consist of two parts - a box and a pin. During the process of screwing box and a pin the contact pressure between their thread surfaces arises, the magnitude of which affects the intensity of wear on these surfaces. For the first time, the authors offer in this work to change the contact area of the thread face by the change in the shape of the cutting edge of the threading turning tool. Previously, in the works of these authors themselves, it was offered to change the shape of the cutting edge exclusively to increase the tightness of the joint. In other works, the justification for changing the geometric parameters of such cutters was offered in order to obtain the possibility of the thread manufacturing from materials with a strength limit of more than 1300 MPa. In this paper an analysis of the influence of the geometric parameters size of the cutter on the value of the contact pressure between the thread surfaces of the box and the pin is carried out. Among above geometric parameters the work height is the most important. As a result of theoretical studies, a certain functional dependence of the contact pressure in the tapered thread joint from the shape of the cutter edge and the magnitude of its work height is obtained.

1. Relevance of the problem

Nowadays tapered threads are very widely used for connection of pipes of oil-and-gas variety in general and for connection of drill pipes in particular, as well as pump-compressor pipes. The quality of these joints is determined by the following parameters: screw-in, tightness and mechanical strength. The last parameter largely depends on the contact pressures that arise between the thread surfaces of the two parts of the joint - a box and a pin. The frictional force between the thread surfaces increases in the screwing up process according to the pressure increasing on these surfaces. The magnitude of the deformation of these surfaces at the moment of their complete screwing up increases too. One and the other leads to the loss of the initial shape of the profile and as a consequence to a decrease in the functional suitability of the joint i.e., reduction of mechanical strength, screwing-ability and tightness. Therefore, the reduction of the contact pressures in the process of screwing those types of pipes between themselves and after their full screwing-up is a very topical issue. Therefore, in the opinion of the authors, it is necessary to investigate the functional effect of certain parameters of the tapered thread on the value of contact pressures. Since the manufacture of the tapered thread of oil and gas pipe assortment occurs with the use of turning tools, an important part of the research should be to obtain functional relationships between the shape of the cutting edge profile of the tool and the contact pressures value on the thread surfaces.

2. Theoretical studies of contact pressures in tapered thread joint in the oil and gas pipes

The contact pressure between the drill string tool joint thread surfaces of the box and the pin is deeply studied at the theoretical level in the work [1]. In this paper, the value of contact pressure is determined for two fundamentally different periods of the joint process. The first period - just when the pin is installed into the box, and the second period when there is a process of screwing the pin into the box already. Figure 1 shows the moment of the installation of the pin in the box and Figure 2 illustrates the layout of the connection parts in the process of the screwing up the pin into the box. On both schemes signs, which correspond to the standard [2], are used:

d_1 - the largest outer diameter of the pin thread on the large base of the taper;

d_3 - the largest outer diameter of the pin thread on a small base of taper.

In addition, in Fig, 1, 2 the following signs are applied:

d_0 - the largest outer diameter of the pin in the place of its contact with the box at the moment before start to screw;

L - the length of the tapered thread of the pin.

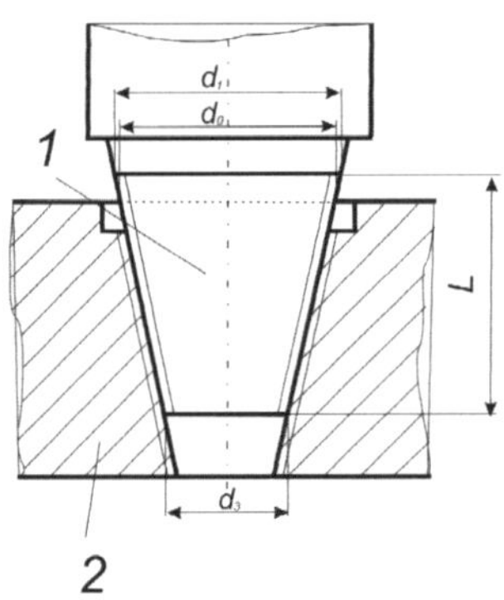

Figure 1. Scheme of the location of the parts of the threaded connection at the time of the pin installation into the box. The numbers indicate: 1-pin, 2-box.

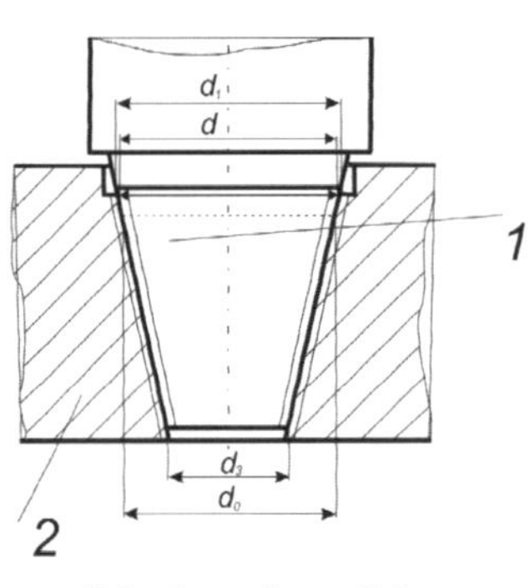

Figure 2. Scheme of the location of the parts of the threaded connection during the screwdriving of the pin into the box. The numbers indicate: 1-pin, 2-box

Figure 3 shows the scheme of the mutual placement of the surfaces of the pin(blue) and the box (black) at the moment the pin is set on the box (position I) and in the process of the screwing the pin into the box (position II). Signs used in Figure 3 are conformed to standard [2]:

h_1 - thread height truncated, P - pitch, φ- taper angle, a - crest flat width, $\alpha/2$ - half profile angle.
In addition, the following notations are applied in Figure 3: x - the value of the mutual displacement of the pin's thread surface onto the box thread surface at the certain moment of screwing, P_1 - the pitch is defined between the larger flanks.

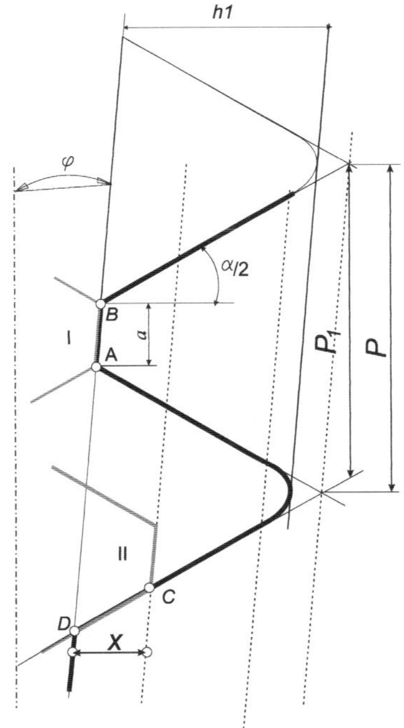

Figure 3. Scheme of the mutual placement of the surfaces of the pin (blue) and the box (black): I - at the moment the pin is set on the box; II- in the process of the screwing the pin into the box

Formula 1 according to the data [1] shows the functional dependence of the area of the mutual contact of the thread surface of the pin and the box in the process of screwing them. This screwing has a certain number of revolutions m corresponding to the displacement of the point C (see figure 3) along the larger flank of the pin in the direction from point D. So distance X depends on number of revolution m.

$$F_k = \frac{\pi m}{4\cos\alpha}\left[\left(d_1 - 2h + KP_1 \cdot m\right)^2 - \left(d_1 - KL\right)^2\right] (\text{mm}^2) \quad (1)$$

where: Fk - the area of mutual contact between the pin and the box in the process of screwing them on the value of X, corresponding to a certain number of revolutions of the pin m;
h - thread height truncated (mm);
d_1 - the largest outer diameter of the pin thread on the large base of the taper;
P_1 - the pitch is defined between the larger flanks
$K = 2tg\varphi$ - taper of thread.
L - the length of the tapered thread.

According to standard [2], the length of the tapered thread of the pin is determined by the equation:

$$L = l_н - 12{,}7 \text{ (mm)}$$

where l_H - the length of the tapered part of the pin.

Formula 2 according to the data [1] shows the functional dependence of the area of mutual contact between the crests of the box and the pin on: parameter a - the crest flat width, the tape angle f and the length of the screw line on the outer surface of the tapered thread *Lb*:

$$F_b = \frac{aL_b}{\cos\phi}\ (\text{mm}^2) \tag{2}$$

where: a - crest flat width;

φ - taper angle;

L_b - length of the screw line on the outer surface of the tapered thread *Lb*. This parameter is determined by formula 3:

$$L_b = \frac{\pi}{2PK}\left[(d_1 - 2h)^2 - (d_1 - KL)^2\right]\ (\text{mm}) \tag{3}$$

Formula 2 shows the calculation of the area of full contact between crests of the pin and the box, that is, the maximum possible area. In the case of point contact, that is, in the vicinity of points *B* or *A* - the area will in theory approach zero.

Figure 4 shows the scheme of the moment of setting the pin into the box (the pin is highlighted in dark gray), but the contact is not on their crests, but on their longer flanks. For a better illustration of this moment, the diagram shows the points *A* and *F* located on a line parallel to the axis of the thread 1. This line is the direction which the pin is setting down into the box along. Unlike the variant of contact I (see figure 3), in this case point *B* on the crest of the pin corresponds to the point *F*, that is, the crest of the pin moved down past the crest of the box and touch to the flank of the box 5. This corresponds to the initial distance X_0 between the crests of the pin and the box.

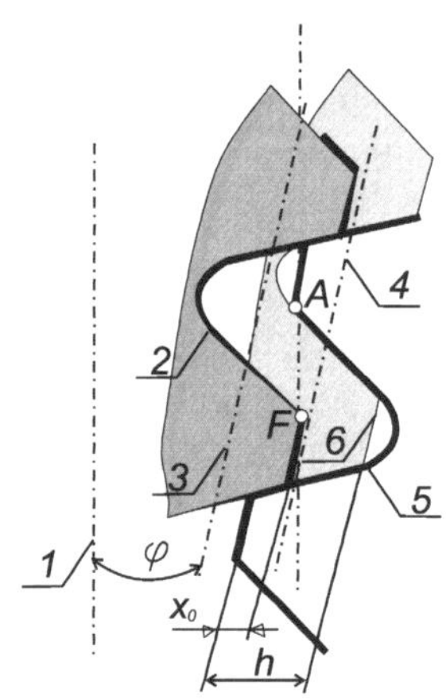

Figure 4. Scheme of the moment of installing pin into the box by use the surface of the thread flanks. Numbers mean: 1- axis of the tapered thread joint, 2- profile of the pin thread, 3- pitch diameter of the pin thread, 4- pitch diameter of the box thread, 5 - profile of the box thread, 6 - profile of the box thread at the moment of the full screwing

Formula 4 in accordance to [1] describes the area of contact between the flank of the pin and the box at the moment of the installation of the pin into the box (see figure 4):

$$F_c = \frac{\pi(P - 2a)}{4P\cos\alpha}\left[\left(d_1 - 2h + K\left(1 - \frac{K}{2}tg\alpha\right)(P - 2a)\right)^2 - \left(d_1 - KL\right)^2\right]\ (\text{mm}^2) \tag{4}$$

The screw-in process begins immediately after the moment the pin is installed into the box and ends when the pitch diameter of the pin coincides with the pitch diameter 4 of the box, and the crest of the pin moves to the value of $(h\text{-}X_0)$. This position of the pin in Figure 4 is shown in light gray and is indicated by the figure 6.

3. Theoretical studies of the functional dependence of the tightness of the tapered thread joint on the value of the working height

The paper [3] presents theoretical and virtual experimental studies aimed at obtaining the dependence of the velocity of the washing liquid moving through the screw technological duct between the thread surfaces of the pin and the box from the geometric parameters of this duct- its height and width. In figure 5 for clarity the cross section of this duct is painted black. Actually it is a gap between the pin and the box thread surfaces and its parameters are: a - crest flat width and $(h_1\text{-}h)$ - the difference between the work height and the thread height truncated of the thread profile. All three parameters according to [2] are optional for the thread and are intended for the design of a cutting tool.

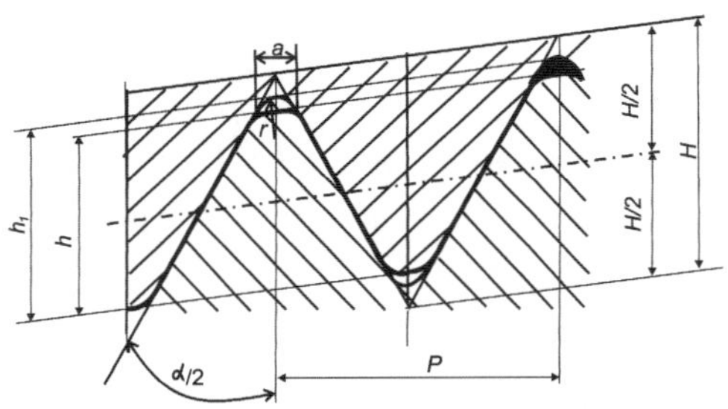

Figure 5. Scheme of the tool joint tapered thread profile according to [2]: P - thread pitch, H - height of fundamental triangle, h1 - work height, h - thread height truncated, a - crest flat width, H/2 - half height of fundamental triangle, r - root flat corner radius, α/2 - half profile angle

Investigations [3] show that with a decrease in the height of the gap $(h_1\text{-}h)$ in the screw channel between the box and the pin from 0,450 mm to 0,134 mm at an input pressure of 10 MPa, the flow rate of the drilling mud decreases from 30 - 55 m / s to 0,56 -1.48 m / s.

4. Theoretical researches of technological possibility of the manufacture of the tapered thread especially designed for low permeability of the clay solution inside the drill pipe tool joint

Studies [4, 5] have been carried out with the aim of identifying the possibility of using the threading cutter tool with negative values of the back rake angle in the range of -3 ° to -5 °. Such a range of the back rake angle value is necessary in case of necessity to ensure an increase in the tool life by the conditions:

- when the requirement for the strength of the material from which the drill pipe thread connection is manufactured increases;
- when a significant reduction of the gap between the contact screw surfaces of the pin and the box is proposed, and this gap is necessary for the design of the cutting tool.

In the paper [4] it is offered:

- according to the technological requirements, to achieve a clearance gap $(h_1\text{-}h)$ by increasing the size h of the profile of the cutting edge of the cutter;
- according to the economic and technological requirements to apply the back rake angle to -5 ° without adjusting the profile angle of the cutting edge of the threading tool since the deviation of this profile angle do not exceed 20% of the tolerance of the value α / 2.

Such offers are positive:

- for use in tools for thread cutting in steel work piece, the ultimate tensile strength of which will exceed 1300 MPa;
- to achieve the technological gap *(h_1-h)*, which is reduced by 2-3 times compared with the standard.

The analysis of primary sources show that the value of the contact pressure between the parts of the pipe connection and the level of its tightness depend on the thread height truncated h, which can be adjusted by correction of the profile of the tool cutting edge in its corresponding part, as shown in Figure 6.

Among the parameters that affect the contact pressure are also the crest flat width a and the profile angle α (equation 1, 2, 3, 4).

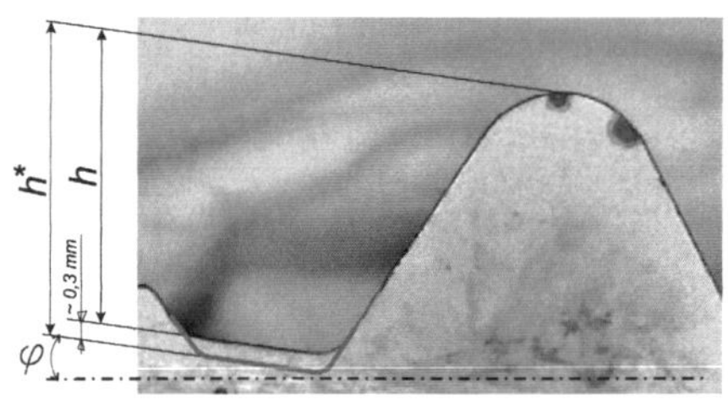

Figure 6. Illustration of changing the profile of the cutting edge of a full-profile threading lathe tool which ensures a reduction of the technological gap between the box and the pin by 0.3 mm

But studies [1] do not show the dependence of the contact pressure of the collected connection between oil and gas pipes or sucker rods on the value of h as the most influential parameter. According to studies [1,2,3,4] it cannot to conclude about direct effect of the change in the shape of the cutter edge of the tool on the values of contact pressures during the installation and screwdriving of the pin into the box and during the operation of the pipe joints.

The purpose of the paper is to determine the effect of changing the profile of the cutting edge and the geometric parameters of the threading lathe cutter tool on the values of the contact pressure that arises at the moments of installation of the pin into the box, during of the process of their screw-in and during operation of the their connection and of the sucker rod connections.

Problem setting: To achieve this goal should

- to conduct a virtual experiment to find out the changes in the contact pressures, stresses and fatigue safety factor (FOS) values that arise during the operation of the pin-box connection of drill pipes and sucker rods, depending on the change in the thread height truncated h;
- to obtain functional dependences of the change of the contact pressures that arise during the installation of the pin into the box and their screw-on by changes in the profile of the cutting edge and certain geometric parameters of the threading lathe tool.

5. Presenting main material. Investigation of the contact pressure in the connection of ZN-80 dependence on the value of the thread height truncated h by means of finite element simulation

Consider the drilling tool joint ZN-80 GOST 5286 with Z-66 GOST 28487 thread (2 3/8 REG API Spec. 7 equivalent). With the help of a finite element simulation, the effect of the major diameter of the pin thread on contact pressures at profile points 1 and 2 was studied (See figure 7). The value of the working height of the profile h is changed by reducing the major diameter of the pin thread, which is possible as a result of the wear during frequent make-up operations. The major diameter of the pin thread in the end face plane of the nipple d varied from 47.674 mm (h = 2.22 mm) to 46.074 mm (h = 1.47 mm). The minor diameter of the box thread is not change.

The material of the parts is steel. Material plasticity and friction are simulated. Two steps of the external axial tensile load were created: $F_{\min}$ =0 N and $F_{\max}$ =1 MN. The joint make-up was simulated

IOP Conf. Series: Materials Science and Engineering **749** (2020) 012007 doi:10.1088/1757-899X/749/1/012007

by axial deformation Δ=0.1 mm of the box shoulder. For the approximate calculation of the fatigue safety factor *D*, the dependence of Sines [6] with a material endurance limit of 207 MPa was used.

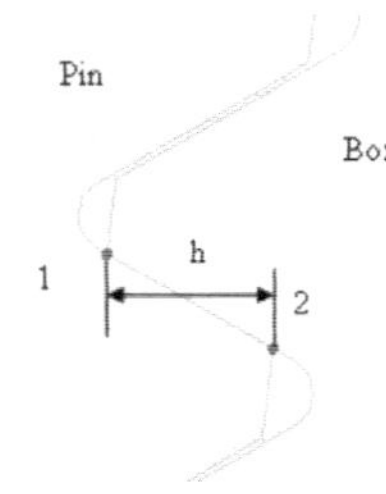

Figure 7. Points 1 and 2 for the calculation of the contact pressure

It is noticeable that a decrease in *d* affects the distribution of equivalent stresses in the connection (see figure 8).

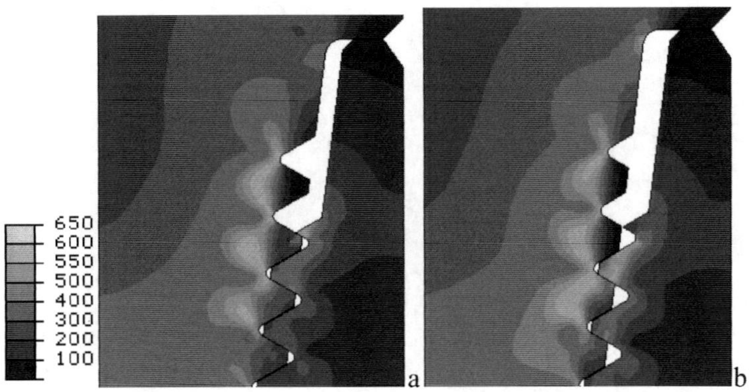

Figure 8. Von Mises stress distribution (MPa) in first threads of the drilling tool joint: d = 47.674 mm (a), d = 46.074 mm (b)

It is also noticeable the reduction in the fatigue strength of the pin with a decrease in the diameter *d* (figure 9).

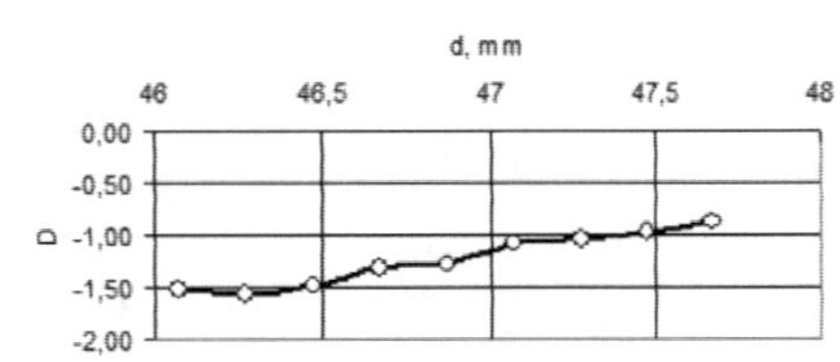

Figure 9. Fatigue safety factor D in the first root of the pin thread as a function of diameter d

Based on the results of the finite element analysis, a regression model of contact pressures *CP* at points 1 and 2 as a function of the thread turn number *N* and the major diameter *d* is found (5). Here a_i - coefficients, p_i - exponents.

$$CP = a_0 + a_1 N^{p_0} + a_2 N^{p_1} + a_3 N^{p_2} + a_4 d^{p_0} + a_5 d^{p_1} + a_6 d^{p_2} + a_7 Nd \tag{5}$$

Values of coefficients and exponents at point 1 (Figure 10): a = -3.77473564e+07, -4.91142995e+03, -2.55477895e+01, 1.00608216e+00, 2.80965184e+07, -4.07911283e+03, 3.66910355e+01, 5.55870891e+00; p = 0.1, 1.9, 2.9. Coefficient of determination $R^2 = 0.71$.

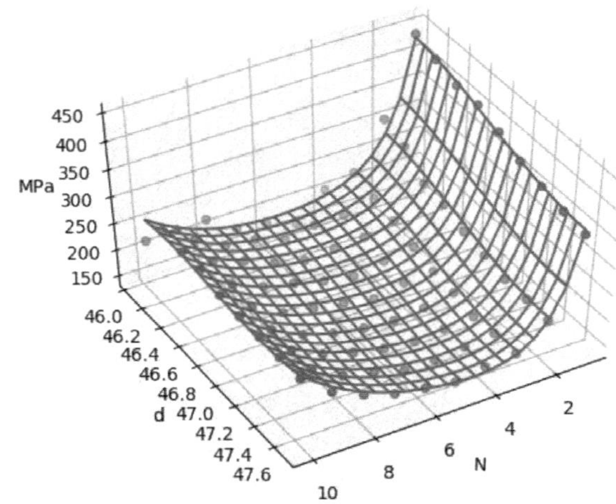

Figure 10. Values of contact pressures at points 1 as a function of N and d for the drilling tool joint

Values of coefficients and exponents at point 2 (see figure 11): a = -3.41589228e+07, -1.03089704e+03, 2.37829334e+01, -1.66910424e+00, 2.94843546e+06, -5.95458890e+04, 4.15737420e+02, 1.48764378e+01; p = 0.9, 1.9, 2.9; $R^2 = 0.97$.

From dependences it can be seen that below $d = 47.2$ mm contact pressures begin to rise sharply.

Consider the axisymmetric FE models of 19 mm GOST 13877 (3/4 in. API Spec 11B equivalent) sucker rod couplings. Material plasticity and friction are simulated. Two steps of the external axial tensile load were created, which form in the rod body the stress P_{min} = 0 MPa and P_{max} = 276 MPa. The joint make-up was simulated by axial deformation Δ=0.1 mm of the box shoulder.

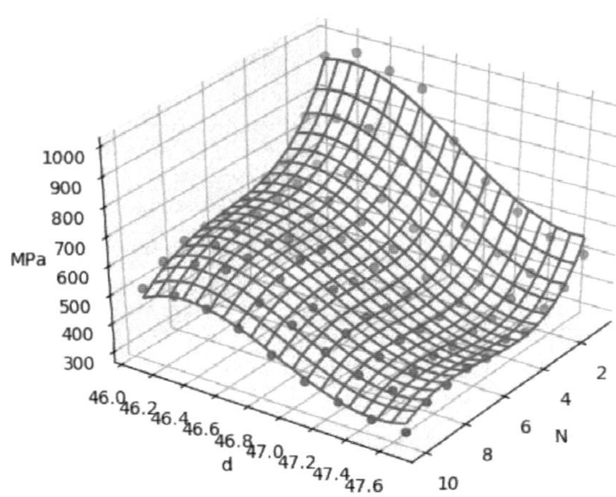

Figure 11. Values of contact pressures at points 2 as a function of N and d for the drilling tool joint

The major diameter d of the nipple thread varies from 26.624 mm (maximum allowable) to 25.624 mm. The minor diameter of the coupling thread is constant - 24.79 mm (maximum allowable). It is noticeable that a decrease in *d* affects the distribution of equivalent stresses in the connection (see figure 12).

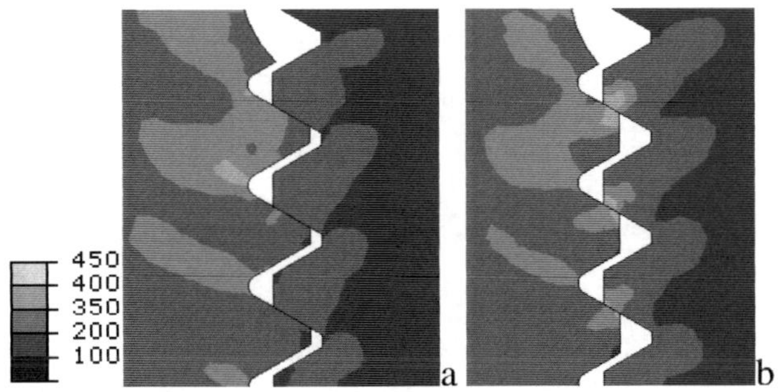

Figure 12. Von Mises stress distribution (MPa) in first threads of the sucker rod coupling: d = 26,624 mm (a), d = 25,624 mm (b)

In the case of a decrease in diameter *d*, the most noticeable decrease in the fatigue strength of the coupling in the zone of the last root of the coupling thread (see figure 13).

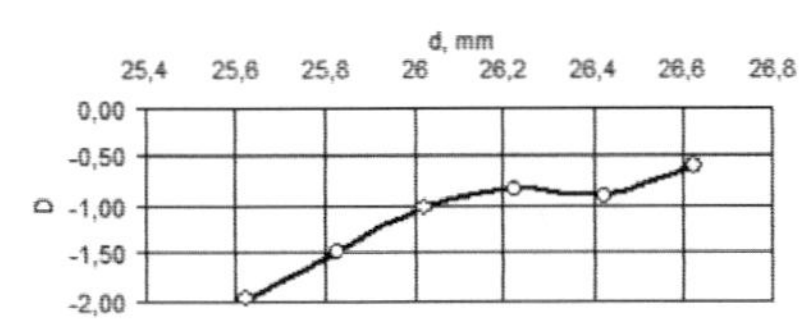

Figure 13. Fatigue safety factor D in the last root of the coupling thread as a function of diameter d

Based on the results of the finite element analysis, the coefficients of the regression model (1) are found. Values of coefficients and exponents at point 1 (see figure 14): a = 1.67302056e+07, 6.12225399e+02, 1.59368037e+01, 2.78069341e-01, -2.43145831e+06, 8.76366121e+04, -1.09226421e+03, -2.32280329e+01; $p = 0.9, 1.9, 2.9$; $R^2 = 0.97$.

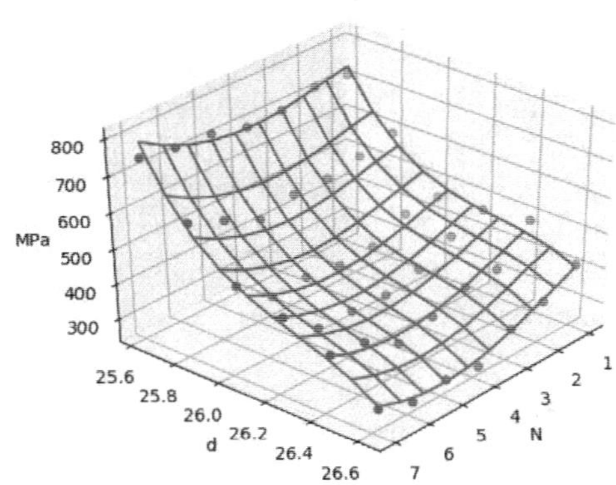

Figure 14. Values of contact pressures at points 1 as a function of N and d for the sucker rod connection

IOP Conf. Series: Materials Science and Engineering **749** (2020) 012007 doi:10.1088/1757-899X/749/1/012007

Values of coefficients and exponents at point 2 (see figure 15): a = -8.52515266e+07, -2.15706639e+03, -1.36373263e+02, 5.62253835e+01, 6.80490010e+07, -7.08061914e+04, 1.97004893e+04, 1.19405460e+01; p = 0.1, 1.9, 2.2; R^2 = 0.98.

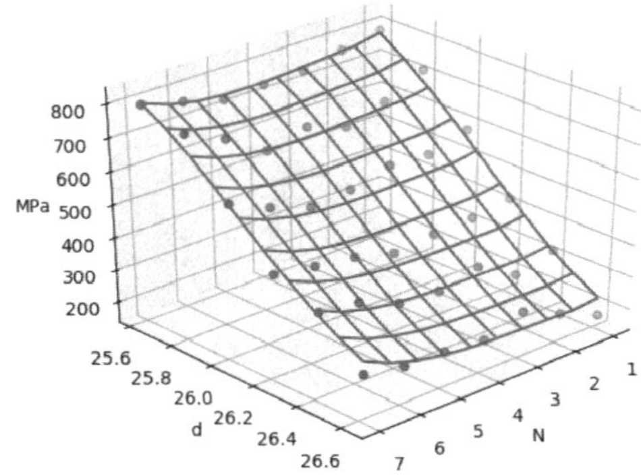

Figure 15. Values of contact pressures at points 2 as a function of N and d for the sucker rod connection

From dependences it can be seen that below d = 26.2 mm contact pressures begin to rise sharply. Functions $CP = f(N, d)$ can be used to justify the allowable value of the major diameter d of the pin thread of the drilling tool and sucker rod joints.

6. Investigation of the functional dependence of the contact pressure on the tool cutter edge profile in the drill string tool joint of the 2 3/8 Reg at the time of the pin installation into the box

Figure 16 shows a schematic on which thick black lines the profile of the tapered thread tool joint 2 3/8 Reg of the drill string are displayed in accordance to standard [2]. It corresponds to Form I [2]. The names of the parameters and their values are shown in table 1. The input parameters and their values for this study are reflected in blue:

- increase of 0,05 mm of the value of h (thread height truncated). This increase is executed five times starting from the nominal h = *2,262* and completing of the size h * = *2,876*. The drawing helps to understand that $h^*=h+5\cdot 0{,}05$ (mm). Because the absence of space, all intermediate dimensions, such as $h + 0{,}05$ (mm), are not shown in the figure, but are presented in Table 1;
- variable width of the crest, which corresponds to the magnitude of the increase in the value of the thread height truncated h; The width a = 1,016 mm according to [2] corresponds to the nominal value h = 2,626 mm. All other correspondences between changes h i a are received by means of geometric construction in the graphic editor. The functional relationship between these variables is shown in Table 2. The red color in Figure 16 shows the limit sizes according to the tolerances [2].

Table 1. Parameters of the profile of the tool joint tapered thread (Form I) according to the data [2]

Legend and parameter name	Parameter size, mm	Tolerance, mm
P pitch	1.143	0.285
H height of the fundamental triangle	1.143	0.285
h thread height truncated *	4.571	0.033
h_l work height *	0.286	0.593
$h/2$ half of thread height truncated*	1.143	0.285

* the sizes and deviations are output for the design of the tool and have an optional dimension for the thread

Table 2. Functional dependence of the value a (variable crest flat width) on the value h (variable thread height truncated)

variable *h mm*	2,626 par	2, 676	2,726	2,776	2,826	2,876
variable *a mm*	1,016 par	0,905	0,841	0,770	0.666	0.537

The results of research on the contact pressure that occurs at the moment of the pin installation into the box is shown in Table 3. Applying the formulas 2, 3, 4 and substituting them with variables *a* (column 7) and *h* (column 6), the corresponding values of the contact area were obtained for the installation of the pin in the flank side - F_b (column 2) and crests F_v (column 3) according to figure 3. In order to obtain the values of the contact pressure, the value of the drill string weight is selected - 10 tons. The calculation scheme of pressure from the load drill string is shown in figure 17.

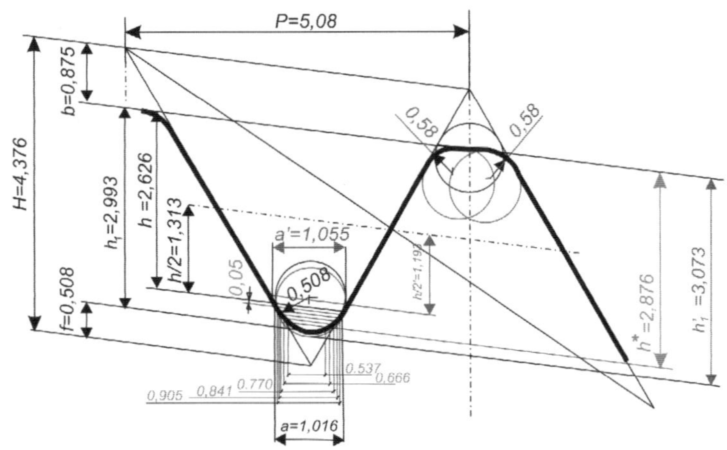

Figure 16. The profile of the tool joint thread profile of the 2 3/8 Reg according to the standard [2]

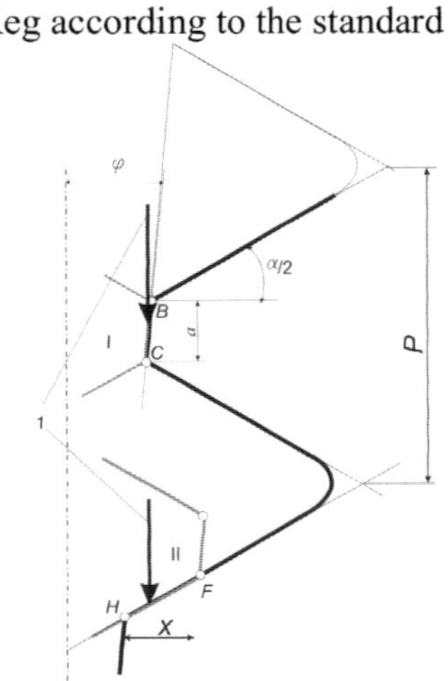

Figure 17. Scheme to calculate the pressure transmitted from the load drill string 1 to the surface of the thread at the time of installation of a pin into the box by the scheme I (crests) or in by the scheme II (flanks)

Using this scheme, the pressure p_b from the weight of the drill string, which is marked by vector *1* is transmitted to the platform of the crest with *BC* width can be calculated by the formula:

$$p_b = \frac{M \cdot 10000 \sin(\varphi)}{F_b} \text{ (MPa)}$$

where: M - weight of the drill sting, t;
F_b - area, which is calculated by the formula 2, mm^2.

According to the same scheme, the pressure p_c from the weight of the drill string, which is indicated by the vector *1*, is transmitted to the side of the flank the width of which is marked *HF* corresponds to the moment of installation of the pin into the box according to scheme II. This pressure can be calculated by the formula:

$$p_c = \frac{M \cdot 10000 \cos\left(\alpha/2\right)}{F_c} \text{ (MPa)}$$

where: F_c - area mm^2, which is calculated by the formula 4;
$\alpha/2$ - half profile angle.

The values of contact pressures at the installation of the flank sides are shown in column 4, and when mounted with crests - in column 5.

Table 3. Results of theoretical studies of contact pressure between the tapered thread surfaces of the tool joint 2 3/8 Reg at the time of the installation of the pin into the box. Load 10 tons

No.	contact area, mm^2		contact pressure, MPa		h, mm	a, mm
	F_c	F_b	between flanks p_c	between crests p_b		
1	*2*	*3*	4	5	6	7
1	649,25	1400,40	133,39	8,86	2,626	1,016
2	693,02	1233,60	124,96	10,05	2,676	0,905
3	714,85	1133,50	121,15	10,94	2,726	0,841
4	739,47	1026,00	117,11	12,09	2,776	0,770
5	778,63	877,32	111,22	14,14	2,826	0,666
6	828,68	699,23	104,51	17,74	2,876	0,537
7	655,22	1512,50	132,17	8,20	2,446	1,055
8			- 28,88	9,54		
9			- 21,7%	116%		

Row 1 of Table 3 corresponds to the standard values of h = 2,626 mm and a = 1,016 mm. Lines 2-6 correspond to the variable h and a according to the data in Table 2. Row 8 shows the difference between pressure values in row 6 (lowest value) and row 1 (highest value) in column 4, as well as between pressure values in row 6 (highest value) and row 7 (lowest value) in column 5. Row 9 shows the relative increase / decrease of pressure in the corresponding columns.

Figure 18 shows the graph where the extreme left point corresponds to the lower limit of tolerance to the size h. Judging by the geometric construction in Figure 16, Table 3 and the graph, the pressure remains virtually unchanged at h = 2,626 mm. The functional dependence of the rest of the array of points is close to the straight linear one, which can be expressed using the following equation:

$$p_c = 132 - 0{,}1(h - 2600), \text{ (MPa)}$$

where h - thread height truncated, mm

IOP Conf. Series: Materials Science and Engineering **749** (2020) 012007 doi:10.1088/1757-899X/749/1/012007

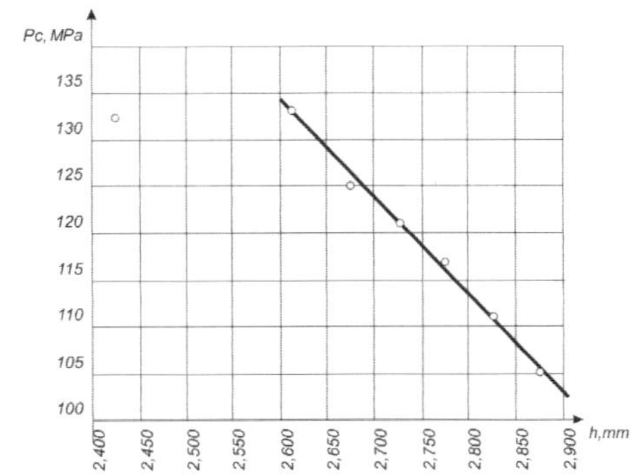

Figure 18. The diagram of the dependence the flank contact pressure of the box on thread height truncated h at the moment of the installation of a pin into the box. Tool joint 2 3/8 Reg. Weight of the drill string is 10 tons

Figure 19 shows the contact pressure dependence on the value of the thread height truncated *h* onto the crest of the box at the time when the pin is installed in it. The extreme left point *A* corresponds to the lower tolerance of the size *h*. Judging by the geometric drawing in Figure 16, Table 3 and the graph, the pressure will remain practically unchanged to the point with coordinate h = 2,626 mm. The functional dependence of the rest of the array of points besides the extreme right point *B* is close to the rectilinear. This dependence can be expressed using the following equation:

$$p_b = 8 + 0{,}0264(h - 2600), \text{ (MPa)}$$

where h - thread height truncated (10^{-3} mm).

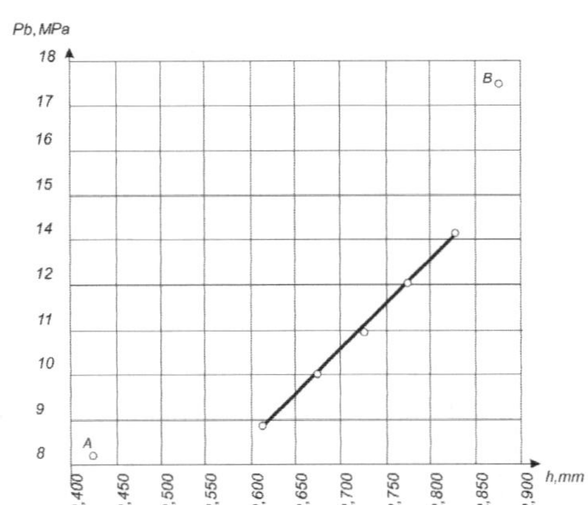

Figure 19. The diagram of the dependence the crest contact pressure of the box on thread height truncated h at the moment of the installation of a pin into the box. Tool joint 2 3/8 Reg. Weight of the drill string is 10 tons

Table 4. Dependence of the values of the contact area F_k and the contact pressure p_k of the thread flank surface on the value of the thread height truncated h under the weight of the drill string of 10 tons at the start of the screw-in process (m = 0.61). Tool joint 2 3/8 Reg. The thread height truncated h varies from the standard h = 2,626 mm (row 1) to the maximum (row 7) in increments of 0.05 mm. The separate row 0 (boundary) shows the value of area and pressure for the minimum permissible value h = 2,446 mm in accordance with standard [2].

No.	F_K, mm^2	P_k, MPa	h
0	6066,32	14,28	2,446
1	5899,58	14,68	2,626
2	5853,42	14,80	2,676
3	5807,33	14,91	2,726
4	5761,32	15,03	2,776
5	5715,37	15,15	2,826
6	5669,49	15,28	2,876
7	5623,69	15,40	2,926

Table 5. Dependence of the values of the contact area F_k and the contact pressure p_k of the thread flank surface on the value of the thread height truncated h under the weight of the drill string of 10 tons at the finish of the screw-in process (m = 3.86). Tool joint 2 3/8 Reg. The thread height truncated h varies from the standard h = 2,626 mm (row 1) to the maximum (row 7) in increments of 0.05 mm. The separate row 0 (boundary) shows the value of area and pressure for the minimum permissible value h = 2,446 mm in accordance with standard [2]

No.	F_K, mm^2	P_k, MPa	h
0	6066,32	14,28	2,446
1	5899,58	14,68	2,626
2	5853,42	14,80	2,676
3	5807,33	14,91	2,726
4	5761,32	15,03	2,776
5	5715,37	15,15	2,826
6	5669,49	15,28	2,876
7	5623,69	15,40	2,926

7. Conclusions

1. The contact pressure between the crests of the drill string tool joint thread surfaces at the time of installation of the pin into the box directly proportional depends on the value of h (thread height truncated), which is formed during the process of cutting the thread.

2.The contact pressure between the larger flanks of the drill string tool joint thread surfaces at the time of installation of the pin into the box inversely proportional depends on the value of the h (thread height truncated), which is formed in the process of cutting the thread.

IOP Conf. Series: Materials Science and Engineering **749** (2020) 012007 doi:10.1088/1757-899X/749/1/012007

3. The contact pressure between the larger flanks of the drill string tool joint thread surfaces the time of the process of screwing the pin into the box directly proportional depends on the value of the working height, which is formed in the process of threading.

4. In the screwed state of the drill tool joint and the sucker rod connection, contact pressures are in certain regression dependencies CP = f (N, d), where d is major diameter of the nipple and N is a thread turn number. Dependencies can be used to justify the permissible value of the major thread diameter of the nipple d. From the dependencies it is noticeable that below d = 47.2 mm for ZN-80 connection (d = 26.2 mm for 19 mm sucker rod connection) contact pressures sharply increase and FOS values sharply decrease.

References

[1] Semyn V Y 2005 *Sovremennye metody proektirovaniya rezbovyh soedinenyj trub neftegazovogo sortamenta dlya stroytelstva skvazhyn: dys. ... dokt. texn. nauk* 25.00.15, 05.02.13 Moskva 344 s

[2] Host P 50864-96 *Mezhhosudarstvennыi standart. Rezba konycheskaia zamkovaia dlia эlementov burylnыkh kolonn,* Profyl. Razmerы. Dopusky (s Yzmenenyem N 1 [Tekst] M.: «Standartynform» 2011 13 s

[3] Borushchak L, Borushchak S, Onysko O 2017 *Influence of the technological gap value of the tool-joint tapered thread on the drilling mud flow rate in its screw coupling* Ukrainian Journal of Mechanical Engineering and Materials Science 2017 Volume 3 No2. pp.24-31

[4] Panchuk V, Onysko O, Medvid I 2018 *Manufacturing technology of the oil and gas drill pipe connector with low-permeability level of the drilling mud in its screw part* Acta Technica Corviniensis Bulletin of Engineering Tom XI Fascicle 4 pp.46-53

[5] Fomin E V 2007 *Povyshenye stojkosty i tochnosti rezbovyx rezczov na osnove modelirovanyya processa rezbonarezaniya* dys.kand.texn.nauk 05.03.01 M 2007 206 s

[6] Sines G 1959 Behavior of metals under complex static and alternating stresses Metal Fatigue New York McGraw-Hill pp.145-69

IOP Conf. Series: Materials Science and Engineering **749** (2020) 012008 doi:10.1088/1757-899X/749/1/012008

Musculoskeletal multibody simulations for the optimal tribological design of human prostheses: the case of the ankle joint

A Ruggiero[1a], **R D'Amato**[2] **and N Ungureanu**[3]

[1] University of Salerno, Department of Industrial Engineering, Salerno, Italy
[2] University of Madrid (UPM), Departamento de Ingeniería Mecánica, Química y Diseño Industrial, Technical, Madrid, Spain
[3] Technical University of Cluj-Napoca, Faculty of Engineering, Department of Engineering and Technology Management, Baia Mare, Romania

E-mail: [a] aruggiero@unisa.it

Abstract. A thorough determination of the loading of the ankle joint is useful both for the optimal design of prostheses and for the preclinical testing in terms of tribological performances. In vivo measurements of joint forces are usually not easy in the in-vivo settings, then non-invasive in-silico methods should be considered. Nowadays resultant joint loads can be reliably estimated by using musculoskeletal modelling in an inverse dynamic approach, starting by motion data obtained in gait analysis laboratories for several human activities. The main goal of this study was to provide a set of dynamical loading curves obtained by the AnyBody Modelling SystemTM (AMS) computer software starting from ground reaction forces and kinematic data obtained by Vaughan et al. in the case of human normal gait. The model accounts for 70 Hill modelled muscles and the muscular recruitment strategy was choose as polynomial criteria. The results are presented in terms of Antero Posterior, Proximo Distal, Medio Lateral Forces and Ankle Eversion, Plantar Flexion, Axial moments, discussing their role on the synovial lubrication phenomena effect in the Total Ankle Arthroplasty (TAR) for the optimal prostheses structural and tribological design.

1. Introduction

In the last year's total joint arthroplasty (TJA) has become a common and a well-established surgical procedure in the case of severe arthritis especially regarding lower limb synovial joints [1]. Total Hip Replacements (THR) [2], Total Knee Replacement (TKR) [3] and Total Ankle Replacement (TAR) [4] are devoted to a substitution respectively of the hip, knee and ankle joint by using prostheses which unfortunately requires, in some cases, revisions and/or substitutions [5-8]. This problem is particularly felt in the case of THA since higher revision rate (about three times more for 100 patients) [9].

Among many others implants failure causes, nowadays particular interest is devoted to a correct tribological design of the implants [10, 11] in order to achieve more performant prostheses and to decrease, in this way, the rate of THA revision.

An optimized prostheses tribological design requires the choice of more and more performants materials in terms of stress, strain but also in wear resistance characteristics [12]; also the optimized geometrical design is necessary to favour the synovial lubrication phenomena especially in terms of full film and/or elasto-hydrodynamic lubrication inside the bio-bearing gap [13,14].

For achieving this, a detailed description of the acting loads on the prostheses and of the kinematical behaviour during the common human daily activities is required. [15]. This is not always simple to obtain in-vivo, since direct force/displacements measurements on joints is in most cases not feasible due to economical and ethical problems [16] and it could be referred only to the body characteristics of the patients under experimental activity. For this purpose the possibility to obtain predicted loads from in-silico simulation is a challenge [17] toward an optimized wear assessment tool necessary for the optimized tribological design of the joint.

This paper aims to give a contribute toward this direction focusing on the theoretical background in which the musculoskeletal multibody algorithms operate, and discussing the obtained results with the use of a musculoskeletal multibody model with a Vaughan gait type.

2. Methods

2.1. The musculoskeletal model

In this study, we used a musculoskeletal modelling software, AnyBody Modelling System™ AMS), to estimate force and moment components acting on the ankle joint during a level walking. The AnyBody Modelling System™ is software based on the human musculoskeletal modelling able to simulate the dynamics of human motion. This environment adopts the inverse dynamics approach and different algorithms allow selecting the appropriate recruitment strategies allowing a complete analysis of the load components acting on the different joints of the human body during a known human body movement. Following the inverse dynamic approach, for achieving the joint forces the kinematic data and ground reaction force must be furnish as input for the simulations. These data are usually obtained in the Gait Analysis laboratories by using special Motion Capture apparatus, which allow to measure by cameras the subject's gait kinematics, monitoring markers fixed in particular points of the body on the person's skin (Figure 1).

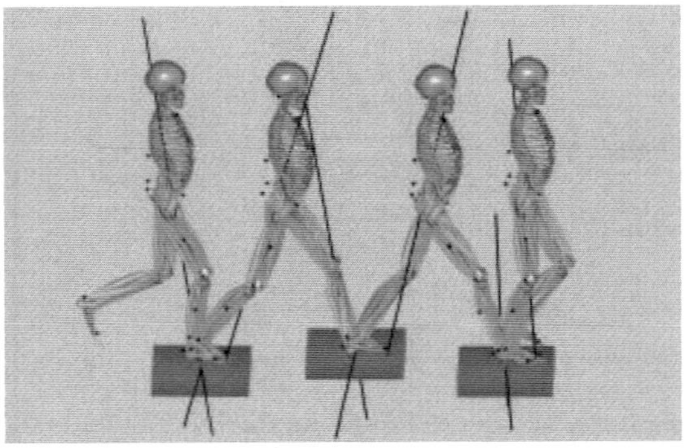

Figure 1. Four frames of gait model driven by kinematic data (The AnyBody System™)

In particular this setup is able to measure an individual gait pattern by collecting the kinematical data of the lower limbs and the pelvis through a walking cycle (gait). Figure 1 shows three frames of the gait model driven by kinematic data. The dark spheres are the skin markers, the black lines representing ground reaction forces. The positions with their first and second derivatives in time, together with knowledge of the ground reaction forces, after a data filtering, represent the software input to predict the net forces in the leg. In fact, the inverse dynamics is based on the knowledge of the motion and the external loads data to determine the unknown internal forces. Following this approach, then the calculation of the behaviour of each muscle force is made possible by solving a redundancy muscular problem. In fact the muscular system is a quite complex system and for each motion many different sets of muscle forces could be involved; chose of the appropriate set is made by the central nervous system (CNS) which instantly chooses one of them in order to produce the assigned kinematics. At moment the selection strategy is still not fully understand, however, the approach used by AnyBody Modelling System is well described in [18]; the software uses an algorithm to determine the activation of each muscle in order to replicate the function of the central nervous system. The list of the considered muscles in the model is reported in table 1.

Table 1. List of muscles included in the lower limb model

Muscle	**Group**	**Attachment**	
Soleus	Quadriceps	Shank	Foot
Gastrocnemius	Quadriceps	Thigh	Foot
Flexor Digitorum Longus	Quadriceps	Shank	Foot
Flexor Hallucis Longus	Quadriceps	Shank	Foot
Tibialis Posterior	Quadriceps	Shank	Foot
Peroneus Brevis	Quadriceps	Shank	Foot
Tibialis Anterior	Quadriceps	Shank	Foot
Extensor Digitorum Longus	Quadriceps	Shank	Foot
Extensor Hallucis Longus	Quadriceps	Shank	Foot
Vastus Lateralis	Quadriceps	Thigh	Shank
Vastus Medialis	Quadriceps	Thigh	Shank
Vastus Intermedius	Quadriceps	Thigh	Shank
Semitendinosus	Hamstrings	Pelvis	Shank
Semimembranosus	Hamstrings	Pelvis	Shank
Rectus Femoris	Hamstrings	Pelvis	Shank
Biceps Femoris Caput Breve	Hamstrings	Thigh	Shank
Sartorius	Hamstrings	Pelvis	Shank
Gracilis	Hamstrings	Pelvis	Shank
Iliopsoas	Hip muscles	Pelvis	Thigh
Gluteus Minimus 1	Hip muscles	Pelvis	Thigh
Gluteus Minimus 2	Hip muscles	Pelvis	Thigh
Gluteus Minimus 3	Hip muscles	Pelvis	Thigh
Gluteus Medius 1	Hip muscles	Pelvis	Thigh
Gluteus Medius 2	Hip muscles	Pelvis	Thigh
Gluteus Medius 3	Hip muscles	Pelvis	Thigh
Gluteus Maximus 1	Hip muscles	Pelvis	Shank
Gluteus Maximus 2	Hip muscles	Pelvis	Shank
Gluteus Maximus 3	Hip muscles	Pelvis	Thigh
Tensor Fasciae Latae	Hip muscles	Pelvis	Shank
Piriformis	Hip muscles	Pelvis	Thigh
Adductor Longus	Hip muscles	Pelvis	Thigh
Adductor Magnus 1	Hip muscles	Pelvis	Thigh
Adductor Magnus 2	Hip muscles	Pelvis	Thigh
Adductor Magnus 3	Hip muscles	Pelvis	Thigh
Quadratus Femoris	Hip muscles	Pelvis	Thigh
Abductor Brevis	Hip muscles	Pelvis	Thigh
Obturatorius Internus	Hip muscles	Pelvis	Thigh
Obturatorius Externus	Hip muscles	Pelvis	Thigh
Pictineus	Hip muscles	Pelvis	Thigh
Gemmelus Inferior	Hip muscles	Pelvis	Thigh
Gemmelus Superior	Hip muscles	Pelvis	Thigh

The used approach for solving the inverse dynamic problem accounting for the muscle recruitment strategy is based on an optimization problem.

Defining an objective function in the form:

$$\min G(f^{(M)}) \tag{1}$$

In which $f^{(M)}$ are the muscular forces for which

$$0 \leq f_i^{(M)} \leq N_i, i \in \{1, \dots, n^{(M)}\} \tag{2}$$

The (2) states the non-negativity constraints on the muscle forces and that muscle can only pull (not push). The upper limit of the i-muscle strength capability is then assumed to N_i.

Once defined the vector of the muscle forces and joint reactions in the form:

$$f = [f^{(M)T}\ f^{(R)T}]^T \tag{3}$$

The dynamic equilibrium equations can be obtained in the form:

$$Cf = d \tag{4}$$

where C is a coefficient matrix for the unknown forces/moments, while d is a vector of the known applied loads and inertia actions.

The most adopted objective function G forms, normalised for each muscle, are the polynomial criteria and the soft saturation criteria [19]:

$$G(f^{(M)}) = \sum_{i=1}^{n^{(M)}} \left(\frac{f_i^{(M)}}{N_i}\right)^p \tag{5}$$

$$G(f^{(M)}) = -\sum_{i=1}^{n^{(M)}} \sqrt{\left(1 - \frac{f_i^{(M)}}{N_i}\right)^p} \tag{6}$$

Both (5) and (6) contain a power variable p and a normalizing function for each muscle. In this study we used the approach (4) in which was settled p = 2.

Ankle joint forces were simulated by using a 18 degrees of freedom lower limb made of 7 rigid members, the pelvis and the thigh, the shank and the foot (for each leg). From a kinematical point of view, the hip joint was assumed in the form of a spherical joint while the knee as a revolute joint and the ankle trochlear joint. In this study we used a set of kinematical input data from Vaughan et al. [20]. Human main parameters here adopted were a weight of 64.9 Kg and a height of 1.75.

3. Results

The output of the model will be presented in the in terms of load components acting on the ankle joint during the gait. With reference to the figure 2, the calculated loads are: the anterior-posterior force (Fx), the proximo-distal force (Fy), the medial-lateral force (Fz), the ankle eversion moment (Mx), the axial moment (My) and the plantar flexion (Mz).

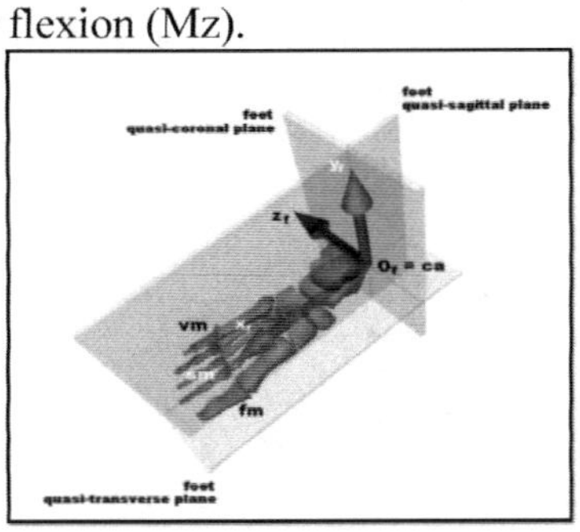

Figure 2. Reference system for ankle joint

IOP Conf. Series: Materials Science and Engineering **749** (2020) 012008 doi:10.1088/1757-899X/749/1/012008

In figures 3, 4, 5 and 6 will be reported the results of the simulations in terms of forces and moment components.

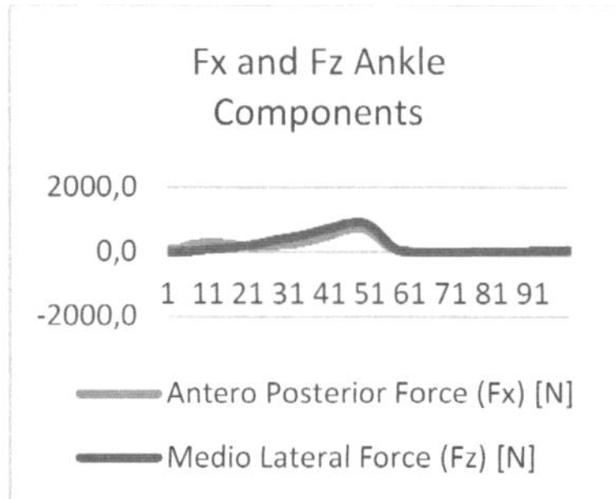

Figure 3. Antero Posterior and Medio Lateral ankle joint components during the gait

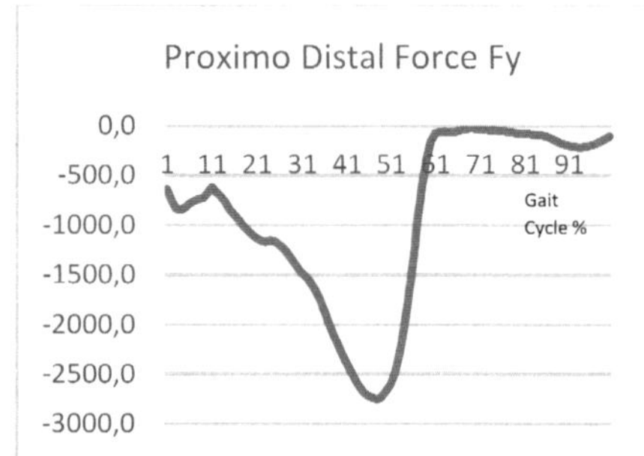

Figure 4. Proximo distal ankle joint force component during the gait.

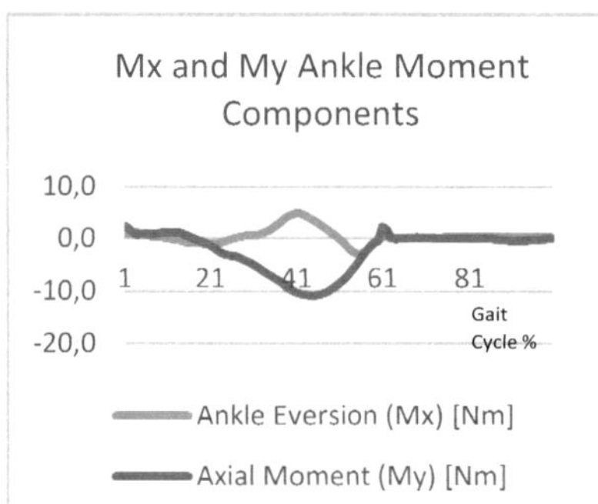

Figure 5. Ankle eversion and axial ankle moment components during the gait

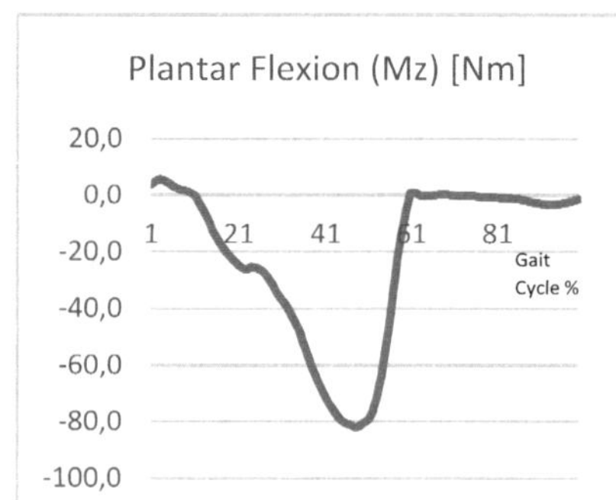

Figure 6. Plantar flexion ankle moment component during the gait

About the muscle recruitment, the simulations allowed to calculate all the forces exerted by all the muscles considered in the lower limb model (Table 2) during the gait. Figure 4 shows a schematic image of the model which highlights the predominance during the toe-off the four muscles just listed. In figure 8 is reported a complete activation during the gait of the muscles involved. From an analysis of the obtained results is possible to observe that, despite the muscles involved during the movement of walking are numerous, more than 60% of the total force exerted during the 50% of gait cycle (toe-off) is provided by four major muscles, which are the Soleus (in the back part of the lower leg), the Gastrocnemius (in the back part of the lower leg), Rectus Femoris (one of the four quadriceps muscles of the human body), and the Iliopsoas (combination of the psoas major and the iliacus at their inferior ends). These muscles are distinct in the abdomen, but usually indistinguishable in the thigh.

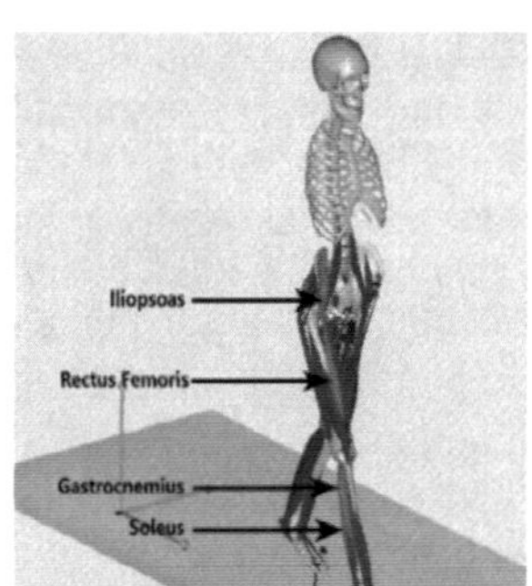

Figure 7. Main active muscles during the gait

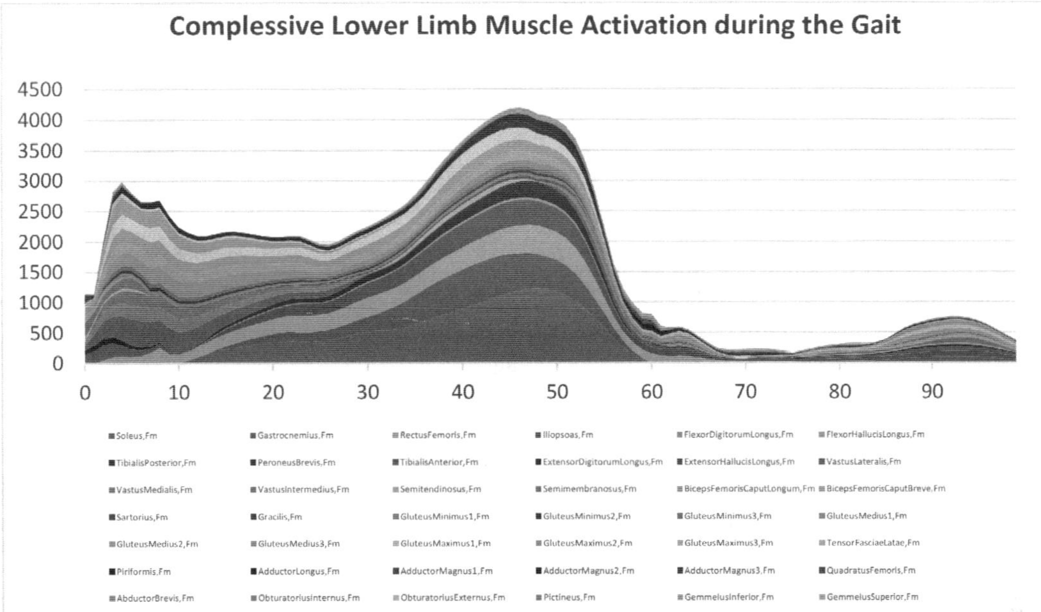

Figure 8. Muscles force activation during the gait

4. Discussion

It is well known that loading of the lower limb joints primarily depends on the physical activity (kinematical data) but they are also influenced by body weight (BW) but, in general, they individually differs greatly, even between subjects with the same BW. The simulations show maximum values of the ankle force and moment components in correspondence of about the 50% of gait cycle (toe-off phase) with a prevalence of the proximo distal force Fy with a value in modulus of 2750 N and of the plantar flexion moment Mz with a value of 82 Nm. The obtained behaviour of the loads, however, show good agreement with the others found in literature for example in [21]. The existing discrepancies however should be attributable to the fact that the considered model assumes limited degree of freedom for the joint with the foot considered a single segment. Decreasing the degrees of freedom in the model allow a reduction in the computational time by reducing the complexity of the calculations to predict the muscle and joint contact forces, but on the other hand, this causes approximations in the force calculation which could accumulate on the whole kinematical chain furnishing more consistent discrepancies in the simulation.

Of course, another cause of alterations of the calculated forces is introduced by the anthropometric differences between the human bodies even if the scaling procedure aims to reduce it. Regarding the muscles activation Figure 8 shows the force exerted by each of the 42 muscles implemented in the model. As can be observed despite the muscles involved during the movement of walking are numerous, more than 60% of the total force exerted during the toe-off (50% of gait cycle) is provided by the four major muscles.

From a TKA design point of view, the obtained results allow the optimized design of the prostheses both from a structural point of view both from a tribological one [22, 23]. In fact the detailed knowledge of the load acting on the joint permits the accurate finite element modelling of the joint [24] to analyse the stability of the implant contributing to improve its stability and structural performances. Moreover the knowledge of the variation of the loads and of the kinematical quantities during the gait is necessary to the geometrical design of the synovial lubricated gap in order to achieve, according to Medley et al. [25] particular lubrication mechanisms (mixed or full-film) [22, 23]. This could be achieved by reaching an optimal value h_{min} (minimum synovial meatus height) divided by the root mean square of roughness values of the prostheses contact surfaces- in order to optimize the prostheses performances in terms of wear resistance.

5. Conclusion

With the purpose of the optimization of the TKA design, a thorough determination of the loading of the ankle joint is necessary both for the its stability and structural resistance design and for its tribological performances improvement, in terms especially of wear resistance. Unfortunately, in-vivo measurements of joint internal forces are not a simple and allowed task, and then non-invasive in-silico approach should furnish a meaning full perspective.

In this paper are presented ankle joint dynamical loading components during the gait, obtained by using the AnyBody Modelling SystemTM (AMS) computer software, adopting kinematical data obtained by Vaughan et al. [20] The obtained results, in terms of Antero Posterior, Proximo Distal, Medio Lateral Forces and Ankle Eversion, Plantar Flexion, Axial moments, shows a satisficing agreement allowing to be used both for detailed prostheses FEM analysis both for the optimized tribological design in terms of synovial lubricating mechanisms. Of course this investigation has limitations regarding the necessary full validation of the proposed model to be executed running several simulations, by varying key parameters of the model and by comparing the results with the ones (few) found in literature from in-vivo testing.

Acknowledgements
The authors wish to thank Jonathan De Mattia for his help with the simulations.

References

[1] Kurtz M et al. 2009 *Future young patient demand for primary and revision joint replacement: national projections from 2010 to 2030* Clinical Orthopaedics and Related Research® 467.10 pp.2606-2612
[2] Saverio A, Ruggiero A, and Merola M 2015 *Advanced biomaterials in hip joint arthroplasty. A review on polymer and ceramics composites as alternative bearings* Composites Part B: Engineering 83 pp.276-283
[3] Ruggiero A, Merola M, and Saverio A 2017 *On the biotribology of total knee replacement: A new roughness measurements protocol on in vivo condyles considering the dynamic loading from musculoskeletal multibody model* Measurement 112 pp.22-28
[4] Vickerstaff J A., Miles A W, and Cunningham J L 2007 *A brief history of total ankle replacement and a review of the current status.* Medical engineering&physics 29.10 pp.1056-1064
[5] Saverio A et al. 2016 *Does metal transfer affect the tribological behaviour of femoral heads? Roughness and phase transformation analyses on retrieved zirconia and Biolox® Delta composites* Composites Part B Engineering 92 pp.290-298
[6] Merola M et al. 2016 *On the tribological behavior of retrieved hip femoral heads affected by metallic debris. A comparative investigation by stylus and optical profilometer for a new roughness measurement protocol* Measurement 90 pp.365-371
[7] Kunutsor S K et al. 2017 *Re-infection outcomes following one-and two-stage surgical revision of infected hip prosthesis: a systematic review and meta-analysis* PloS one 10.9 e0139166
[8] Saverio A, et al. 2017 *Does metal transfer differ on retrieved Biolox® Delta composites femoral heads? Surface investigation on three Biolox® generations from a biotribological point of view.* Composites Part B: Engineering 113 pp.164-173
[9] Terrier A et al. 2017 *Fixed and mobile-bearing total ankle prostheses: Effect on tibial bone strain* Clinical Biomechanics 48 pp.57-62
[10] Ruggiero A, Merola M and Saverio A 2018 *Finite element simulations of hard-on-soft hip joint prosthesis accounting for dynamic loads calculated from a musculoskeletal model during walking* Materials 11.4 pp.574

[11] Saverio A, Merola M, and Ruggiero A 2019 *Tribological performances of total knee prostheses: Roughness measurements on medial and lateral compartments of retrieved femoral components* Measurement 135 pp.341-347

[12] Saverio A, Merola M, and Ruggiero A 2018 *Development of a Novel in Silico Model to Investigate the Influence of Radial Clearance on the Acetabular Cup Contact Pressure in Hip Implants* Materials 11.8 pp.1282

[13] Ruggiero A, Emilio G, and D'Amato R 2011 *Approximate analytical model for the squeeze-film lubrication of the human ankle joint with synovial fluid filtrated by articular cartilage* Tribology letters 41.2 pp.337-343

[14] Ruggiero A, Gómez E and Roberto D 2013 *Approximate closed-form solution of the synovial fluid film force in the human ankle joint with non-Newtonian lubricant* Tribology International 57 pp.156-161

[15] Jaber S A et al. 2015 *In vitro effects on mobile polyethylene insert under highly demanding daily activities: stair climbing.* International orthopaedics 39.7 pp.1433-1440

[16] Saverio A and Ruggiero A 2019 *A Critical Analysis of TKR In Vitro Wear Tests Considering Predicted Knee Joint Loads* Materials 12.10 pp.1597

[17] Ruggiero A et al. 2017 *FEM analysis of metal on UHMWPE total hip prosthesis during normal walking cycle* Proceedings of the XXIII conference of the Italian Association of Theoretical and Applied Mechanics AIMETA

[18] Damsgaard M R, Rasmussen J, Tørholm S et al., 2006 *Analysis of musculoskeletal systems in the AnyBody Modeling System* Simulation Modelling Practice and Theory 14 pp.1100–1111

[19] Rasmussen J, Damsgaard M, and Voigt M 2001 *Muscle recruitment by the min/max criterion-a comparative numerical study* Journal of biomechanics 34.3 pp.409-415

[20] Vaughan C L, Davis B L, O'Connor J C 1992 *Dynamics of gait* 1st ed. Human Kinetics Publishers The Man data set provided at the website of the International Society of Biomechanics http://isbweb.org

[21] Hlavácek M 2005 *Squeeze-film lubrication of the human ankle joint subjected to the cyclic loading encountered in walking*. Journal of tribology 127.1 pp.141-148

[22] Ruggiero, A, D'Amato R, and Gómez E 2015 *Experimental analysis of tribological behavior of UHMWPE against AISI420C and against TiAl6V4 alloy under dry and lubricated conditions* Tribology International 92 pp.154-161

[23] Ruggiero A, Gómez E and Merola M 2016 *Experimental comparison on tribological pairs UHMWPE/TIAL6V4 alloy, UHMWPE/AISI316L austenitic stainless and UHMWPE/AL2O3 ceramic, under dry and lubricated conditions*. Tribology International 96 pp.349-360

[24] Ruggiero A, Merola M and Saverio A 2018 *Finite element simulations of hard-on-soft hip joint prosthesis accounting for dynamic loads calculated from a musculoskeletal model during walking*. Materials 11.4 pp.574

[25] Medley J B, Dowson D, Wright V 1984 *Transient elastohydrodynamic lubrication models for the human ankle joint* Eng. Med. 13 (1984) 137–151 (accessed November 15 2018)

IOP Conf. Series: Materials Science and Engineering **749** (2020) 012009 doi:10.1088/1757-899X/749/1/012009

Innovative approaches to the formation of environmental safety at the objects of oil and gas production

L Shkitsa[1,a], T Yatsyshyn[2], M Lyakh[3] and O Sydorenko[4]

[1] Ivano-Frankivsk National Technical University of Oil and Gas, Department of Engineering and Computer Graphics, Ivano-Frankivsk, Ukraine
[2] Ivano-Frankivsk National Technical University of Oil and Gas, Department of Ecology, Ivano-Frankivsk, Ukraine
[3] Ivano-Frankivsk National Technical University of Oil and Gas, Department of Oil and Gas Equipment, Ivano-Frankivsk, Ukraine
[4] Research and Design Institute PJSC "Ukrnafta", Ecological Researches Laboratory, Ivano-Frankivsk, Ukraine

E-mail: [a] lshkitsa@nung.edu.ua

Abstract. The innovative approach to environmental safety upgrade of pumping-circulatory equipment of drilling rigs based on the analysis of life history and continuous perfection principles. Examples of implementation of innovative approach in environmental safety investigation of oil and gas wells in order to define environmentally hazardous units of equipment and technological processes are given. In order to solve the problem of environmental safety improvement and the effectiveness of flushing out of well that are being drilled a modernized scheme of drilling rig pumping-circulatory system has been designed. System approach of modernizing separate elements of pumping-circulatory system has been used. The maximum air-tight pumping-circulatory drilling rig system has been designed. The proposed model takes into account the necessity of hazardous evaporations outlet with the aim of their further utilization. Such an approach will avoid the entry harmful substances into the environment and prevent harmful effects of aggressive substances on health personnel of drilling rig, environment and on state of the technological equipment.

1. Introduction

Oil and gas industry complex is one of the objects of high ecological hazard. It uses ignitable, highly flammable, dangerously explosive and toxic substances. The process of drilling oil and gas wells is the source of air pollution. The pollution takes place during scheduled technological processes at the cost of evaporations and emitting hazardous substances.

An important of the strategic direction, according to 2008/50/EC Directive of the European parliament and council [1], is preventing hazardous substances emission from particular sources. At the same time the most effective measures of defining and implementation of air pollution reduction are taken at the local level.

Pumping-circulatory system (PCS) of drilling rigs is mainly open. The mud circulates there together with different hazardous substances; some of them are used in order to improve technological processes: 55% - gel powder, chalk, barytes; 16% - acids; 11% - surface-active substances; 7% -

water-soluble salts, polymers - 5%, dissolvent - 4%, caustic and different chemical reagents 1% each. Hazardous substances may create in the process of mud circulation and it contact with drilling layers, saturated for example with hydrogen disulphide and other dangerous substances [2].

The open equipment containing drilling fluid cause evaporations and distribution of hazardous substances saturated with toxic components. This may have following negative consequences:

- Health effect for staff and population of adjacent territories covered by hazardous evaporations distribution;
- Negative effect for flora and fauna;
- Negative effect for operating conditions and useful time of all main equipment that comes into contact with evaporations.

That is why the investigations aimed at upgrade the environmental safety of oil and gas extraction process, namely pumping-circulatory equipment of drilling rigs (DR), for the time being remain up-to-date and require complex approach based on evaluation of the present state and finding faults of main equipment and processes with their further elimination.

2. Analysis of present designs and defining ways of improvement

The evaporation of liquid substances used during well construction creates the risk of some ailments origination with drilling rig staff, aggressive evaporations may damage the equipment and at favorable climate conditions there happens intensive dispersal of pollutants onto adjacent territories. The origination of evaporations during drilling fluid circulation has been described bin works by Steinsvag, K., Galea, K.S., Krüger, K., & Peikli, V. 2011, 7. [3].

At the time being all the regions of the world possess a great number of oil wells. These objects are quite dangerous for the environment as in the course of regular operating processes and also at states of emergency. A great amount of scientific papers in the sphere of ecological safety are about oil and gas industry complex problems. However, the analysis of the whole state found the problem of defining separate problematic fields. Famous investigations of the state of atmospheric air of DR and approved methodologies do not take into account the evaporations of chemical substances that originate during DF circulation. In order to evaluate and forecast the state of atmospheric air the important role is played by the analysis of evaporations from evaporation surfaces of PCS taking into account real conditions and operating functions of DR equipment.

During technological drilling operations especially after used drilling fluid lifting, its passing the chutes and its purification, as well as keeping bore mud in sludge pits the pollution of surface layer by hazardous substances that constitute DF takes place. These substances come to atmosphere in the process of evaporation that takes place quite intensively due to comparatively high temperature of the flushing fluid [4]. The results of investigation of evaporation processes at different units of PCS DR are given in work [5]. Gaseous substances that penetrate ground layers of atmosphere as the result of eddy diffusion and by the action of wind expands over drilling rig territory and beyond its borders which poses risk of toxic effect for staff and population of adjacent territories. The forecast of prevalence of pollution of ground layers of atmosphere caused by evaporations is given in work [6].

The existing constructions of pumping-circulatory systems and their separate components studied in the works of [2, 7] that state about the negative impact of evaporations that appear in the process of primary purification of the mud on the mud screen.

The above-mentioned investigations prove the necessity of design modernized circulatory system which will make it possible to perform flushing process at the rig effectively and will ensure utilization of hazardous evaporations in order to increase environmental safety of oil and gas wells drilling.

3. Innovative approaches to the increase of the level of environmental safety of pumping-circulatory system of drilling rigs

Academic specialists of Ivano-Frankivsk National University of Oil and Gas conducted investigations and estimated life history of the objects of wells construction in compliance with ISO 14000 standards which helped identify the most powerful environmental effects. The decrease of environmental pollution at oil and gas extraction works demands thorough investigation of every stage from the ecological safety viewpoint. It will make it possible to establish faults in operation of the equipment, technological processes and management decisions that need to be disposed of. An offer has been made to use systematic investigation approach that will be based on the analysis of each life cycle stage of the well. The above-mentioned approach will make it possible to trace the connections of the combination of factors and the results of the environmental impact as the whole set of elements. The result of implementation of this approach will be the definition of real ecologically hazardous processes and their delimitation [8].

The sections of PCS DR that are more likely to emit hazardous substances into ground layer of the atmosphere have been defined. Guided by the principle of life history management concerning continuous improvement several stages of modernizing operating machinery and processes on order to reduce emissions of pollutants into ground layers of atmosphere have been conducted. Fig. 1 shows the chart of the stages of the increase of levels of environmental safety of pumping-circulatory system of drilling rig.

The first step in order to reduce the amount of emissions was to modernize the set of operating machinery, namely: the mud screen model with insulating screen for primary purification of drilling fluid has been improved which prevents penetration of hazardous substances to the environment that emerge intensively during mud screen vibration [5]; a new model of mud cone for fine drilling fluid purification has been designed which makes it possible to operate as mud cleaner and desliter which respectively reduces the amount of equipment that is necessary to install in order to obtain required result of drilling fluid purification [9]; there has also been designed constructive solution purification equipment block of drilling mud and device for cleaning the downhole tool [10].

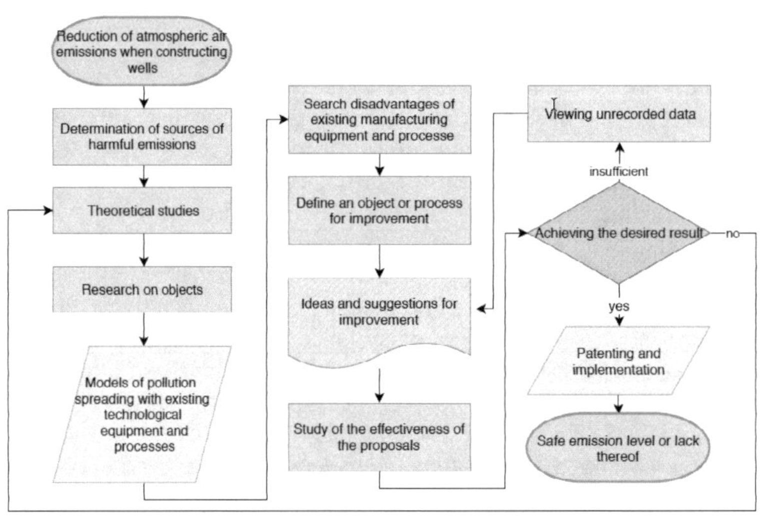

Figure 1. The chart of basic stages of the increase of environmental safety level of pumping-circulatory system of drilling rig

Using the proposed equipment will avoid harmful substances into the environment, prevent harmful effects of aggressive substances on health personnel and increase fire safety rig.

At the second stage modernized pumping-circulatory system from the viewpoint of obtaining efficient operation results, meets the following requirements: quickly readjusts for the drilling mode in certain conditions; takes minimum time for transportation to the operating site, assembling and disassembling; is compact for mobility; has convenient and safe labor conditions for mans; has high resistivity to damages [2].

During further analysis of the chart of closed off PCS it has been defined that vapors in isolated sections can accumulate and create unfavorable conditions for pumps operation and later penetrate ground layers of the atmosphere thus endangering environment. That is why a decision has been made to improve the design of closed off PCS by means of vapor disposal from definite sections of PCS with their further utilization in order to minimize the risk of poisoning by DF vapors.

4. Recommendations for improvement of closed off PCS

In order to ensure safe, reliable operation of closed off PCS with separate modernized equipment it is necessary to conduct additional analysis and take into account the specifications of operating conditions. The work [2, 10] contains the chart and operating peculiarities of closed off PCS where it describes the operating peculiarities and design features of separate units of equipment from flexible material that make it possible ensure air tightness, mobility, compactness and a set of other positive factors that help increase the level of environmental safety of PCS. Fig. 2 shows the chart of improved closed off PCS with vapor disposal for utilization.

Modernized pumping-circulatory system consists of: 1 air-tight accept tanks are made of elastic material; 2, 3, 4, 24 - intake line; 5, 8, 11 - drilling mud pumps; 6 - drilling pumps; 7 - tanks with chemical reagents; 9, 23 - section of channel-circulatory line; 10 - purification unit; 12 - back valves; 13, 25, 27 - lock valves; 14 - the drilling mud preparation unit; 15, 16, 17, 18 - the sections of discharge line; 19 - standpipe; 20 - equipment of the mouth of the well; 21 - rotary swivel; 22 - safety switch units; 26 - fill-up line.

The offered PCS in chart 2 contains separated units where vapors accumulate. Though PCS is closed off separate units accumulate vapors. The most favorable conditions for evaporation rate are created by high temperatures of the fluid coming out of the well and in summer when the temperature of the environment is high. These sections are marked in the chart as: A1, A2, A3, A4.

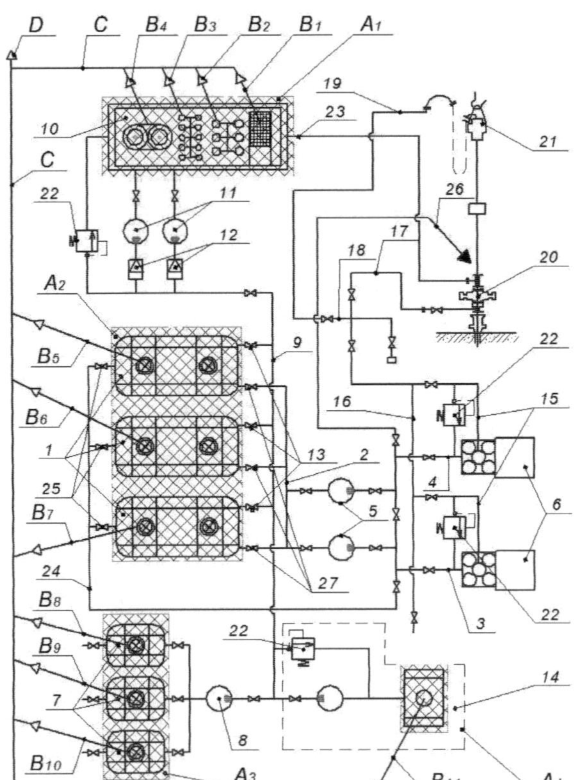

Figure 2. Improved closed off PCS with vapor disposals for utilization

A1 - the unit of drilling fluid purification that comes out of the well. The given unit contains mud screen for primary drilling fluid purification designed with insulating screen [8], under which saturated vapor microclimate is created; for fine purification - hydroclone batteries, desanders and desilters, skimming machines and also degasser. All these technical means are sources of intensive evaporation. And it is recommended to install at every above-mentioned objects vapor disposals at B1, B1, B1, B1 to common manifold C.

A2 - the section of elastic, airtight reservoirs that contain and store spare drilling fluid. It is recommended to install vapor collectors in the upper part. If the vapors are not disposed of their volume increases and they may penetrate pumps 5, 6, 8, 11, and they as a rule at the presence of gas in the fluid stop acting as they should. This may cause the failure of closed off PCS DF. That is why there has been offered to install disposals B5, B6, B7 to common manifold C.

A3 - the section of elastic, airtight reservoirs for containing chemical substances meant for drilling fluid processing. It is also recommended to install disposals analogic to A2. This section is equipped with disposals B8, B9, B10 and connected to common manifold C.

A4 - the section of drilling mud mixing unit. It contains operating machinery for drilling mud mixing, the mining itself takes place, dispergation of fluid that is being mixed. These processes facilitate intensive evaporation. At the above mentioned objects, it is recommended to install vapor collectors and with the aid of disposal B11 move vapors to common manifold C.

If the drilling rig equipped by diesel engines the reservoirs and tanks for fuel storage can also be the sources of pollution. That is why this machinery should be equipped by analogic vapor disposals to fuse carriers that prevent vapors from penetrating the atmosphere (utilization section).

From common manifold C vapors enter line D and flow to safety area. This area can be discharge to atmosphere at the absence of hazardous substances. If hazardous substances are present it is necessary to utilize them. There exist designs for vapor recovery of volatile matters at big and small breathings of reservoirs [11]. The principle of the given design is based on two processes: sorbate for adsorptive light carbohydrates entrapping, thermoelectric cooling for denuding and condensation of recuperated vapors and sorbents regeneration. The design [12] is also to our knowledge, where the vapors are scavenged out of the reservoir by the pump, pressed to 0,2 - 0,4 MPa and components similar to amyl hydride and hexane become liquid and collect at holding tank and propane and iso-butane come to cold storage room where they condense at appropriate temperature.

5. Conclusions

The innovative approach to the increase of environmental safety of the equipment of pumping-circulatory system of the drilling rig, which is based on the analysis of life history and constant perfection principles. The given methodological approach can be used to reduce pollution of atmosphere and for other objects where liquid hazardous substances are kept.

The existing charts of PCS have been analyzed and the necessity to dispose vapors for utilization has been defined. The directions of equipment improvement have been outlined and the modernized scheme of pumping-circulatory system has been offered. The given chart takes into account all evaporation sources. This design of pumping-circulatory system makes it possible to ensure high environmental safety level when high-toxic substances are used in the system.

References

[1] *Directive 2008/50/EC of the European Parliament and of the Council of 21 May 2008 on ambient air quality and cleaner air for Europe* OJ L 152 11.6.2008 pp.1-44

[2] Skitsa L, Yatsyshyn T, Lyakh M and Sydorenko O 2018 *Ways of increasing the safety of pumping-circulatory system of the drilling rig* Mining of Mineral Deposits Vol. 12 Issue 3 pp.71-79

[3] Steinsvag K, Galea K S, Krüger K and Peikli V 2011 *Effect of drilling fluid systems and temperature on oil mist and vapour levels generated from shale shaker* Annals of Occupational Hygiene Vol. 55 Issue 4 DOI: 10.1093/annhyg/meq097 pp. 47-356

IOP Conf. Series: Materials Science and Engineering **749** (2020) 012009 doi:10.1088/1757-899X/749/1/012009

[4] Shkitsa L Y, Yatsyshyn T M and Popov O O 2012 *Matematuchne modeluvannya pryzemnogo sharu atmosfery paramy burovogo rozchuny* Naukovy visnuk IFNTUNG 3(33) pp.56-62

[5] Shkitsa L Y and Yatsyshyn T M 2014 *Pidvucshunya rivnya ekologichnoyi bezpeku nasosno-tsurkulyatsiynoyi systemu burovoyi ustanovku Exploration and development of oil and gas fields* 3(52) pp.7-16

[6] Shkitsa L Y, Yatsyshyn T M, Popov O O and Artemchuk V A 2013 *Prognozirovaniye rasprostraneniya zagryaznyayushchikh veshchestv v atmosfere na territorii burovoy ustanovki* Neftyanoye khazyaystvo No.11. pp.136-140

[7] Shvydkyi O 2010 *Pro kompleksnyi pidkhid do modernizatsii nasosno-tsyrkuliatsiinoi systemy burovoi ustanovky. Zbirnyk naukovykh prats NAK «Naftohaz Ukrainy»* Problemy naftohazovoi promyslovosti 8 pp.197-200

[8] Shkitsa L, Yatsyshyn T, Liakh M et.al. 2018 *Innovative development of resourse-saving technologies for mining* Multi-authored monograph Sofia Publishing House "St.Ivan Rilski"

[9] Lyakh M M, Vakalyuk V M, Yatsyshyn T M, Solonychnyy YA V, Lyakh YU M and Vil'chyk O H 2010 *Pat. 89267 Ukrayina, MPK (2009)* N 04 S 5/00 Hidrotsyklon Byul No1

[10] Shkitsa L, Yatsyshyn T, Lyakh M and Sydorenko O 2016 *Means of atmospheric air pollution reduction during drilling wells* IOP Conference Series: Materials Science and Engineering, Vol. 144 Issue 1 pp.012009

[11] Boychenko O V, Aks'onov O F, Boychenko S V and Berehovyy O M 2001 *Aprobatsiya laboratornoyi modeli «dykhal'noho» klapana novoho pokolinnya dlya rezervuariv* Visnyk NAU No. 3

[12] Kryk H V, Lavrenko O M and Stadnyk O D 2007 *UA27877 U MPK (2006) B01D 53/14 Ustanovka dlya utylizatsiyi benzynovykh pariv*

IOP Conf. Series: Materials Science and Engineering **749** (2020) 012010 doi:10.1088/1757-899X/749/1/012010

The influence of the angles of the asymmetrical gear rack on the geometrical and functional parameters for the pinion and gear with asymmetric teeth

F Suciu[a]

Technical University of Cluj-Napoca, Faculty of Engineering, Department of Engineering and Technology Management, Baia Mare, Romania

E-mail: [a] flavia.suciu@cunbm.utcluj.ro

Abstract. The paper presents some research on the design of the asymmetric gears, special gears with involute asymmetrical teeth. The aim of these researches is to establish the influence of the angles of the asymmetrical gear rack upon some geometrical and functional parameters of the gears. We have used, as design method, the direct gear design, followed by the gear rack design, considering that asymmetric gears are not conventional gears, so the design method it is not the standard one. The results that are presented was obtained by using a computer modelling program, for asymmetric gears, in MatLab, and for one can see the asymmetric gears and the zone of the tooth action, a computer program in AutoLISP is developed. Based on the computer applications developed for determinate the geometrical parameters and for representing the teeth and gears models, the asymmetric teeth pinion and gear can be manufactured. Between the design parameters established at starting the described process there are the angles of the asymmetrical gear rack.

1. Introduction

Involute gearing, with involute curves as teeth profiles, is the most common because it has a lot of advantages as related to other types of gearing.

The opposite flanks of an involute tooth function in a different way, for the majority of gearings. The load on the active flank is significantly larger and is applied over longer periods of time than on the inactive flank. These aspects led to the idea of using gears called asymmetric gears [3], for which in order to build the tooth profile two involutes are drawn from two different base circles, involutes that are at a certain distance one from another, in order to create the tooth body, which this time is no longer symmetrical as related to the radius that crosses the intersection point of the two involutes, radius which for the common teething was the tooth symmetry axis. Thus, an asymmetrical involute flanks tooth results with supplementary advantages.

Designing gears with asymmetrical teeth, is aimed at improving the performance of active flanks, by reducing the performance of inactive flanks, taking into consideration the fact that these are not loaded or rarely loaded and usually for short periods of time. Improving the performances can mean an increase in the loading capacity and weight, noise and vibration reduction.

The degree of asymmetry, determined by the two different pressure angles (α_a, α_i), on the active and inactive profile:

$$k = \cos\alpha_i / \cos\alpha_a \tag{1}$$

can be chosen in accordance with the requirements of the particular application for which the gearing is designed.

The conventional gear design is generally based on the standard tools, in these cases dimensions are determined with the parameters of standard gear rack. This makes gearing quite simple, economical, reducing the costs with tooling and design, but in the same time it is known that standard design can't obtain the optimum performance.

For some particular application, in many industries, it is necessary to reduce the noise, the vibrations, the weight and dimensions, by developing design method for more compact and higher capacity gear pairs. The direct asymmetrical gear design, used in the computer program, a part of which results are presented in this paper, is one of these methods.

The computer program, in MatLab, developed for determining the geometrical and functional parameters of asymmetric gears, makes possible to study the performance of many gear pairs with different coefficient of asymmetry and to establish for the initial data, the number of teeth and centre distance, which is the best one.

The program can also be used for analysing the performance of asymmetrical gears in relation with the symmetrical gears, considering those as asymmetrical gears with the coefficient of asymmetry equal to the unit. Thus, one can see how the computer design makes production of nonstandard gears as easy as production of standard ones.

2. Asymmetrical generating gear rack

Considering, in figure 1, that one of the gears of an asymmetrical gear pair, for example gear 2 with center O_2, has the base circles infinitely large, its active and inactive profiles will degenerate into lines tangential to the active and inactive gear 1 profiles respectively, the gear 2 rolling circle will also have an infinite radius and becomes a line tangential to the gear 1 rolling circle. Gear 2 thus obtained is a gear rack with asymmetrical flanks.

While generating the symmetrical teeth the opposite profiles are in pairs symmetrical to a line perpendicular on the reference line.

While generating asymmetrical teeth, the opposite profiles generating the active and inactive flanks respectively, are no longer symmetrical, having different angles in relation to the gear rack's reference line [5]. The pressure angles for the active and inactive flanks at generation of the gear with asymmetrical rack are the generating rack angles α_{ac} and α_{ic}, the angles of the cutting edges in relation with the perpendicular to the reference line of the rack.

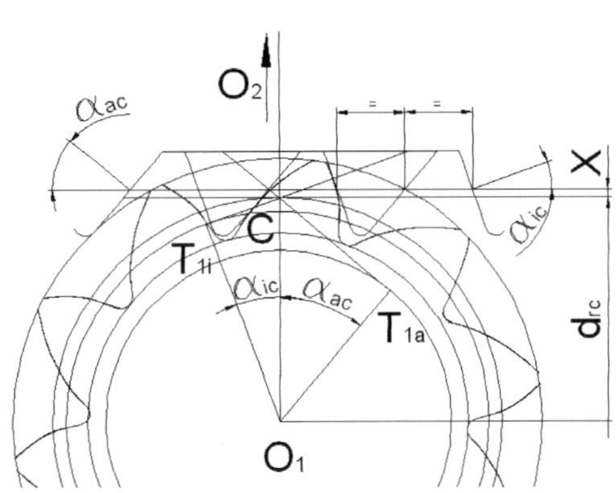

Figure 1. Generating with the gear rack with asymmetrical flanks

If the numbers of teeth, for the gear pair, and the center distance are initial date of the design program, one can determine the sum of the pinion rack shift and gear rack shift, by the generating rack angles with the following relation:

$$(X_1 + X_2) = \frac{p_c(z_1 + z_2)}{2\pi \cdot (\tan\alpha_{ac} + \tan\alpha_{ic})} \cdot \\ \cdot (\text{inv}\,\alpha_a + \text{inv}\,\alpha_i - \text{inv}\,\alpha_{ac} - \text{inv}\,\alpha_{ic}). \tag{2}$$

where p_c is the generating rack pitch.

Thus, the pinion's teeth's thickness and the gear's one, is determined by the generating rack angleα_{ac}, from which the other α_{ic} results.

If the generating angle α_{ac} is equal with the pressure angle α_a, the values of the pinion rack shift and gear rack shift are equal, but with different signs. If the generating rack angles are chosen different from the pressure angles of the gear pair, the rack shifts are different.

In the booth case the shifts are established in the computer program, finding the values corresponding to the condition to avoid the interference, to avoid the undercutting, to ensure the contact ratio greater than one.

For a given centre distance, number of teeth and pressure angles, the sum of the shift rack of the pinion and the shift rack of the gear is determined by the angle of the rack, in one of Matlab [3] developed application. It is necessary to study how this angle influences the geometrical and functional parameters of the gears.

The maximum value for the active profile angle of the rack is equal to the active pressure angle and the minimum value is:

$$\alpha_{ac\_min} = \arccos(1/k) \tag{3}$$

corresponding to $\cos\alpha_{ic}=1;\ \alpha_{ic}=0.$

The α_{ic} angle of the rack results from the coefficient of asymmetry:

$$\alpha_{ic} = \arccos(k \cdot \cos\alpha_{ac}) \tag{4}$$

The design and modelling program make possible also study the influence of generating the pinion and the gear with different racks, or with the same rack.

Thus, with the computer program, one can obtain from a great number of possibilities of design, for the given initial date, which is the optimum set of design parameters for a particular product [1], [2].

3. The variation of some geometrical and functional parameters of the gears in relation with the asymmetrical rack's angles

Because the rack's angle for the active profile is considered as an initial date that one can choose between minimum and maximum values, and the rack's angle for the inactive profile results in relation with the coefficient of asymmetry, we have obtained the diagrams of variation of gears parameters depending on the gear rack angle α_{ac}, corresponding to the generating profile of the generated gear active profile.

The initial data of the analysed gears are chosen with the following values:

$z_1=16,\ z_2=16,\ z_2=57,\ a=120mm,\ \alpha_a=40^o,\ \alpha_i=20^o,\ P=18kW,\ n=1000\ rot/min$

The gears can be used with greater pressure angle on the active profile, in which case the coefficient of asymmetry is bigger than one, or with smaller pressure angle on the active profile, in which case the coefficient of asymmetry is smaller than one. We have analysed both of these cases and we have taken into consideration also the generation with one or two racks. The variations of studied parameters, for the coefficient of asymmetry k>1, are represented in the following diagrams:

IOP Conf. Series: Materials Science and Engineering **749** (2020) 012010 doi:10.1088/1757-899X/749/1/012010

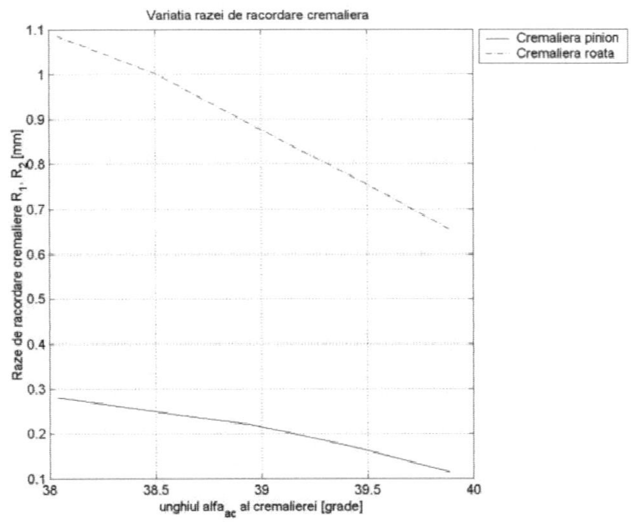

Figure 2. The variation of the fillet radius R_1, R_2

Figure 2 presents the variations of radius of the fillet of the racks in relation with α_{ac} angle. One can see that the increase of α_{ac} determines the decrease of booth radius R_1, R_2 of the pinion's and gear's racks.

The 2D gears models were obtained with two computer programs in AutoLisp [6], developed using the parametrical equations of the asymmetrical involute profiles and of the two joint profiles at the tooth base, and also an algorithm for determine the dimensions of the gear racks.

The zone of the tooth action represented for different generating angles α_{ac} pairs of gears (figures 3, figure 4), show the variations of the fillet radius.

The zone of the tooth action for an example of designed and manufactured gears, based on the developed applications, is presented in figure 5.

The dimension of the maximum cross section, at the bottom of the gear tooth, increases by increasing the α_{ac} angle and the maximum cross section of the pinion tooth decreases by increasing the α_{ac} angle (figure 6).

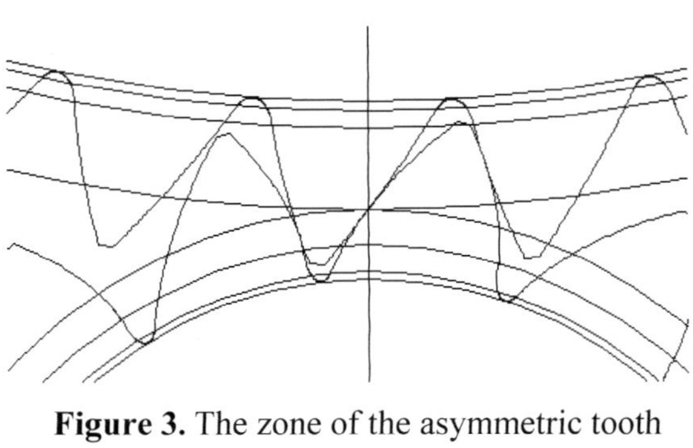

Figure 3. The zone of the asymmetric tooth action $\alpha_{ac} = 0.6960$, $R_1 = 0{,}1138$, $R_2 = 0{,}6564$

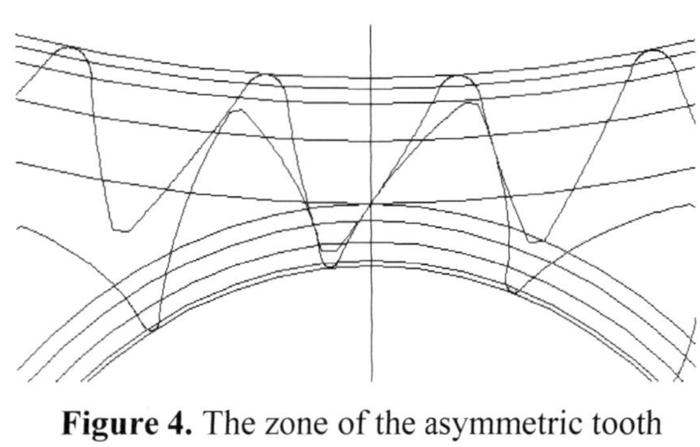

Figure 4. The zone of the asymmetric tooth action $\alpha_{ac} = 0.6640$, $R_1 = 0{,}2803$, $R_2 = 1{,}0847$

IOP Conf. Series: Materials Science and Engineering **749** (2020) 012010 doi:10.1088/1757-899X/749/1/012010

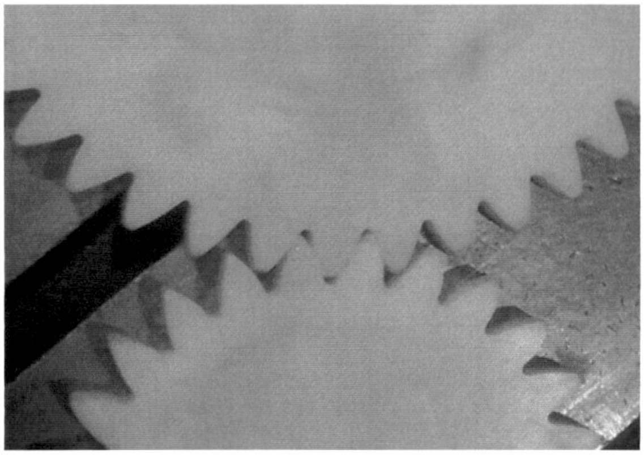

Figure 5. The zone of asymmetric tooth action for the manufactured gears

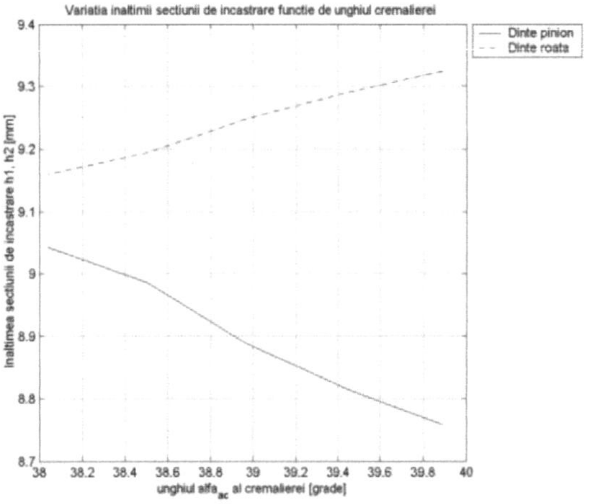

Figure 6. Variation of the dimension of cross section

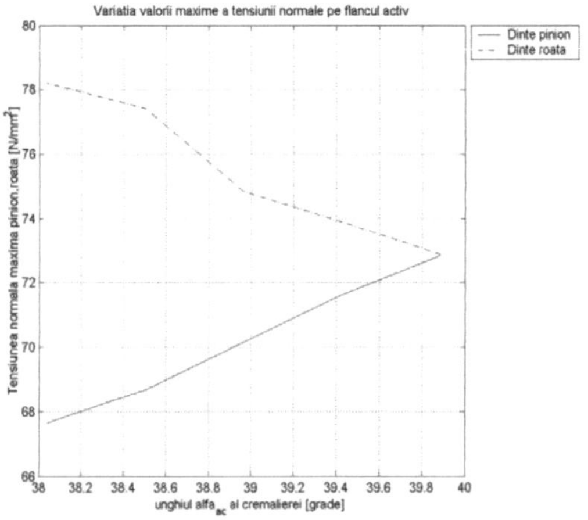

Figure 7. The variation of maximum tensile bending stress in relation with α_{ac}

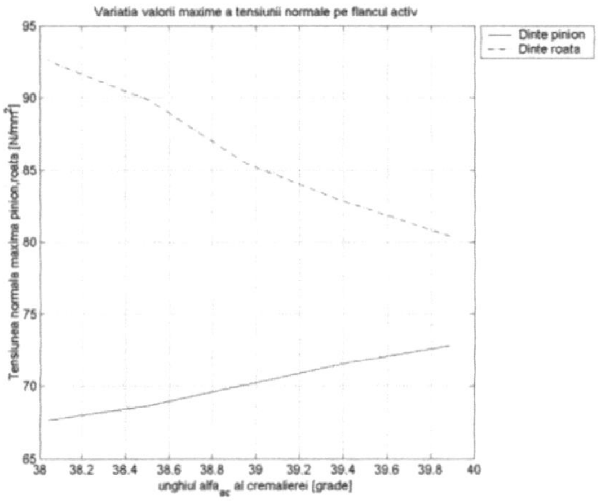

Figure 8. The variation of the maximum bending tensile stress, k>1, using the same one rack

Therefore, the cross-sectional area and the modulus of tooth section will be also significantly increased at the gear and decreased at the pinion. That has a direct influence on the bending stress of the tooth. The dimension of the maximum cross section was determined with a method established using the equation of the joint profiles of asymmetric tooth. The equations of the joint profiles have been obtained considering those profiles as curves generated by the filet profile of the gear rack. Increasing the α_{ac} angle, the maximum tensile bending stress, at the bottom of the active profile, increases for the pinion and decreases for the gear, in such a manner that the two very different values of the maximum stress, for smaller values of the generating gear rack angle α_{ac}, may become equal for greater values for α_{ac} (figure 7). In the case of generating the pinion and the gear with the same one rack, the difference between the maximum bending stress values at the pinion's tooth and at the gear's tooth are larger in comparison with the case of generating with two racks, because the dimensions of the rack are calculated on the base of corresponding gear's parameters (figure 8).

In this paper have been presented diagrams of variations of geometrical and functional parameters for gears having the coefficient of asymmetry k>1, with the pressure angle on the active profile bigger than the pressure angle on the inactive profile, but it have been developed also computer applications for analysing the behaviour of the asymmetric gears with coefficient of asymmetry smaller that one

4. Conclusions

A small generating rack angle α_{ac}, chosen as the initial date in the design program for asymmetric gears, ensures greater fillet radius, decreases the contact stress, decreases the maximum tensile bending stress at pinion tooth but increases the maximum tensile stress at the gear's tooth in comparison to the pinion's one. If the aim of the design is to obtain equal values for the bending stress at pinion and gear one, there must be used the generating with two different racks and as an initial date, a large α_{ac} angle for the active profile of the rack.

References

[1] Banica M 2004 *The Optimization of the Relief Parameters of the Involute Spur Gearing with the View of Decrease the Noise Emission* Annals of the Oradea University Fascicle of Management and Technological Engineering Vol. III (XIII) ISSN 1583-0691 6p

[2] Banica M 2006 *Optimizarea Dinamicii Angrenajelor (Optimization of Gear Dynamics)* Publisher Risoprint Cluj-Napoca ISBN 978-973-751-385-4

[3] Ghinea M and Fireteanu V 1995 *MATLAB Calcul numeric - Grafica - Aplicatii* ISBN 973-601-275-1

[4] Kapelevich A L 2000 *Geometry and design of involute spur gears with asymmetric teeth* Mechanism and Machine Theory 35 (Retrived on 1.06.2015 from www.akgears.com)

[5] Kapelevich A L and Kleiss R E 2002 *Direct Gear Design for Spur and Helical Gears* Gear Technology, September/October 2002 pp.29-35 (Retrived on 1.06.2015 from www.akgears.com)

[6] Tisan V 1997 *Graphical representation of gears using the AutoLisp language* Proceedings of the MicroCAD 97 Miskolc, 26.02.1997

Managing the suitability of cost calculation for manufacturing technology in the mechanical processing industry

V Birle

Technical University of Cluj-Napoca, Faculty of Engineering, Department of Engineering and Technology Management, Baia Mare, Romania

E-mail: vasile.birle@cunbm.utcluj.ro

Abstract. Maximizing returns and increasing efficiency in manufacturing can also be achieved by organizing, managing and controlling production costs. Moving from the independent approach of costing to a systemic approach, by considering dependence and conditioning correlations with manufacturing technologies means reconsidering the concept of cost management. Both subsystems involve specific and common instruments and procedures, which allows a conceptual summation leading to a new system of approaching the goal of higher resource utilization in economic processes. The adaptation of calculation methods to manufacturing technologies is circumscribed to managerial measures undertaken to increase efficiency and internal yields as fundamental objectives. Reorganizing the enterprise through segmentation into Responsibility Centres can provide viable solutions for achieving management goals. Segmenting the enterprise and implementing integrated procedures requires significant costs that can't be borne by all businesses, especially small and medium-sized enterprises that are still in the early stages of development.

The proposed solution responds to the need of directing the activity through costs in SMEs in early developmental stages, which hold modest resources that do not allow the acquisition of integrated software. The application provides information with a high degree of utility, both in the pre-calculation and post-calculating phase, and has a high capacity to adapt to changes in technology. The results obtained from the analysis of the deviations require corrective or improvement operational measures, shortening the time gap between the moment of consumption and the corrections.

1. Introduction

Industry is the most dynamic sector in modern economies that contributes most to added value (45.5%) and to Gross Domestic Product (23.6%). Within the industry, enterprises carrying out mechanical processing (machining) activities are SMEs that are in early stages of financial accumulation. Machining means several specific operations, such as milling, turning, splicing, grinding, sanding, etc. which require significant consumption of material, human and informational resources. The limited financial resources from these activities make it impossible to apply technology management principles, methods and tools as an interdisciplinary field: *"engineering, science and management disciplines for the purpose of planning, developing and implementing technological capabilities to form and achieve the strategic and operational objectives of an organization."*

Considering the need to increase return and efficiency in the management of mechanical processing enterprises, this paper proposes an application for adapting cost-calculation methods to manufacturing technologies. Implementation of this application has as a fundamental requirement the process of segmentation of activity by Responsibility Centres: *"...a component of the enterprise management system ..."*

2. Technical and economic information, linking the system of suitability of cost calculation for the manufacturing technologies

Management in the mechanical processing industry is subordinated to the concepts of: planning, designing, drawing up technical documentation, organizing technological flows and executing specific operations. These directions of action aim to increase efficiency and internal returns, indicators that support the fierce competition in specific markets. Measurement of efficiency indicators is also done through the cost of production, which reflects internal efficiency and returns. The production cost economic indicator represents *"the numerical expression of the quantitative side of economic phenomena and processes..."* and is obtained from an organized and scientifically applied calculation. The size of this indicator shows the quality, efficiency and yields recorded in the production process. Cost calculation is the duty of management.

The cost of production, as *"the monetary expression of the consumption of production factors",* is determined in a distinct process called cost calculation. This phrase refers to all the mathematical operations that are carried out under the authority of technical and economic thinking in order to determine specific indicators that reflect the returns and the efficiency of resource consumption. The significance of the cost has two valences: economic, found in internal returns and financial, measuring the degree of recovery of advanced resources.

The syntax manufacturing technology is the summation factor of the methods, processes, operations performed or applied to resources such as labour force, raw materials and materials and information to produce an industrial or commercial product for the market. The concept of system means several interconnected elements or components that influence and coordinate each other, aiming at reaching certain technical - economic parameters or performing certain functions.

Cost calculation is indissolubly linked to the specifics of the manufacturing technologies. The manufacturing technology validates its efficiency through the size of the cost. The combination of the two technologies with the cost calculation results in an independent system with specific functions, methods and tools.

The tools, procedures and working methods used by the costing process are associated with the economic sciences, and the methods and operations defining the manufacturing technologies belong to the engineering field. Both the technologies and the cost calculation underpin the same objective: increasing economic and financial efficiency in the resource management process. Both have the capacity to measure returns: technologies can measure economic return while calculations can measure financial return.

It can be argued that the two subsystems are mutually conditional and measure the objectives of the economic activities, i.e. to generate added value and positive financial results. The combination of the two concepts is done through technical and economic information. Information combines the two subsystems as a new concept, which we can define as a mechanism for the suitability of costing methods for manufacturing technologies.

Technical and economic information, as a *"process research tool"* shows cause-effect objective links between system components. At the same time, information is the decisional support of the act of conducting and controlling economic activities. The connection between technology and calculation is made by managerial accounting, understood *"as a set of procedures for the identification, quantification, collection, analysis and reporting of information"*. This data processing and analysis laboratory is not a regulated one. The organization of managerial accounting is related to the company's abilities, requirements, guidelines and needs. From this perspective, the technical and economic information will be built, processed and provided on the management-driven flows to serve them in substantiating decisions.

Cost calculation adapts to manufacturing technologies, not vice versa. The choice of calculation methods is based on the managerial decision in accordance with the applied technology. The result is a comprehensive perception of the higher use of available resources in relation to the proposed goals.

3. Directions for action for the suitability of costing for manufacturing technologies

The suitability of costing methods, respectively, *"the way or method to carry out research and interpret phenomena"*, for manufacturing technology has as its starting point the segmentation of the enterprise on responsibility centres. These centres are relatively independent structures designed to collect costs by a particular method, a numerical quantification model of production costs, variable and fixed, as determinable economic dimensions. At the same time, the responsibility centre also means *"a hierarchical sequence of enterprise management"* capable of self-government on the basis of correlative relationships with other centres, generating new functional relationships. By applying the principle of self-management, the centres can independently measure returns and efficiency based on cost.

Enterprise segmentation can be performed on the basis of functional or spatial criteria and their choice depends on: the field of activity, the technology used, the way of organizing the production process, the diversity and complexity of the manufactured production, the level of equipment, social or psychological factors. The segmentation criteria may be: production activity (execution of a product or service); technological process (machining, assembly, finishing, etc.); field of activity (design, supply, sale, etc.) or related and auxiliary activities (repairs, energy production, maintenance, SDVs etc.)

The enterprise segmentation decision will be substantiated by the quality and content of the answers to a set of questions, such as:

1. *How to translate expenditures into cost elements?*
2. *What are the possibilities for variable cost planning?*
3. *How can indirect expenditures be allocated to the responsibility centre and the product?*
4. *How do we determine actual costs?*
5. *Are there preconditions for determining deviations between planned and actual costs?*

Obtaining answers to the above questions means just as many stages of work to be done so that, finally, it is possible to connect the most suitable calculation method to the manufacturing technology.

4. Steps to suit the calculation of production costs for manufacturing technologies

Once the enterprise segmentation stage has been completed on responsibility centres, which can also be referred to as production centres, the next stage is the organization of the "laboratory" for collecting, processing and supplying the technical - economic information, called cost management.

Through the five steps below, management in the mechanical processing industry can organize a personalized cost management system, capable of providing the required high-quality technical and economic information. For any corrections it is possible to intervene in real-time.

4.1. Limiting expenditures and translating them into cost elements

Responsibility Centres (CR) are represented as Production Centres. In this category, it is recommended to include the machines used for the mechanical processing operations, i.e. for production.

From this perspective, the construction of the production cost takes into account the presentation structure. It will be determined by the following calculation formula:

$$Cp = Cmd + Csd + Cind \tag{1}$$

where: C_p - the cost of production, which will be obtained as a sum total of:

C_{md} - direct material cost which means all specific inputs of identifiable raw materials and materials for each product,

C_{sd} - the direct wage cost is calculated according to the consumption norm (Ncs) and the norm of the machine's operating time (Nut). The consumption norm means the time it takes

IOP Conf. Series: Materials Science and Engineering **749** (2020) 012011 doi:10.1088/1757-899X/749/1/012011

an employee to perform an operation in the technological flow. The machine norm is the time it takes to operate the machine within the normal work norm. For a correct determination, several times are taken into account: preparation, operation, maintenance or accidental stops,

C_{ind} - are all indirect costs that will be allocated to the product based on allocation keys, determined a priori:

$$Cind = \sum_{r=1}^{n}(Chc + Chg + Cha) \tag{2}$$

r = the category to which the indirect expenditure belongs,

C_{hc} - the indirect common costs of responsibility centres, such as those incurred for repairing work equipment and adjacent areas, the cost of electricity consumed in the responsibility centre which cannot be allocated directly to the product, spare parts, maintenance materials other than direct ones etc.,

C_{hg} - the general indirect costs, which consist of: the salary costs of engineers, designers, supply and sales staff, foremen and other staff involved indirectly in the production centre's work. In the category of general indirect costs it is also possible to include those for lighting and heating of production sites, cleaning, maintenance of buildings related to the production area etc.,

C_{ha} - the indirect management costs related to the TESA department, including the maintenance and repair of the buildings in which they operate.

4.2. Planning variable cost elements by Responsibility Centers

In the mechanical processing industry, variable cost is determinant. The variable cost elements are directly proportional to the volume of production. This is the reason why they are pursued with priority and the fixed costs are left in the secondary plan. From the value point of view, the variable cost has the majority share in the total cost of production.

The choice of cost collection processes is made by adapting to manufacturing technologies to best meet the cost control requirements. In practice, there are a number of cost-accounting procedures for cost carriers, among which: weighing quantities with prices or purchase rates, applying proportionate percentages to absolute amounts, *pro rata temporis*, applying flat rates etc.

When calculating the material expenses (raw materials, auxiliary materials, semi-finished products, fuel, spare parts, electricity, water, steam, other utilities, inventory items, protective equipment etc.) that can be expressed quantitatively, the mathematics of weighing quantities with prices or tariffs is widely used.

Transforming a material expense into a cost element is done by applying the generic formula:

$$C_{md} = \sum_{i=1}^{n}(q \cdot P)_j \tag{3}$$

where: C_{md} - the monetary expression of material consumption,

j - the product in relation to which the expense is identified and for which it will become a cost element,

i - raw materials, materials etc.,

q - the amount consumed according to the consumption norm,

P - the unit price of the element of material cost.

For the assessment of actual consumption of materials and utilities, in the post-calculation phase we will use the evaluation based on the actual cost of acquisition of the cost element, by applying one of the stock evaluation methods: FIFO, LIFO or Weighted Average Cost.

The following expenditure components are considered in the calculation of costs for direct salaries to the bearer, before they become calculation elements.

IOP Conf. Series: Materials Science and Engineering **749** (2020) 012011 doi:10.1088/1757-899X/749/1/012011

The salary rate that takes into account the categories of personnel, working time and the forms of pay used, respectively, the agreement: individual or global or the time-based pay system. The form of pay in the agreement is common for directly productive workers who are directly involved in the production process.

The established formula for determining wage costs is the following:

$$C_{sd} = \sum_{s=1}^{n} (N_t \cdot T_s)_j \tag{4}$$

where: N_t - is the time norm obtained as a result of normalization of work,
T_s - pay rate,
j - product, cost bearer.

Once the theoretical model of variable expenses planning has been defined, the manufacturing technology can be outlined to the smallest details, respectively, the operations flow from the technological flow, in their logical sequence, starting from the reality which states that *"manufacturing flexibility is aimed at increasing productivity"*. Once these terms have been obtained, the specific consumption of raw materials and materials for each product is calculated, as well as the wage costs according to the rates applied to the personnel involved in the production activity. At this stage, both the raw material costs and those resulting from the remuneration of staff for the next production cycle are predicted, depending on the volume of production. Thus, we are able to know the working times of the machines involved in the technological flow.

The running time of the machine will become a basis for calculating the amount of indirect costs for a product.

4.3. Distribution of indirect costs for the Responsibility Center and Product

Indirect costs (C_{ind}) are collected at enterprise level, in each period, according to the requirements of the organizational and managerial system. In most cases, in the mechanical processing industry, indirect costs are collected monthly. Where control and tracking of production costs is done online, indirect costs are collected at shorter intervals, sometimes daily.

In the case of the mechanical processing industry, the number of hours of operation of the machines for the next month is planned from the design phase. In the production cost pre-calculation phase we can determine the amount of indirect costs with a high degree of accuracy for one hour of operation of a particular machine. For the purposes of calculating indirect cost elements we need to choose the basis of allocation, which typically is a relatively stable cost element, such as the *"direct payroll expense"*. The choice of the allocation criterion is based on the stability and complexity of the operations performed on a machine, and the direct payroll cost directly meets these requirements.

Once the allocation base has been defined, it is possible to determine the indirect costs of a product out of their total. The indirect cost allocation operation is made by using the following calculation algorithm:

$$K_{ind} = \frac{C_{ind}}{\sum_{c=1}^{n} C_{hs}} \tag{5}$$

IOP Conf. Series: Materials Science and Engineering **749** (2020) 012011 doi:10.1088/1757-899X/749/1/012011

Knowing the normal running times of all machines (N_{fu}) in the responsibility centres to which the indirect costs will be allocated, it is necessary to determine the share of indirect costs for each machine using the calculation formula:

$$THM_n = \frac{k \cdot C_{hsc}}{N_{fu}} \tag{6}$$

where: $k \cdot C_{hsc}$ - is the sum of the indirect costs of a machine (CR),
THM_N - is the indirect cost of one hour of operation of the machine (CR) to which the indirect costs have been allocated,
N_{fu} - the number of hours of operation of the machine,
n - represents the machine (CR).

The dispersion of indirect costs of the Responsibility Center on each product obtained with the respective machine is determined by the sum of the working times counted per machine (ΣTfun) using the calculation formula:

$$C_{ind\,i} = \frac{THM_n}{\sum T_{fui}} \tag{7}$$

where: i is the product.

The planned production cost is obtained from the sum of the above-determined cost structure elements, respectively:

$$Cp_{0i} = Cmd_{0i} + Csd_{0i} + Cind_{0i} \tag{8}$$

4.4. *Calculation of the actual production cost*

The same steps will be taken in the post-calculation phase. The difference between pre-calculation and post-calculation is the nature of the expenditures. For the pre-calculation stage we use the planned expenditures, which are determined based on measurements or norms, while the post-calculation takes into account the actual expenditures incurred, provided by the management accounting.

$$Cp_{1i} = Cmd_{1i} + Csd_{1i} + Cind_{1i} \tag{9}$$

In most cases there are differences between the planned production cost (C_{p0}) and the actual one (C_{p1}), differences which are called deviations.

4.5. *Determining the deviations between actual and planned costs*

One of the significant cost functions relates to ensuring the conditions for measuring the efficiency of resource utilization, the return recorded in the manufacturing process. Comparing actual and planned consumption results in positive differences when actual costs are below those planned or negative in the reverse situation. The analysis of deviations is of great importance for management, from both perspectives:

a) in the case of positive deviations, they may result from streamlining the technological flow, the intensive use of working time or from a poor planning mechanism;

b) the causes for negative deviations can be given by: deviations from the technological flow, communication and relationship deficiencies between the Responsibility Centres, exceeding consumption norms or the manufacturing times (negligence or misuse of the time base), increases in actual purchase prices, mistakes in planning activity etc.

Deviations are calculated at the level of each structural component of the production cost, for the entire responsibility centre and for each product, and the results are subject to independent analysis.

Once the causes are identified, management may order the appropriate measures for each situation that derives from the economic activity carried out.

IOP Conf. Series: Materials Science and Engineering **749** (2020) 012011 doi:10.1088/1757-899X/749/1/012011

5. Conclusions

The application that can be used to efficiently suit production costs calculation for manufacturing technologies can find its usefulness in the managing process of SMEs that do not have sufficient resources to acquire integrated accounting and computing systems. The process of directing activity through costs allows management to have at least the following directions of action:

- *building a planning and budgeting platform on sub-units up to the machine level that allows the activity to be broken down to unit cost per product;*
- *the possibility of keeping track of the internal returns, caused by the internal consumption of material, human and informational resources;*
- *measuring individual efficiency at the level of each segment (equipment) on economic and financial coordinates, but also from the perspective of valorisation of the human potential;*
- *deviation analysis is an efficient and pragmatic internal control tool that allows for the operative implementation of corrections but also the undertaking of measures to increase returns and efficiency;*
- *the application is flexible and can be easily adapted to any changes in the manufacturing technology.*

References

[1] Albu N and Albu C 2003 *Instrumente de management al performantei (Performance management tools)* vol. I and II Economic Publishing House, Bucharest

[2] Briciu S et.al. 2010 *Contabilitatea si controlul de gestiune, instrumente pentru evaluarea performantei entității (Accounting and management control, tools for assessing entity performance)* Aeternitas Publishing House, Alba Iulia

[3] Caraiani C and Dumitrana M (coordinators) 2008 *Contabilitate de gestiune&Control de gestiune (Management accounting & Management control)* University Publishing House, Bucharest

[4] Alazard Claude and Sabine Separi 2001 *Contrôle de gestion*-5th edition Dunod Paris

[5] Cojocaru C 2002 *Contabilitatea de gestiune (Management accounting)* Moldavia Publishing House Bacău

[6] Nita Dobrota (coordinator) 1999 *Economics Dictionary*, Economic Publishing House Bucharest

[7] Oprea C and Cârstea Gh 2000 *Contabilitatea de gestiune si calculatia costurilor (Management accounting and costing)* Ginecod Publishing House

[8] Rusu C 1995 *Analiza si reglarea firmei prin costuri (Analyzing and adjusting the company by cost)*, Asachi Publishing House Iasi

IOP Conf. Series: Materials Science and Engineering **749** (2020) 012012 doi:10.1088/1757-899X/749/1/012012

Relevant aspects of the responsibilities in the international project management

A Dascalescu[a] and M Ungureanu

Technical University of Cluj-Napoca, Faculty of Engineering, Department of Engineering and Technology Management, Baia Mare, Romania

E-mail: [a] anamaria.dascalescu@cunbm.utcluj.ro

Abstract. Usually, the international Projects are of greater importance and with greater founding than national projects, involve multiple partners from different countries and are subject of nationals and international controls. The paper approach the case of international projects with EU funding, including cross-border projects and underline responsibilities of the management team both to the Project Contract and to the national laws requirements, responsibilities in managing human resources, time and costs. For a successful Project implementation, the same importance should be given to communication, visibility and dissemination rules as well as reports quality. Major importance is granted to the quality of Project results.

1. Introduction

Facilitating the global interconnection of human society, the increase of the volume of international affairs implies the need for appearance of international management. International management is the management of organizations involved in activities and transactions that take place across the borders of two or more states.

The financial resources used by the international projects have a growing role in economic development and their area of applicability is growing.

The managers are involved in unprecedented international activities this fact requiring specific personal abilities to overcome the language and culture barriers between partners.

International management is more complex than domestic because of economic, political, legislative and cultural diversity.

Management theories are transferable and managers perform the same functions regardless of the organization structure, project type, income differences or cultural differences.

Efficient management styles are similar in culturally related countries (languages, Religion), forms of government, attitude towards work, [1].

As a result of the experience in managing of four cross-border projects, in this paper we will highlight the profile required by a project manager to successfully operate in international projects and the responsibilities of the project management team.

2. The characteristics of the project management

The project represents a sum of activities that lead to a common goal and require an important consumption of resources (human, material, financial, equipment, documentary information and time).

IOP Conf. Series: Materials Science and Engineering **749** (2020) 012012 doi:10.1088/1757-899X/749/1/012012

The consumption of all resources is done during project implementation with the purpose of achieving the objectives and obtaining the results to the required quality.

Project management consists of *tasks* and *resources* planning, organizing and managing (controlling) to achieve a particular *goal*, given the constraints of *time*, *resources* and *costs*.

The areas covered by project management is presented in figure 1.

Figure 1. Areas covered by project management

3. The characteristics of international project

In the second half of the 20th century the evolution of management at the international scale involve the transition from internationalization to the globalization of a series of aspects of social-economic life: global nature of the science and technology, global financial system, global network for transmitting and receiving information (communication infrastructure), the world institutional framework (governmental and nongovernmental organization) who are debating actions on global issues [3].

Compared to the national projects management, a number of peculiarities are deduced in the management of international projects.

The study of international management revealed the *influence of the culture* of the participant countries. The most important influences on international management have the following elements of culture:

- language differences requiring the use of a "project language" known by members of the management team at such level as to allow direct communication and avoid confusion of any nature (it is not recommended to use translators in the joint activities and it is more difficult to learn the languages of the partners) - translator is recommended to be use in official document translation,
- different beliefs and behavior of the project team members reflected in different ability to make decisions, responsiveness to difficult situations or risk, ability to collaborate or competitiveness, resistance to change or to new, attitude towards work.

Planning activities in the international management is more complex due to the number of external elements involved:

- multilingual, multinational and multicultural factors;
- political factors;
- financial climate;

- different legislation and rules, changing, sometimes unclear;
- coins with different stability and real value;
- different salaries for similar positions in the Project team, in accordance with national legislation.

These differences increase the difficulty of plans drafting and implementing.

The *implementation team selection* has a main contribution to achieve the specific objectives and obtain the results of the Project. The selected experts must cover all activities with necessary expertise in order to obtain the required quality of the Project results. Also, the selected team members must have the ability to work together in common activities, eventually previously proven in other projects and to accept the authority of the manager.

It is necessary for each team member to have real implication and determination, positive attitude and strong filing in order that the Project to succeed.

4. The characteristics of the international manger's profile

An efficient international management need to adapt to the management style of the involved countries and understanding the cultures of these countries.

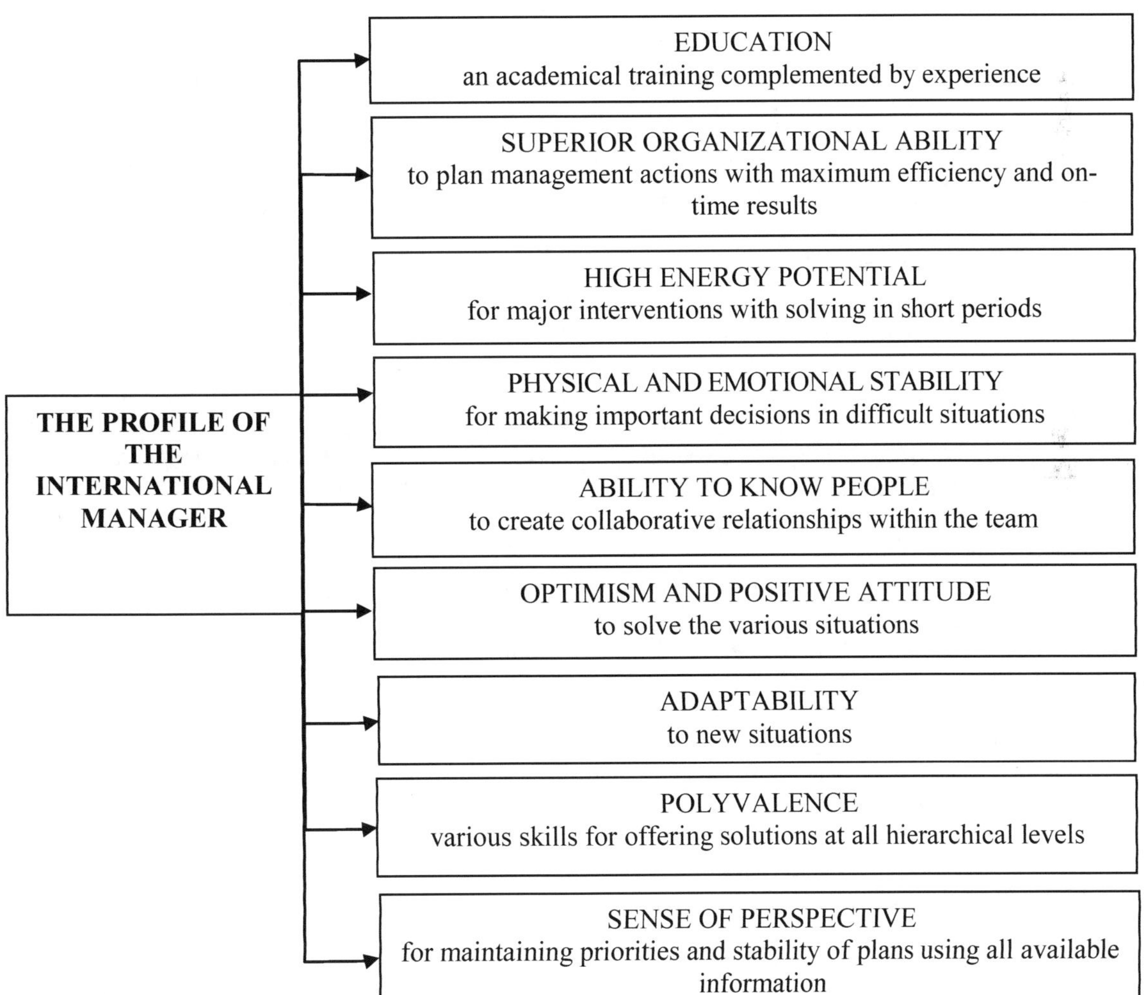

Figure 2. The characteristics of the international manager's profile

IOP Conf. Series: Materials Science and Engineering **749** (2020) 012012 doi:10.1088/1757-899X/749/1/012012

To be efficient and to perform in domain, the international manager must have the profile features presented in figure 2.

The international manager decisions are based on information, listening, intuition and observation, has flexibility and adaptability to situations. Does not look for behavioral patterns, for general rules or guidelines for different situations, [2].

It is important for the manager to have the last decision on the implementation team members' selection. The partners' countries managers must be certain that their implementation team are the best to cover all activities in the benefit of the quality of Project's results.

5. Responsibilities of the management team in the international project

In the international project, the entire Project management is coordinated and overseen by *Project Manager Coordinator* and applied by project management team composed by *each partner country management team* led by *Partner Country Manager*.

Project Manager Coordinator, as lider of project managenet team, assumes the responsibility for ensuring implementation of the entire project.

Ensures that the expenditure presented by the Partners has been incurred for the purpose of implementing the project and correspond to activities set in the grant contract and Partnership agreement;

Verifies that the expenditure presented by the Partners has been examined by an auditor or by a first level controller according to the procedures.

The managers respond of implementing the *activities* of the project *in time*, using the *designed resources* and *costs* and obtaining the *results* to the required quality.

All partners managers shall actively cooperate for the implementation of the project, they shall ensure staffing and financing of the project. Each partner manager shall be legally and financially responsible for the activities that is implementing and for the share of financing that it receives.

The Partnership Agreement provisions must guarantee the sound financial management of the funds and shall include the arrangements for recovery of funds unduly paid to the Partners [4].

A rigorous communication system links the project manager coordinator to all Partners countries managers and also to the Director of the Project representing JTS (Join Technical Secretariat) of the Financing Programme. The communication strategy between decision factors allows quick and highly efficient solution to the encountered problems. It is maximal importance that to not exist language barrier in this internal communication system.

The Project Manager Coordinator and Partners Countries Managers respond to all requirements stipulated in main documents:

- Financing Contract (including *Project Action* and *Project Budget)* signed between Contracting Authority and Partners,
- Partnership Agreement signed between Project Partners from a different country,
- National legislation and Financing Contract procurement rules,
- National legislation for human resources management,
- Visibility Guide.

Partners Countries Managers respond to all requirements of own national controllers (First Level Control) and the Project Manager Coordinator respond to international controllers (Second Level Control).

The Project Manager Coordinator must have the ability to link experts from different countries' implementation team in joint activities of work packages and has the responsibility to brought closer the team members through professional experience exchange and mutual understanding of cultural values.

The manager should detect in time the changes occurred in project implementation and to apply quick measures to solve problems. That requires the elaboration of a complex control system which determines all deviation from established standards.

IOP Conf. Series: Materials Science and Engineering **749** (2020) 012012 doi:10.1088/1757-899X/749/1/012012

The manager commits himself in doing everything in his power to foster the implementation of the Action and ensure its performance in time and in compliance with obligations to the JMA (Joint Management Authority).

6. The control issues in international management

The control process of international management activity involves a series of stages starting with setting objectives, setting results and outputs. The *objectives* must clearly outline the goals of the project. The *results* must be clear, effective, quantifiable and verifiable.

For an *efficient control system,* it is necessary to choose the *control methods*, to establish the *performance standards*, the *measurement and the evaluation of the performances*:

- the control alternatives in international management are direct control and indirect control,
- the performance standard derives from the purpose of the project and must be clearly defined, accepted and understood. If the operational standard is to general, must be established intermediary standards to achieve the general goal,
- the managers must have the ability to supervise the achievement of current performance expressed by the quality of the results guaranteed by the experts of the implementation team.

Also, the control system includes establishing *responsibilities* and establishing *communications system:*

- responsibilities are set both to the management team and to the implementation team and must be clearly stipulated in the personal attribution sheet (job description),
- the communication system is connected with reporting system and represent an information system needed to ensure a continuous flow of data necessary for analysis and decision making. The communication system must be fast and efficient, without unnecessary information, to ensure proper control of the project's activities,
- the communication system links the Manager Coordinator both with the Partners' Managers and with financing Programme organizations and the information flows in both directions.

Each Partner manager coordinates the activities of the own implementation team, has to inform and to be informed about the performance of the activities and the degree of achievement of the results.

If in performance evaluation is detected deviations from standards, methods corrections are applied and revisions of responsibilities and communications if it is necessary.

7. Conclusions

The international Project has notable differences of management toward national Project. A main importance has the Project human resource's capability of adapting to the several challenges due to cultural, economic, political, linguistic differences between Partners countries' implementation teams.

The international manager must have several extra capabilities and to assume extra responsibilities to succeed the implementation of the Project.

The manager should detect in time the changes occurred in project implementation and to apply quick measures to solve problems. That requires the elaboration of a complex control system which determines all deviation from established standards.

References

[1] *** 2010 *Manager de proiect* EUROCOR ISBN 973-7929-15-2 Modul 16
[2] Thomsett M 1990 *The Little Black Book of Project Management* AMACOM Books
[3] Popa I, Filip R 1999 *Managementul international* Bucuresti Ed. Economica
[4] Ministry of Regional Development Public Administration and European Funds ROMANIA *Guidelines for grant Applicants* Joint Operational Programme Romania-Ukraine ENI CBC 2014-2020

IOP Conf. Series: Materials Science and Engineering **749** (2020) 012013 doi:10.1088/1757-899X/749/1/012013

Experimental research based on statistical control, aimed to improve the quality of the parts obtained through machining, at some consumer goods

L Dragan

Technical University of Cluj-Napoca, Faculty of Engineering, Department of Engineering and Technology Management, Baia Mare, Romania

E-mail: liliana.dragan@cunbm.utcluj.ro

Abstract. The paper aims to carry out experimental research meant to increase the quality of some mass-produced furniture items. In order for that to happen, a great number of size measurements have been done on beech and birch wood parts obtained through milling and drilling with numerically controlled machines. The experiment lasted for a week, with work being done in two shifts and with a volume of 300 parts per shift. The experimental database was then processed and interpreted in line with the statistical analysis methods. Finally, conclusions regarding the measures that need to be taken in order to assure an accurate adjustment of the machine as well and an optimisation of the production process were formulated.

1. Introduction

The quality of an industrial product must satisfy the customer's needs. These must be specified, controlled and observed during the manufacturing process. In this context quality is a complex concept that includes issues related to conception, manufacturing and use.

To solve this problem, we rely on statistical control. This is a modern mean to monitor the manufacturing process and to assess the quality, the manufactured parts having to fulfil a number of conditions: to be manufactured in large scale; to have a significant share in the company's production; to be manufactured using an automatic or semi-automatic technological process.

Attention is directed to three levels of statistical control [1]:

- Quality planning (planning level) setting the preliminary parameters: monitored features, the number of units contained in the survey sample, type of control card, evaluation;
- Quality verification (operational level): determining the measured values, documentation on verification data, documentation of the process development, process control based on control cards;
- Quality adjustment (administrative level): long-term forecasts are issued based on verification reports and the process is optimized.

2. Data on the manufacturing process

The paper aims to analyse the stability of the process of drilling parts (figure 1) used to make sofas, in a wooden plate cutting workshop. The process is done on multiple drilling machines (figure 2) which are simultaneously using 6 drills, whose adjustment is essential. This study is required following the observation that some components of the sofa's structure show a higher percentage of rejects.

Figure 1. The monitored part

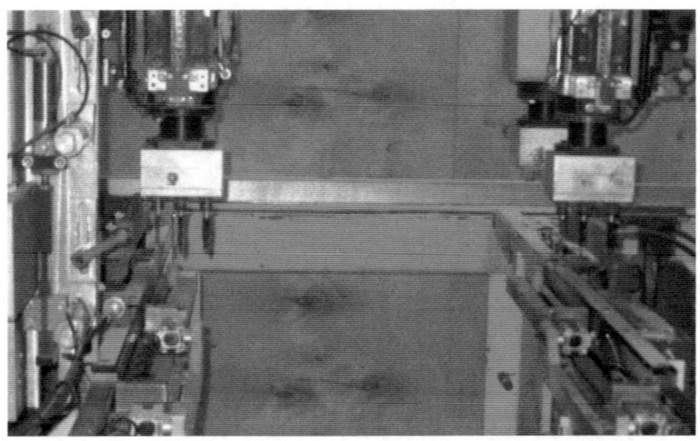

Figure 2. The multiple drilling machine used for processing

Manufacturing process control [2] is done by sampling, through surveys and calculations based on probability theory and mathematical statistics. In this regard, a preliminary statistical analysis was done, setting the dispersion parameters (mean square deviation σ), then the quality control cards were issued, meant to monitor the capacity of the process to be kept under control.

3. Data on statistical control

Manufacturing quality can be estimated/ assessed by the measurement of some discrete variable features of the product in accordance with the requirements of the technical documentation. For the samples taken successively, for values of the features in the survey sample, two parameters were monitored, one related to the machine adjustment - by the position of the field scattering (median), and the other one concerning the accuracy of processing - by the size of the field dispersion (amplitude). Depending on their position against the limits calculated and plotted for each one, given a significance level $\alpha = 0.002$, the manufacturing process is assessed and measures to correct possible occurrences are provided: optimum / excessive / insufficient precision and centered/ decentered process. Each time it is assessed if there are prerequisites for worsening the current situation over time and whether stoppage and correction of the process is economically justified.

The size of the survey sample is $n = 5$ units. The sample consists of the last pieces taken from the machine in their production order, complying with the recommendation that the proportion of controlled parts represent 5÷10% of the batch if normal series. The proportion of controlled pieces can be estimated using the formula [3]:

IOP Conf. Series: Materials Science and Engineering **749** (2020) 012013 doi:10.1088/1757-899X/749/1/012013

$$k = 100 \cdot \sqrt{\frac{n}{Q}} = 100 \cdot \sqrt{\frac{5}{600}} \cong 9\% \tag{1}$$

n is the size of the survey sample;
Q is the average number of pieces drilled between two successive samplings or dis-adjustments (600);
Δt, the time interval between two samplings, determined by the formula:

$$\Delta t = \frac{60}{q} \cdot \sqrt{n \cdot Q} = \frac{60}{70} \cdot \sqrt{5 \cdot 600} \sim 1\, ora \tag{2}$$

q is the hourly production of the machine (70/h).

4. Processing the results of measurements. Preliminary statistical analysis

Considering the standard tolerances specified for the monitored size, the distance between the hole and the edge of the part was measured using a calliper with 0.01 mm precision.

After the collection of the measured results over several 8 hour shifts, the data was analysed and one important dimension of the product were determined: 185 ± 0.5 mm. For this size, the sequence of measured values was divided into 10 groups, for each group the followings were set: absolute frequencies, relative frequencies, the distribution function, and finally, the mean square deviation and data dispersion for the measured values (figure 3). The data of the preliminary analysis were entered in Excel program [4] and were the basis for drawing the histogram (figure 4), the frequency polygon and the graph of the distribution function (figure 5).

Nr clasei	Limitele claselor		Frecventa absoluta, n_i	Frecventa relativa, $f_i = \frac{n_i}{n}$	F(x)	x_i	$f_i * x_i$	σ	σ^2
1	184,00	184,216	14	0,029166667	0,029167	184,108	5,3698167	0,265365504	0,0704189
2	184,217	184,433	23	0,047916667	0,077083	184,325	8,8322396		
3	184,434	184,65	45	0,09375	0,170833	184,542	17,300813		
4	184,651	184,67	7	0,014583333	0,185417	184,6605	2,6929656		
5	184,868	185,084	237	0,49375	0,679167	184,976	91,3319		
6	185,085	185,301	63	0,13125	0,810417	185,193	24,306581		
7	185,302	185,518	45	0,09375	0,904167	185,41	17,382188		
8	185,519	185,735	8	0,016666667	0,920833	185,627	3,0937833		
9	185,736	185,952	24	0,05	0,970833	185,844	9,2922		
10	185,953	186,17	14	0,029166667	1	186,0615	5,4267938		
							185,02928		

Figure 3. Summarization of data from the preliminary analysis

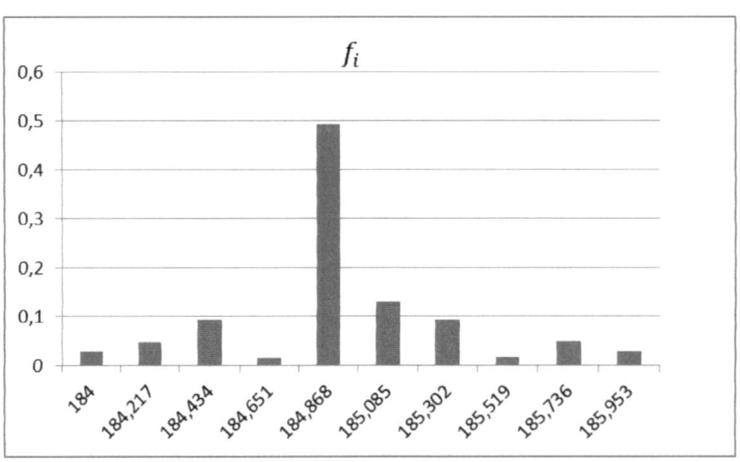

Figure 4. The histogram

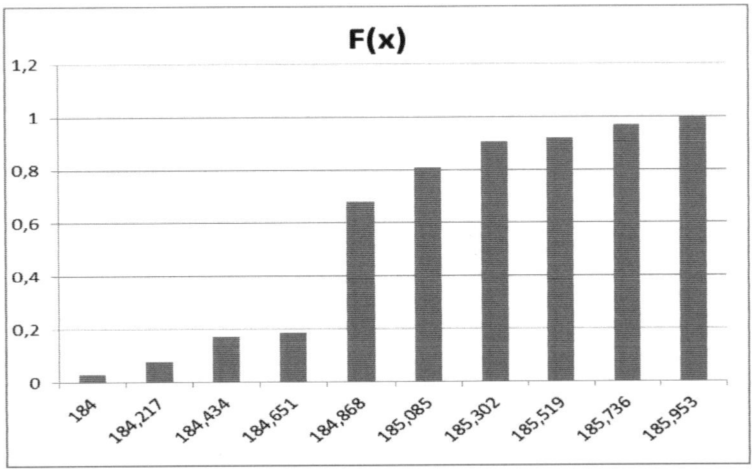

Figure 5. The field dispersion

5. Drawing up the control cards

The classic control cards allow continuous monitoring of the process, enabling a constant and predictable production both in terms of cost and quality. There is also a reduction of products dispersion, a cost reduction and an increase in the effective capacity of the process to be kept under control. To allow the identification of the causes of errors, all the changes and influences on production must be mentioned in the sheet accompanying the process [5].

The control cards [2] contain the following: general information (department, machine, product, operation, monitored feature, limits of the tolerance interval, size of the sample, the interval between two samplings); date and time of sampling; diagrams of the monitored parameters; the measured values and the accuracy of the measuring instrument used; the values of the parameters (average / standard deviation, median / amplitude); conclusion of the survey and decision.

The diagrams of the location parameter (average / median) contain the control limits (upper Lcs and lower Lci) and tolerance limits (upper Ls and lower Li), while the diagrams of the variation parameter (standard deviation / amplitude) need only the upper control limit.

If a location parameter (median ME) with bilateral risk is monitored, the values of the control limits are calculated as follows:

$$L_{CS} = T_C + 1{,}2533 \cdot z_{\alpha} \cdot \frac{\sigma}{\sqrt{n}} \tag{3}$$

$$L_{CI} = T_C - 1{,}2533 \cdot z_{\alpha} \cdot \frac{\sigma}{\sqrt{n}} \tag{4}$$

$$ME = X_{\frac{n+1}{2}} \tag{5}$$

Tc - middle of the tolerance interval.

If a variation parameter like amplitude R is monitored, it requires a single control limit Lc. Because only positive deviations are of interest, the unilateral risk is adopted.

$$L_C = t_{\frac{\alpha}{2}} \cdot \sigma \tag{6}$$

$$R = x_{max} - x_{min} \tag{7}$$

x_{max} and x_{min} are the limit values of the feature.

It was noted that the frequency distribution has a single maximum, meaning that the measured values observed the normal distribution. The preliminary analysis is used to set the mean square deviation and then the control- cards for median and amplitude are drawn up (figure 6 and figure 7).

Only two control cards are presented here, covering four consecutive shifts, in which the most significant developments of the monitored statistical parameters were observed.

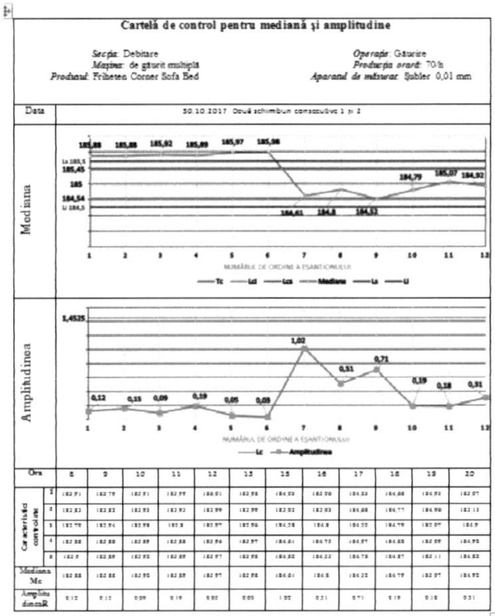

Figure 6. The control card for median and amplitude, drawn up for 31.10.17

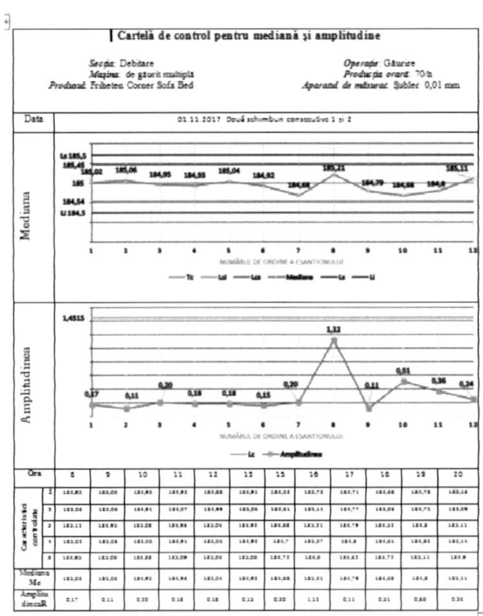

Figure 7. The control card for median and amplitude, drawn up for 1.11.17

IOP Conf. Series: Materials Science and Engineering **749** (2020) 012013 doi:10.1088/1757-899X/749/1/012013

6. Conclusion

The analysis of the drilling process during 16 shifts revealed the following:

For dimension ϕ 10.5 ± 0.2 mm:

- for the survey samples of 30 October / shift 1 the successive parameters of the median are on the same side of the mean value, towards the upper values, and they are located outside the control and tolerance limits; also for shift 2 there are parameters outside the control limits;
- for the survey samples of 1 November / shift 1 there are successive parameters of the median on the same side of the mean value, but within the control limits, and for shift 2 there are also parameters outside the control limits.

For dimension 185 ± 0.5 mm:

- On 30 October / shift 1 the median parameters are also on the same side of the mean value, towards the upper values, and they are also located outside the control and tolerance limits; for shift 2 the median parameters were located within the control limits;
- For samples of 1 November / shift 1 and 2 the median parameters fall on both sides of the mean value, within the control limits.

In all analysed cases the amplitude parameters were located within the control limits.

It could be concluded that on 30 October the adjustment of the machine was faulty, and then the process parameters were reset so that on 1 November the results were acceptable. However, the fact that on 1 November / shift 1, more than 2/3 of the parameters were situated in the middle third of the interval may raise a suspicion that the recorded data were "doctored".

Nevertheless, an amplitude jump should be noted at the beginning of shift 2, which may be associated with changing the operator on the machine. The explanation could be frequent machine adjustments, which would result in lower productivity.

References

[1] Bulgaru M, Bolboacă L 2001 *Ingineria calității. Managementul calitatii, statistica si control masurari in 3D* Ed. Alma Mater ISBN 973-85153-0-0 Cluj-Napoca

[2] Ivan M, Antonescu N, Dumitraş C, Rusan Gh, Bădescu Gh, Popescu I 1980 *Masini-unelte si control dimensional* Ed. Didactica si Pedagogica Bucuresti

[3] Banica M and Medan N 2015 *Statistical Analysis of the Experimental Data Obtained in Studying of Water Jet Cleaning* Academic Journal of Manufacturing Engineering ISSN 1583-7904 Vol. 13 Issue 1 pp.6-11

[4] Cathy K 1999 *Excel pentru Windows* Ed. Teora ISBN 973-601-457-6 Bucuresti

[5] Tănăsescu I 1987 *Controlul statistic al produselor si proceselor* Ed. Didactica si Pedagogica Bucuresti

IOP Conf. Series: Materials Science and Engineering **749** (2020) 012014 doi:10.1088/1757-899X/749/1/012014

Determination of the cutting parameters and forces for the rolling surfaces machining in the re-manufacturing of the wheelset wheels: some research and results

A L Ghionea[1,a] and I G Ghionea[2]

[1] University Politehnica of Bucharest, Department of Robots and Production Systems, Bucharest, Romania
[2] University Politehnica of Bucharest, Department of Manufacturing Engineering, Bucharest, Romania

E-mail: [a] ionut.ghionea@imst.pub.ro

Abstract. The paper presents theoretical and applicative aspects regarding the turning machining of the rolling surfaces of the wheelset wheels. The surface profile (generatrix) is defined by 9 areas of points whose coordinates are determined according to the UIC 510 relations. The directrix of the rolling surface must also meet the circularity precision requirements. For processing, it is used a specialized re-manufactured machine tool that has two working units. Each unit is provided with 2 axes numerically controlled. The cutting tools used for turning are based on catalogs, as recommended by the manufacturing companies. The authors present results and recommendations of the theoretical and experimental researches on the main parameters of the cutting process, requirements of conformity and of quality for the profile and the rolling surface. The cutting parameters and the number of passes are established in the possible range of speeds and feed rates provided by the kinematic and control structure of the lathe. The working parameters can be adjusted corresponding to the values of the cutting forces and torques. The data and results of the application of the machining process must meet the requirements regarding the dimensional and shape precision and roughness of the machined surfaces, but also those considering the dynamic behaviour of the machine during the processing of wheels, minimum costs, low energy consumption, optimum machining times. These are important for evaluating kinematic and dynamic precision, but also for different analysis simulations and CAD modelling.

1. Introduction

Wheelsets are the most loaded components of railway vehicles and present a continuous process of wearing due to different and difficult running conditions: varying loads of the wagons, modifications of the rails surfaces and of the wheels profiles, sudden temperature variations due to brakes, curved segments of rails, breakings, etc. At regularly occurring intervals, the wheelsets are checked and when they reach a critical level of wear, they should be reshaped or replaced [3]. The complete replacement is difficult and it is applied only if the material allowance to be removed surpass a certain limit.

Figure 1 presents two samples of worn wheels, the first one has a light wear, but the other is significantly worn. If the wear includes also deep cracks, the reshaping process cannot be achieved and the wheelset will be replaced for the train safety. Re-profiling is normally performed to restore the geometry of the wheel running surface for purpose of maintenance, but experience showed that the quantity of useful metal cut during this reshaping is greater than the abrasion-induced metal loss during operation. There are many studies on wheel reshaping strategy (machine tool, turning tools, NC program, control methods before and after the reshaping process etc.) and profile optimization design.

Following the research project [1], several phases of machine re-manufacturing were conducted, consisting of kinematic improvements, of actuation and control of the processed surfaces. Thus, experimental kinematic solutions and series of constructive and functional parameters were determined and applied to determine the optimal processing options.

Figure 1. Examples of wheels with worn rolling surfaces

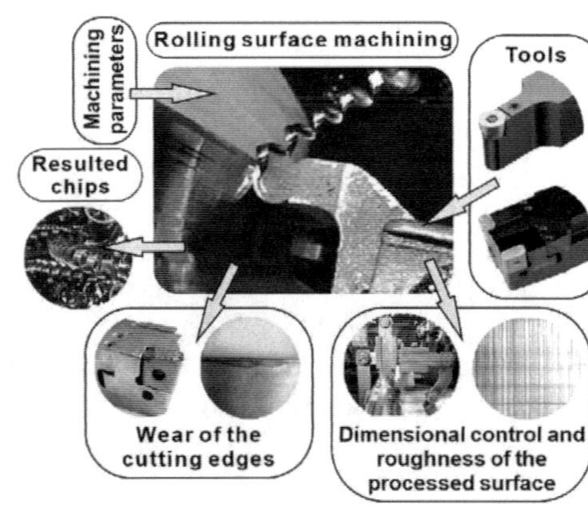

Figure 2. Schematic representation of the experimental setup

Based on these parameters, the remanufacturing solution of the Rafamet UBC 150 lathe machine was chosen to increase the generation precision of the rolling surface profile. Two feed - positioning kinematic chains have been adapted to each working unit, on X and Z axes of the machine.

The schematic representation in figure 2 shows the main aspects of the study carried out in the research project [1], with reference to: wheel surface analysis, choice of cutting tools, setting of the cutting parameters, accomplish the shape conditions, dimensions and roughness of the processed surfaces, limitation of the cutting forces, maintaining an optimal durability of cutting tools, etc.

Figure 3. Main view of the machine after remanufacturing

Figure 3 shows the final assembly of the machine having two main shafts coaxial disposed, provided with plates and clamping devices. On each shaft it is placed a fixed headstock on the machine's bed frame. Also, there can be observed the two working units of the re-manufactured machine. Between them is located the operator's workplace and the command panel.

Each working unit has in its structure two radial slides, each contained in a casing mounted on a longitudinal slide. The working units ensure simultaneous machining of both wheels of the wheelset.

The re-manufactured solution, in terms of driving, consists of adapting 2 + 2 electric actuators and 2+2 steering ball screws with precision intermediate elements.

2. Rolling profiles and materials for wheels

The rolling profile of the railway vehicles wheels must fit within the geometrical characteristics imposed by internal and international instructions [11], [13], being an important factor for the passengers' comfort and decrease of the environmental impact by reducing the noise caused by the wheel-rail contact [10], according to the current standards.

In the last years many research projects studied the development of modernized or new advanced equipment, in order to enable precise machining and accurate measurement of the wheel and rolling profiles, both in static and dynamic mode [5].

The surface and profile of the railway vehicles wheels have a complex geometry that must fit within certain sizes imposed by national and international standards [7].

The classification and indication of the geometric parameter limits for the railway vehicle wheels are indicated by EN 15313/2010 standard. Each wheel is provided with a flange that is designed to provide the vehicle guidance inside the two rails of the railway, with a rolling surface which runs on the rail.

The wheel profile is given by its periphery contour in a median plane of the wheelset. This profile is complex, being composed of a multitude of spline curves [13], with connection zones having a defined geometry, in order to form a continuous curve from a geometrical point of view in any point.

Both the profile of the rail and of the wheel are governed by UIC recommendations, in the case of the wheel profile existing versions derived from the profile recommended determined by each railway authority depending on the characteristics of the railway network.

Usually, the wheels have the diameter in certain dimensional domains, which are established and used for international transport vehicles intended for passengers.

Diameter values may vary from minimum to maximum, as follows: 330-390 mm, 390-470 mm, 470-550 mm, 550-630 mm, 630-680 mm, 680-760 mm, 760-840 mm, 840-920 mm and 920-1000 mm, according to standards [11], [13]. These dimensions are imposed for both wagons and locomotives.

The steel materials used for wheels and rails across Europe have predominantly pearlitic structures containing hard cementite lamellae for a high wear resistance. A pearlitic microstructure, ensures a better resistance than bainitic or martensitic structures [5], [14].

Although the standard UIC 812-3 [14] for solid wheels lists seven types of steel, which mainly differ in carbon content, heat treatment state and strength, the actual SR EN 13262 [11] contains only four types (ER6, ER7, ER8 and ER9) and this represents the current state of technology in Europe.

The maximum percentages of the various specific elements are presented in table 1.

Table 1. Steel categories for wheels and chemical composition

	Maximum content, in %										
Steel category	C	Si	Mn	P	S	Cr	Cu	Mo	Ni	V	Cr+Mo+Ni
ER6	0.48	0.40	0.75	0.020	0.020	0.30	0.30	0.08	0.30	0.06	0.50
ER7	0.52	0.40	0.80	0.020	0.020	0.30	0.30	0.08	0.30	0.06	0.50
ER8	0.56	0.40	0.80	0.020	0.020	0.30	0.30	0.08	0.30	0.06	0.50

Regarding these percentages, [5] considers that low carbon material has a better thermal crack resistance and fracture toughness but has less resistance against wear.

In selecting the wheel material, an appropriate level of carbon content is mandatory to be selected on the basis of the braking condition in service and the actual situation of maintenance.

3. Stages of the study and calculus of the process parameters

3.1. Initial setup

The wheelset is first subjected to a visual analysis of the rolling surfaces of the wheels and then there were determined by measuring the dimensions of the worn rolling surfaces.

An evaluation document is created based on measurements mainly referring to: the diameter of the wheel in the plane of the rolling circle, the width of the wheel, the distance between the inner front faces of the two wheels, the distance between the rolling circles, the flange gradient, the wheel condition of wear etc.

Depending on the wear of the two wheels, there are indicated in [9] these states: low wear, surface wear with flattening, exfoliation, thermal cracks.

The results of the measurements made for each wheel are used to determine the reprofilation data and the number of passes. Based on these, there are set the values for technological parameters, as well as other data necessary for writing the program. It is then uploaded in the numerically controlled equipment (two axes X1, Z1 and X2, Z2 for each wheel) for positioning and profiling movements.

After the fixture of the wheelset between the lathe struts, it is checked at low speed if the grip is correct. It is also important to analyse the state of wear (within the limits of durability, about 45 minutes) for each one of the two cutting tools positioned in the seats of each radial slides. The parameters required for machining are chosen from those indicated in the machine's technical documentation, thus: the range of D_{nc} spindle speeds: 9 ... 35.5 rpm in 7 steps (9, 11.2, 14, 18, 22.4, 28 and 35.5); longitudinal feed range D_{fL}: 0.2 ... 1.25 mm/rev. The range of feed speeds is within the D_{vf} values: 1.8 ... 17.5 mm/min.

3.2. Components of the cutting force

In the cutting process, in the contact area of the cutting tool's edge with the machining allowance (figure 4), the resulting forces and torque involve and load all the components of the technological system. Determining them by calculation and measurement is important to establish the load of the electric motors and of the main kinematic feed / positioning mechanisms, of the other assemblies and structural elements of the lathe.

Figure 4. Functions of quality assurance in a broad sense

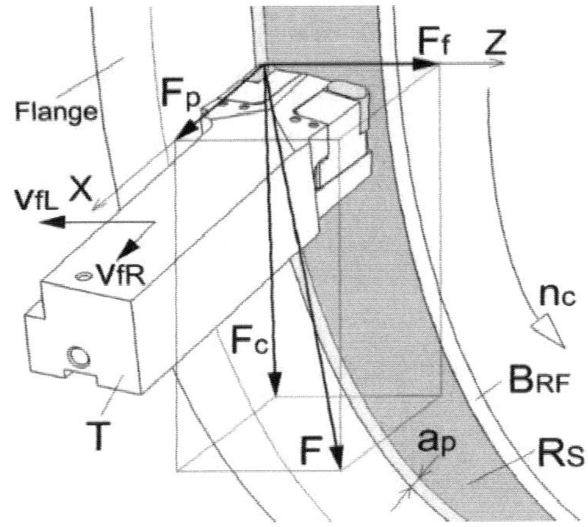

Figure 5. Total cutting force components

Thus, it is very important to determine the forces and moments of cutting, to optimize the cutting parameters [6] and to evaluate the dynamic behavior of the machine. The components of the total cutting force F (figure 5), denoted F_c, F_f and F_p, are defined according to the ZX horizontal working plane. These are obtained by projecting the total force F on the directions of the kinematic reference systems axes for the two working units, as follows:
• F_c - the cutting force is important for determining the loads of the main shaft; • F_f - the feed force is important for dimensioning the mechanisms in the longitudinal feed kinematic chain; • F_p - the passive force determines the constructive solution of the screw-nut mechanism with intermediate elements from the kinematic chain for feed/radial positioning.

The tool T has a feed move with the v_{fL} speed in the longitudinal direction and with the v_{fR} speed in the radial direction. The result of the two moves lies on a trajectory tangent to the RS rolling surface profile. The force components have the application point in the contact spot of the tool T cutting edge with the machining allowance a_p. The main cutting movement is performed by the wheelset with the nc speed which is adjustable step by step. The total machining allowance for the re-profiling of the wheelset wheels is determined by the amount of wear of the two rolling surfaces. The difference between the processed surfaces of the wheels regarding the diameters of the rolling circles should not be greater than 1 mm [11]. To obtain the roughness and precision conditions of the rolling surface, it is recommended that the machining allowance to be removed in two or three passes.

In the literature, there are some differences regarding the calculus relations of these forces. The results can be of different values and fit within certain limits. Thus, according to [8], the cutting force F_c is determined by the calculus relation (1) according to [1]:

$$F_c = b \cdot h \cdot k_c \cdot K_{\gamma o} \cdot K_v \cdot K_{ms} \cdot K_{uz} \cdot K_{prd} = b \cdot h^{1-m} \cdot k_{c1.1} \cdot K_{\gamma o} \cdot K_v \cdot K_{ms} \cdot K_{uz} \cdot K_{prd}, \text{ [N]} \quad (1)$$

In this relation, the following values are considered: $m = 0.20$ is a polytropic exponent; $K_{\gamma o} = 0.81$ - a correction factor according to the orthogonal departure angle γo; $K_v = 1.11$ - a correction factor according to the cutting speed; $K_{ms} = 0.95 \ldots 0.9$ - a correction factor depending on the material of the cutting tool insert; $K_{uz} = 1.3 \ldots 1.5$ - a correction factor according to the wear of the cutting tool edge; $K_{prd} = 1.15$ - a correction factor according to the machining process. These values correspond best to the real cutting conditions.

Between the specific cutting force k_c and the specific cutting resistance $k_{c1.1}$ exists the calculus relation, according to [2]:

$$k_c = \frac{k_{c1.1}}{h^m}, \text{ [N/mm2]} \quad (2)$$

The specific cutting force k_c is a material characteristic from which the wheels of the wheelset are manufactured and takes values according to the thickness of the chips (eg: $h = 0.05$ mm, $k_c = 3590$ N/mm2; $h = 0.32$ mm, $k_c = 2760$ N/mm2). The feed force F_f and the passive force F_p are determined with similar calculations as that of the cutting force F_c:

$$F_f = b \cdot k_{f1.1} \cdot h^{1-x}, \text{ [N]} \quad (3)$$

$$F_p = b \cdot k_{p1.1} \cdot h^{1-y}, \text{ [N]} \quad (4)$$

The values for constants and polytropic exponents are considered: $k_{f1.1} = 355$ N/mm2, $1\text{-}x = 0.38$, $k_{p1.1} = 255$ N/mm2, $1\text{-}y = 0.57$ (for steel categories ER6 and ER7) [2]. The thickness h and the width b of the nominal chip are determined according to the approach angle Ƙr, to the longitudinal feed rate f_L and, respectively, to the cutting depth a_p: $h = f_L \cdot \sin \kappa_r$ (5), $b = a_p / \sin \kappa_r$ (6).

Figure 6 shows the shape and orientation of the active part of a rectangular insert [9], according to figures 4 and 5, in two successive positions corresponding to the longitudinal feed motion with the parameters f_L and a_p.

The insert is oriented in the tool body at a position angle of 84^0 [4], [9]. In contact with the wheel surface, the insert has a main side cutting edge (approach) angle Ƙr, which determines a chip section defined by the parameters *h* and *b*.

The shape of the chip is also determined by the radius *r* of the active part of the insert's cutting edge. According to the catalogue data [9], the insert has the following dimensions: $r = 4$ mm, $l = 19.05$ mm and $iW = 10$ mm, values that also influence the shape of the chip and the size $R\varepsilon$ of the surface roughness.

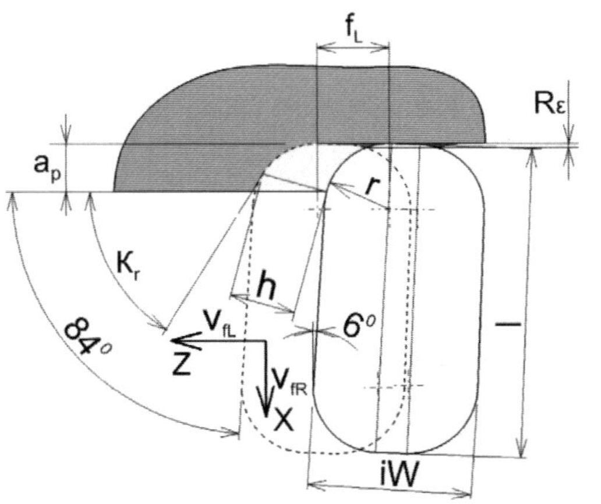

Figure 6. Geometric elements of the cutting edges, cutting parameters

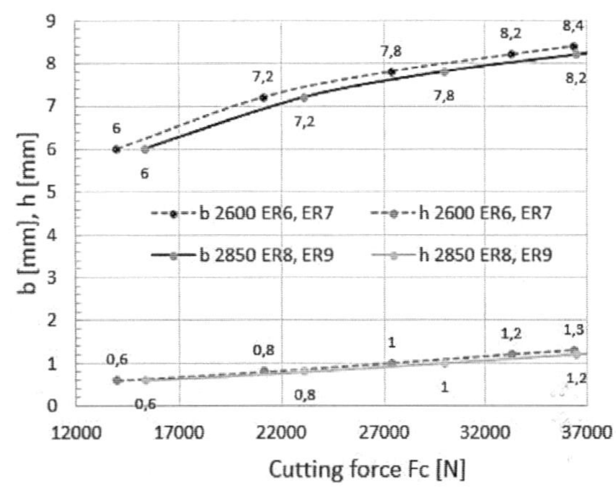

Figure 7. Cutting force Fc variation

On the basis of the forces determination relations, using as parameters the width and the thickness of the chips (determined experimentally after some processing sets with feeds $f_L = 0.4 \ldots 1.2$ mm/rev and cutting depth $a_p = 2 \ldots 9$ mm) their values are calculated, some of them are presented in Table 2.

As can be seen from the values calculated in Table 2, the main component of the force *F* (figure 5) is the F_c force.

In figure 7 it was represented the variation of this force according to the geometric parameters of the chips. The $k_{c1.1}$ value was considered for two categories of recommended steels [7], [14].

Table 2. Values of the cutting force components

No.	Steel category	b [mm]	h [mm]	$k_{c1.1}$ [N/mm^2]	F_c [N]	F_f [N]	F_p [N]
1	ER6, ER7	6	0.6	2600	13988.2	1754.2	1143.5
2		7.2	0.8		21129.74	2348.2	1616.7
3		7.8	1		27364.3	2769	1989
4		8.2	1.2		33285	3119.8	2320
5		8.4	1.3		36351.61	3294.6	2487.5
6	ER8, ER9	6	0.6	2850	15333.22	1803.6	1188.3
7		7.2	0.8		23161.45	2414.3	1680.1
8		7.8	1		29995.48	2847	2067
9		8.2	1.2		36485.48	3207.7	2411
10		8.4	1.3		39846.96	3387.4	2585.1

Determination of the total force *F* components (F_c, F_f and F_p) is also important for establishing the clamping forces to drive the wheelset into the main cutting movement.

IOP Conf. Series: Materials Science and Engineering **749** (2020) 012014 doi:10.1088/1757-899X/749/1/012014

The F_c component determines the moment of cutting for the two wheels simultaneously processed. The resulted torque must be taken over by the grip and driving system elements. Fixing the wheelset must be secure. The clamping systems are controlled by conditioning in the operation of the lathe's command system.

4. Considerations and recommendations on tools wear

The durability of the tools cutting edge for turning the wheels rolling surfaces is within 25 ... 48 minutes. This corresponds to the admissible limit of wear on the rake face VB_{max} = 0.3 ... 0.4 mm. At this size, the superficial layer of the plate's super hard material is already removed on the worn cutting edge, both on the rake face and on the clearance face.

The cutting edge becomes worn (figure 8), with a sharpening radius much larger than the initial state, and increases the frictions and the cutting temperature. The quality of the processed surface is less affected by additional hardening. It is recommended to change the insert at a maximum wear VB_{max} = 0.4 ... 0.5 mm. Above this wear, the insert can still work, but the cutting edge resistance decreases and the cutting temperature increases. Also, the risk of breaking the cutting tool increases.

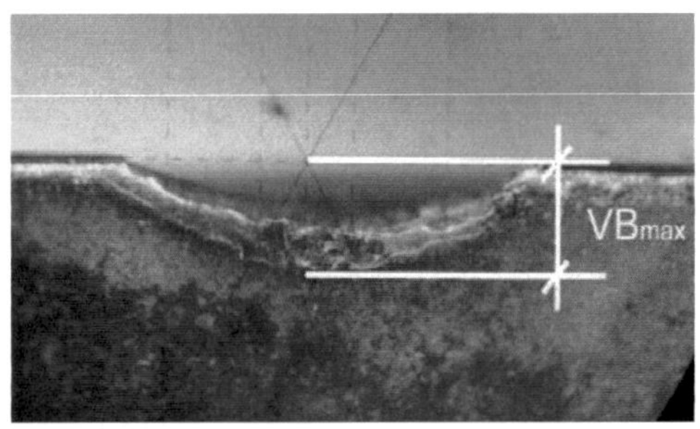

Figure 8. Detail of worn area on the insert's cutting edge

The wear of the cutting edge suddenly increases to 0.7 ... 1.5 mm, cutting forces also increase much. As a result, the processing process must be stopped.

5. Conclusions

The simultaneous machining of the rolling surface of the wheels of the wheelset is processed on a specialized lathe in conditions of dimensional accuracy, of shape conformity and quality. The characteristics of the material to be machined, the surface condition and the unevenness of the machining allowance make the cutting process to produce vibrations, wear of the cutting edge and some unevenness of the surface quality. Taking into account the theoretical and experimental researches done in the frame of the project for the remanufacturing and NC equipment of the Rafamet UBC 150 lathe, some specific data resulted for its optimal use, necessary in the running phases of this lathe.

The research team has carried out a wide range of studies, proposed and applied technical solutions. Many results of these researches are published and already in use in the daily activity of the Wagons Factory of Aiud, Romania.

Acknowledgement

The paper presents researches in the frame of Partnerships in Priority Areas Programme-PNII supported by MEN-UEFISCDI, in project PN II-PT-PCCA-2013-4-1681, *Mechatronic system for measuring the wheel profile of the rail transport vehicles, in order to optimize the reshaping on CNC machine tools and increase the traffic safety*, 2014-2017.

IOP Conf. Series: Materials Science and Engineering **749** (2020) 012014 doi:10.1088/1757-899X/749/1/012014

References

[1] Contract 250/2014, 2015-2017, PN-II-PT-PCCA-2013-4-1681 Scientific reports 1-4
[2] Degner W, Lutze H, Smejkal E 2009 *Spanende Formung. Theorie, Berechnung, Richtwerte* Carl Hanser Verlag Münche-Wien
[3] Ghionea I, Ghionea A, Cioboată D, Ćuković S. 2016 *Lathe machining in the era of Industry 4.0: remanufactured lathe with integrated measurement system for CNC generation of the rolling surfaces for railway wheels* 13th IFIP WG 5.1 International Conference on Product Lifecycle Management chapter 27 vol. 492 Springer University of South Carolina Columbia USA pp.296-308
[4] Ghionea I, Ghionea A, Predincea N. 2018 *Cutting parameters and analysis by FEA simulation of their influence on the re-profilation of the wheels of the railway wheelset* Acta Technica Napocensis Vol. 61 Issue I Series Applied Mathematics, Mechanics and Engineering Technical University of Cluj-Napoca Romania ISSN 1221-5872 WOS: 000428903500013 pp.73-84
[5] Jianxi W, Xin X, Yongjie L. 2017 *Study on cutting form and surface machining quality of wheel tread under reprofiling* Advances in Materials Science and Engineering Vol. 2017 article ID 6950351 Hindawi
[6] Oberg E, Jones F. 2012 *Machinery's Handbook* 29th Edition pp.2120-2136
[7] Okagata Y. 2016 *Design technologies for railway wheels and future prospects* Nippon Steel&Sumitomo Metal Technical Report
[8] Perovic B. 2006 *Machine tools guide. calculation and construction* Carl Hanser Verlag Munchen-Wien
[9] Sandvik 2014 *Railway turning, re-turning and new wheel turning* Application Guide
[10] Shevtso, I. Y. 2008, *Wheel/rail interface optimisation* Phd. Thesis Delft University of Technology Faculty of Civil Engineering and Geosciences Netherland
[11] SR EN 13262+A2 2011 *Romanian standard, Railway applications-Wheelsets and bogies-wheels, product requirements*
[12] SR EN 13715+A1:2011 *Romanian standard, Railway applications-Wheelsets and bogies-Wheels-Tread profile*
[13] UIC 510-2 2004 *Trailing stock: Wheels and wheelsets. Condition concerning the use of wheels of various diameters* 4th edition
[14] UIC 812-3 1984 *Technical specification for the supply of rolled solid wheels of non-alloy steel for traction and rolling stock* 5th edition

IOP Conf. Series: Materials Science and Engineering **749** (2020) 012015 doi:10.1088/1757-899X/749/1/012015

Sliding surface structure comparison of 3D printed polymers using FDM and DLP technologies

M M Hanon[1,2a] and L Zsidai[1]

[1] Institute for Mechanical Engineering Technology, Mechanical Engineering Doctoral School, Szent István University, H-2100 Gödöllő, Hungary
[2] Baquba Technical Institute, Middle Technical University (MTU), Baghdad, Iraq

E-mail: [a] harba.Muammel.M.Hanon@phd.uni-szie.hu

Abstract. This study aims to review the surface structure of the parts that are manufactured using various techniques of 3D printing. Particularly, two different methods were employed in the current work: Fused Deposition Modelling (FDM) and digital light projection (DLP). The output of both technologies regarding the surface construction aspects was evaluated and compared. For determining the features of the surfaces, an optical microscope was utilised for taking the pictures from the cross-sectional area (upper and lower edges) and the outer face of the samples. Various magnification values have been investigated to find the best sights at the surface structure of the printed bodies. Surface roughness was examined due to its relevance to the texture properties of the fabricated pieces. The tribology of the workpieces has been tested as well. The results show that the products of the FDM technique have been characterised with a rough surface and anisotropic properties which were reflected on the examined measurements. In contrast, the DLP method has revealed objects with a very smooth surface and approximately homogeneous structure.

1. Introduction

Three-dimensional (3D) printing technology has become very popular in the recent decade. It turned in to a robust alternative for building many parts which are involved in attractive applications [1], [2]. The products in the 3D printing are created as a layer by layer. Thus, manufacturing of components with a complex structure is enabled [3], [4]. Fused Deposition Modeling (FDM) is one of the 3D printing techniques based on extrusion principle. With FDM, thermoplastic material is extruded. The molten material layer is printed on top of the previous layer (see figure 1), then almost merges and hardens immediately when the printing nozzle leaves.

The FDM technique is an inexpensive 3D printing method and the most commonly utilised in home 3D printers [5], [6]. Nowadays, the most often materials used are ABS-Acrylonitrile butadiene styrene (oil-based plastic) [7], [8] and PLA-Polylactic acid (bio-based plastic) [9], [10].

Digital Light Processing (DLP) is a 3D printing technology based on the Photo-polymerization mechanism. This technique uses a projector UV-light for hardening of liquid photo-curable resins as layer by layer. The whole layer will be exposed with UV-light at once through a mask, which makes this technology basically faster than FDM. The process contains a vat filled with a light-curable liquid resin. The platform rises, and a layer of the resin distributes evenly as shown in figure 2. This process is iterated up to the object is created [12].

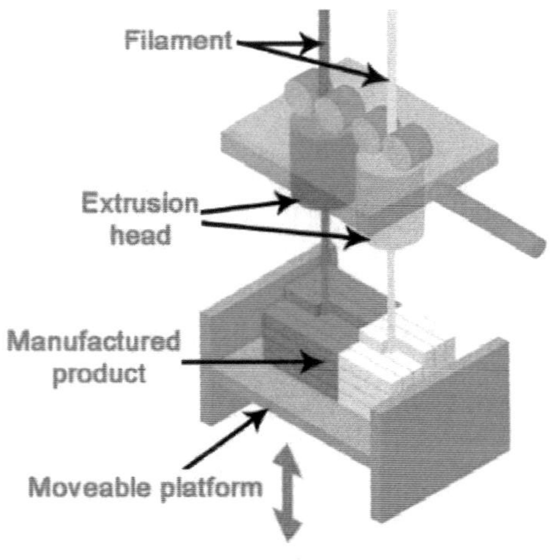

Figure 1. FDM technique mechanism [11]

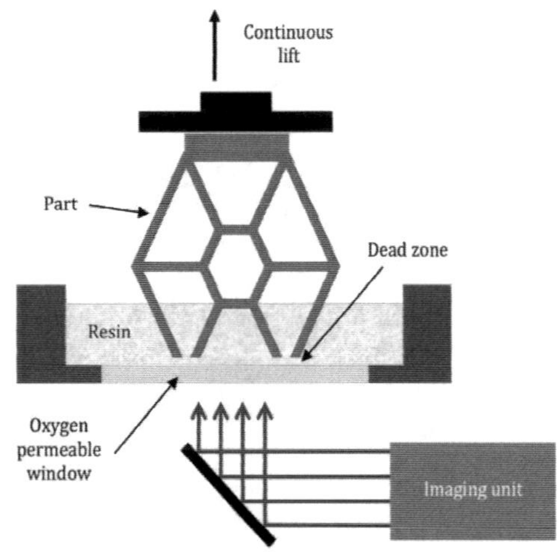

Figure 2. Schematic of a DLP device [13]

The surface properties are considered a critical issue for certain implementations. Hence, studying the surface structure of the produced objects is an essential matter. The surface morphology of the printed parts has a key role in enhancing their performance in terms of lowering friction and wear. There are many researchers studied the microstructure of 3D printed bodies, e.g. glass for bone scaffolds [14], tissue engineering [15], microfluidic chips [16], and fibre-reinforced polymer composites [17]. However, still there is a lack in the literature about the surface construction of the printed bulks of polymers with different methods.

In the present work, Fused Deposition Modelling (FDM) and Digital Light Projection (DLP) techniques of 3D printing have been used. Numerous cylindrical tribology test specimens were fabricated with diverse printing options and prepared for the experiment. The resulting of both technologies (FDM and DLP) concerning the surface construction aspects was examined and compared. An optical microscope was employed to determine the features of the surfaces of the samples. The tribological behaviour and surface roughness of the fabricated pieces were evaluated. This research forms the basis for a prospective series of tribological studies with basic knowledge of surface morphology.

2. Materials and methods

2.1. The employed FDM and DLP 3D printers

The printed test pieces were manufactured using two different 3D printers work according to the FDM and DLP technologies. The models of the used 3D printers are WANHAO duplicator 6 for the FDM and WANHAO D7 for the DLP, which are exhibited in Figures 3 and 4 respectively.

For the FDM 3D printer WANHAO duplicator 6, the G-Code of the 3D model can be incorporated directly from the SD card. The model could be designed using any CAD software, which is in this case Solid work software. The file must be saved as “stl” format due to it is compatible with most slicing programs. Cura (Wanhao edition v16.01) software was used as a slicer for this work. The FDM printing technology utilises the materials in a filament form.

Regarding the DLP 3D printer WANHAO D7, all parameters and commands could be given from a supplement controller supplied with the printer (Shown in figure 4). For the slicing purpose, CreationWorkshop software used which is the recommended and prepared for the same printer. “cws” format is the file extension that can be recognisable by the controller. The model file can be uploaded to the controller using a pen drive. The material that employed for the DLP printing is the so-called resin which is a chemical liquid curable by light. The Technical specifications of both 3D printers (FDM WANHAO duplicator 6 and DLP WANHAO D7) are listed in table 1.

IOP Conf. Series: Materials Science and Engineering **749** (2020) 012015 doi:10.1088/1757-899X/749/1/012015

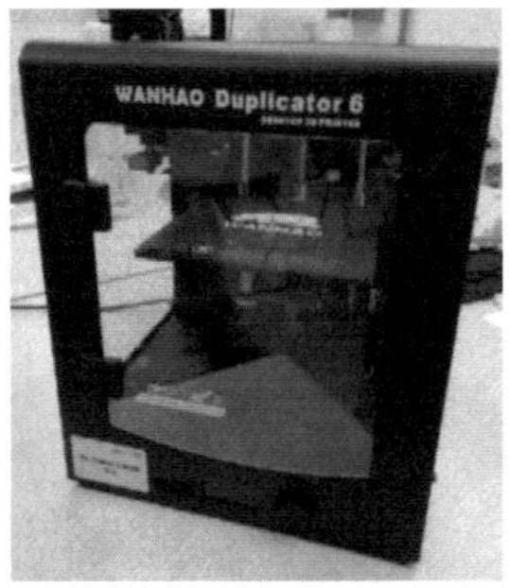

Figure 3. The FDM 3D printer WANHAO duplicator 6

Figure 4. The DLP 3D printer WANHAO D7

Table 1. Technical specifications of the used printers

Specifications	WANHAO duplicator 6 (FDM)	WANHAO D7 (DLP)
Technology	FFF (Fused Filament Fabrication)	DLP LCD printer
Printing materials	PLA, ABS, PVA, PEVA and HIPS	405nm Resin
Printable volume	200x200x175mm	120.96*68.5*180mm
Layer thickness	50-400µm	35-100µm
Material form	Filament diameter of 1.75mm	Resin
Heating or curing source	Extrusion nozzle (0.4mm)	5.5inch LCD, 405nm UV lamp
Print speed	30-150mm/s	30mm/hour (in height)
Operating temperature	180-260°C	23 °C
Position precision	X: 12.5µm, Y: 125µm and Z: 5µm	X: 0.01µm, Y: 0.01µm, Z: 4µm

2.2. Printing options

Once the STL file of the model completed, it is imported to the previously mentioned slicing programs. Hence, different settings can be examined when printing the test pieces using the FDM and DLP printers. These settings are expected to affect the properties of the finished object. The primary investigated option is the print orientation (horizontally "X", 45° angle, and vertically "Z") which are shown in figure 5 (a). In the FDM, raster direction means, the orientation of the extruder head when filling the printing material within the contour of the printed part. The angle of the raster directions of the test piece with the longitudinal axis can be adjusted to any value. Also, successive layers may have different orientations. So, there is a possibility of variation, e.g. if the raster direction angle of the first layer is 45° the second could be built at 135° (see figure 5 (b)). For the present work, while varying the print orientation, the raster direction was set as crossed 45/135°. Layer thickness is also one of the parameters that can be varied in the printing process. These parameters are expected to influence the results of the subsequent tests (Tensile, tribology, bending, impact, and surface roughness). Table 2 summarises all the examined parameters of the current study.

Table 2. The employed options for printing the tribology test specimens

Parameter	**FDM**	**DLP**
Raster angle	Crossed 45/135°	**-
‡Orientation	*X, 45°, Z	*X, 45°, Z
Thickness	200µm	100µm

*The variables of each examined set.
‡Where X, 45°, Z refer that the orientations of printing are: horizontally, 45° angle, and vertically consecutively.
**No raster direction in the DLP technology, since every layer is built once after it exposes to the UV light.

IOP Conf. Series: Materials Science and Engineering **749** (2020) 012015 doi:10.1088/1757-899X/749/1/012015

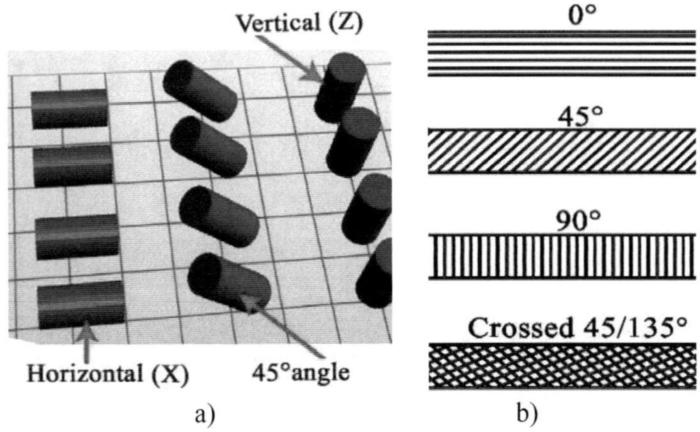

Figure 5. Print settings: a) orientations, b) raster direction

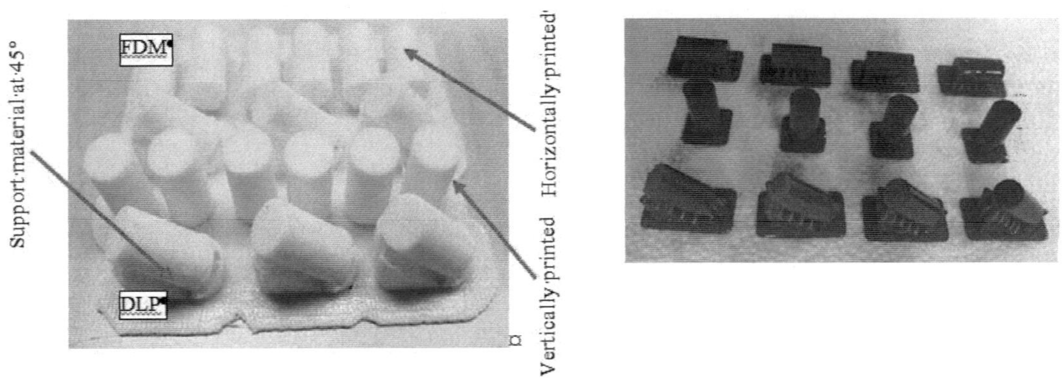

Figure 6. Sets of 3D printed tribology specimens

The 3D printed specimens for tribology test have a diameter of 8 mm and a length of 15.2 mm. Some sets of tribology test pieces printed in various orientations using FDM and DLP are exhibited in figure 6.

2.3. The achieved tests

The printed specimens were examined by three tests which are tribology, surface structure (optical microscope) and surface roughness. The tribological measurement was performed at tribology laboratory in Szent István University under the laboratory conditions. The first step in the measurement process is to connect the measuring circuit, which consists of a computer, Spider 8 measuring converter, a tribotester, and an inverter. Spider 8 is a strain gauge measurement device for measuring load, force, and wear. The indispensable part of the measuring system is the tribotester, which is in this case compatible with polymers. Figure 7 displays the structure of the used PLINT TE 77 tribotester (High-Frequency Tribotest), employed by Zsidai and Kalácska in previous research [18]. Before starting the measurement, some parameters must be defined which are given in Table 3. A computerised microscope (ZEISS brand) with a high quality camera and four magnification lenses (10x, 20x, 50x and 100x) was used to study the surface structure of the samples before and after the tribology test (displayed in figure 8).

Table 3. Parameters of tribology test

Surface roughness of steel counterpart, Ra [μm]	**0.8**
Load, F [N]	**200**
Alternating motion frequency, f [Hz]	**10**
Stroke length, [mm]	**6**
Relative humidity, Rh [%]	**50**
Ambient temperature (°C)	**23**

Figure 7. Tribotester

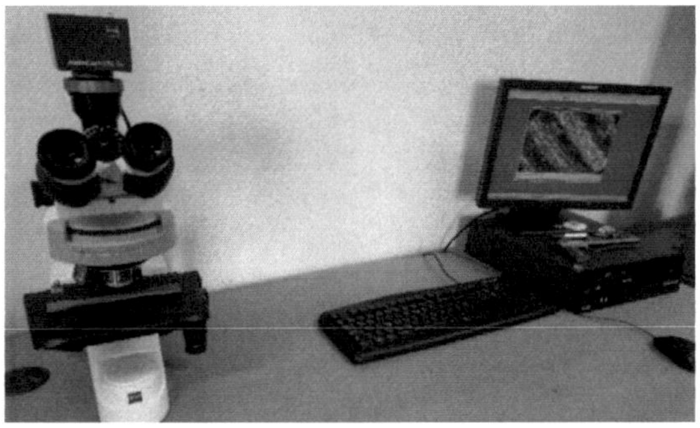

Figure 8. Optical microscope

3. Results and discussion

3.1. Tribology measurements

The friction and wear are the most critical values to be obtained for the tribological measurements. The attitude of the coefficient of friction as a function of the sliding distance has been represented in Figure 9. During the test, both static and dynamic friction coefficient were evaluated. The value of the static friction coefficient (blue line) is always higher than the relative dynamics (red line). The behaviour of wear versus the sliding distance is shown in Figure 10 for one of the inspected test pieces (PLA printed in 45° orientation by FDM method).

A number of measurements with the given materials have been carried out. For each material (printed with FDM and DLP technologies) and print orientation (horizontal, 45°, and vertical), three identical specimens were prepared and subjected to the same tribology test parameters. A comparison among the average results of the examined samples is exhibited in figures 11 and 12 concerning the static and dynamic friction coefficient, and the wear depth respectively.

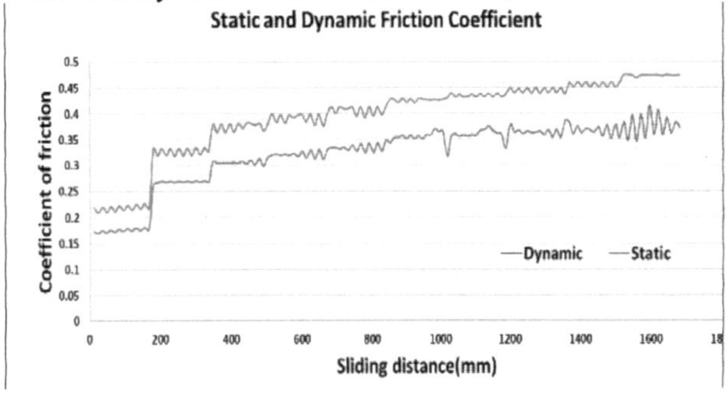

Figure 9. Static and dynamic friction coefficient of PLA printed in 45° orientation by FDM method

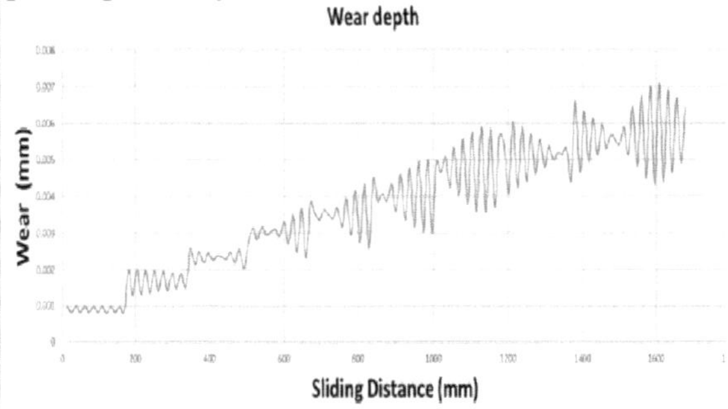

Figure 10. Wear behaviour of PLA printed in 45° orientation by FDM method

In the current work, the materials that were used in the FDM and DLP 3D printing methods are the PLA filament (white and grey colours) and WANHAO resin (red colour) consecutively. In figure 11, the red column represents the maximum value of the dynamic friction coefficient, while the blue column is complemented by the maximum static friction coefficient measured during the experiment. It can be clearly seen that the static max value of the specimens printed in the X orientation (Horizontally) is greater than those of the other orientations. The highest value of the dynamic max friction coefficient was observed at the X oriented sample (printed with the FDM method) on an average about 0.79. Whereas the lowest is noticed at the test pieces printed with FDM in 45° angle orientation, averaging 0.49, representing a 38% difference in comparison with the greatest value. Regarding the static friction coefficient, the maximum value was 0.86 for DLP technique specimens in X orientation. While the minimum value was measured at FDM printed in 45° orientation on an average of 0.53, which also brings 38% difference between the highest and lowest static max values. The biggest variation between static and dynamic friction coefficient has been measured in DLP test specimens, which indicates the high tendency of these materials for the stick-slip phenomenon.

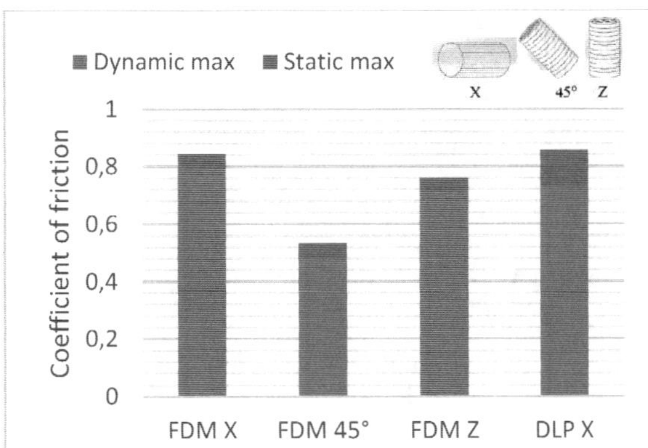

Figure 11. Comparison of friction coefficient among tested specimens regarding print method and orientation

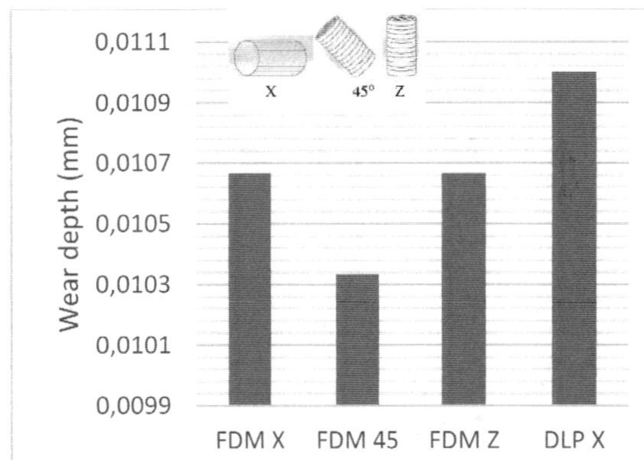

Figure 12. Comparison among the average wear values of specimens concerning print method and orientation

The comparison among the average wear for the specimens printed with various methods and orientations is illustrated in figure 12. The values of average wear have been calculated from three identical specimens' measurement. The maximum wear values belong to the specimens printed with DLP method. The highest measured wear is 0.0110 mm at the DLP X oriented samples. The smallest value was measured at the FDM specimens printed in 45° orientation with an average of 0.0103 mm. It is clear from Figures 11 and 12 that similar attitudes are observed between the behaviour of the coefficient of friction and wear among the tested samples regarding the rise and drop of the values.

3.2. Surface roughness measurements

Taking a glance by the naked eye on the samples printed using FDM and DLP methods, it is easy to realise the obvious difference in their surface roughness (see Figure 13). To study it accurately, surface roughness test was accomplished for the specimens manufactured in the three orientations (X, 45°, and Z) by both mentioned printing technologies. The measurements were examined from the upper side and the perimeter of each tested cylinder.

Figure 14 illustrates the values of the surface roughness (R_a). The surface roughness of the worn area after the tribology test (at the cylinder perimeter) was checked. It is apparent from the overall results that surface roughness of the worn area after the tribology test is much lower than other surfaces due to abrading the rough layers from the previous surface. Figure 15 (a) displays the R profile diagram of the most distinguished rough area which was recognised in the upper side of the 45°

oriented specimens fabricated by the FDM technology. While figure 15 (b) presents the R profile of the lowest surface roughness value that was measured in the upper side of the DLP samples printed at Z orientation.

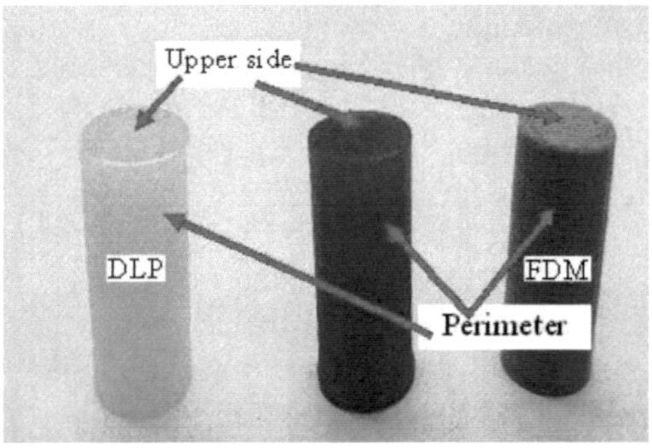

Figure 13. Examined areas in surface roughness test

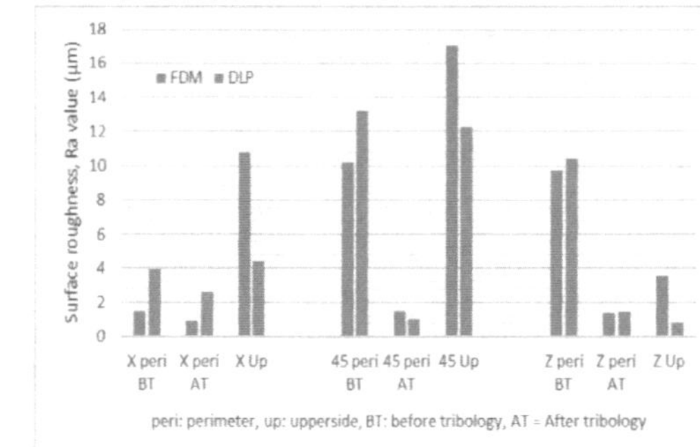

Figure 14. Comparison among the average surface roughness values in the tested specimens

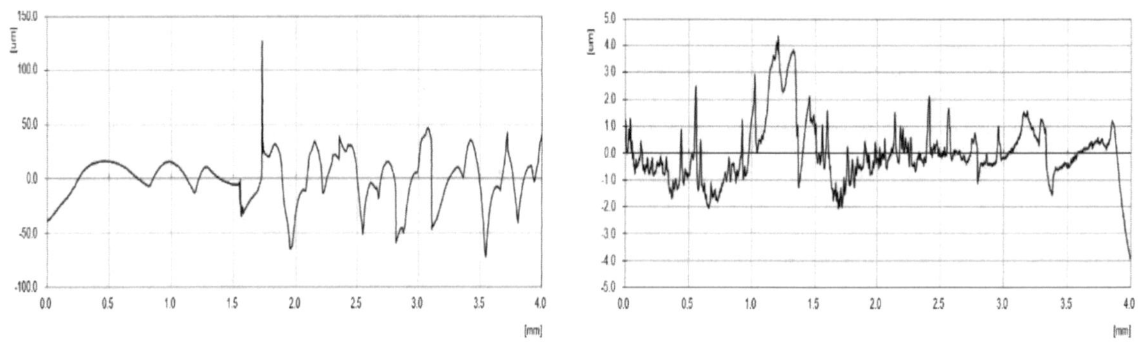

Figure 15. R profile diagram of the highest and lowest surface roughness values

3.3. *Surface morphology*

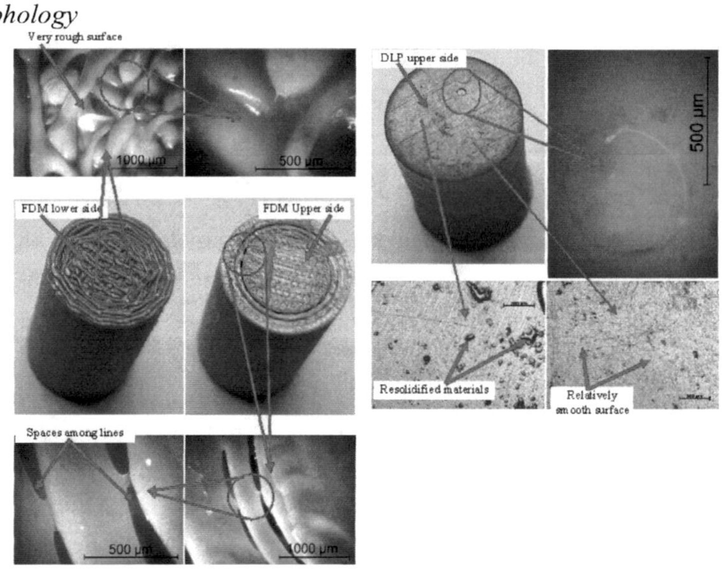

Figure 16. Surface texture of the upper and lower faces in samples printed using FDM and DLP methods

The surface structure of the lower (the face that is in contact with the platform) and upper faces of the cylindrical test pieces which printed in FDM and DLP methods has been demonstrated in figure 16. In general, it can be clearly seen that DLP technology has displayed much more smooth surfaces than the FDM. The lower side of the FDM exhibited very rough texture due to the contact with the raft layer (the layer between the built sample and the platform of the 3D printer). Spaces among the lines have been revealed at the upper side which indicates the anisotropy of the FDM method. On the other hand, it is apparent that DLP specimens have demonstrated relatively smooth surfaces in comparison with the FDM. Resolidified materials were noticed with the DLP products which arise from the unremoved resin spots when cleaning by alcohol after the printing and then cured and resolidified during the postprocessing with UV light. The perimeter surface morphology of the FDM and DLP samples has been shown in Figures 17 and 18 respectively. The essential points related to the surface morphology of the perimeter were illustrated at the mentioned Figures. These Figures represent the surface structure before and after the tribology test for different oriented specimens (X, Z, and 45°).

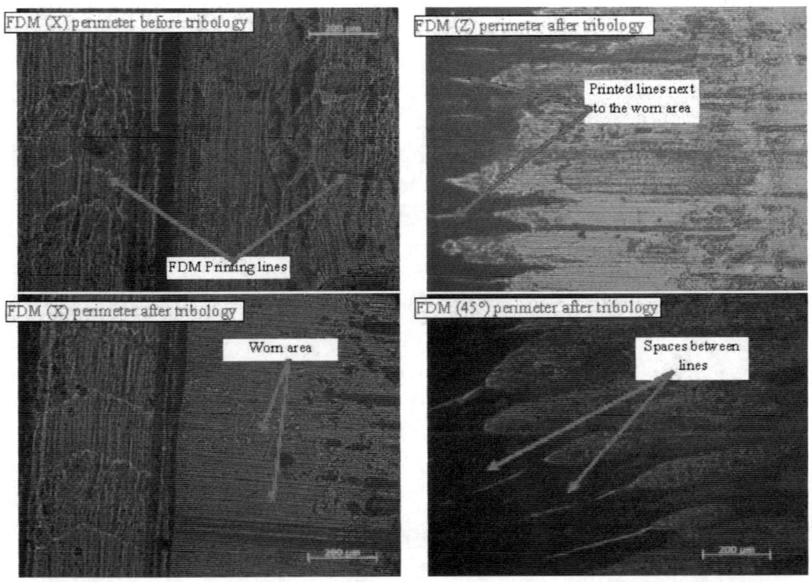

Figure 17. Surface morphology of FDM specimens printed in X, 45°, and Z orientation before and after tribology test

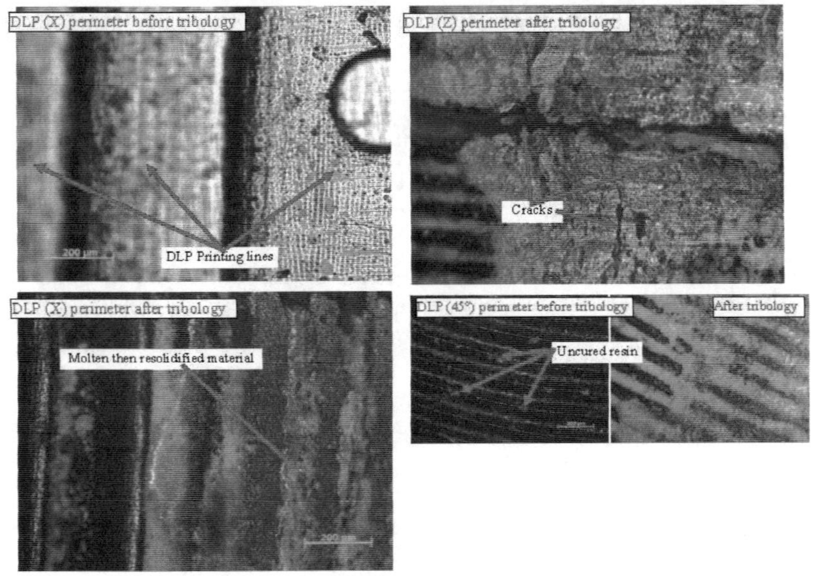

Figure 18. Surface morphology of DLP specimens printed in X, 45°, and Z orientation before and after tribology test

4. Conclusions

In this research, the surface structure, tribological properties, and surface roughness of specimens fabricated using two of the 3D printing technologies (FDM and DLP) have been examined. With each method, three different orientations (X, 45°, and Z) as a print parameter were investigated. The tribological measurements wear carried out under dry condition utilising cylindrical plane in the reciprocating tribotest system. The measured wear as well as the static and dynamic friction coefficient among the tested samples was compared. After evaluating the results, the following can be concluded:

- Altering the 3D printing method and the print options (particularly print orientation) has revealed a significant effect on the surface structure and the tribological properties of the manufactured objects;
- Concerning the surface roughness test, very smooth surfaces were observed for the bodies' printed utilising DLP technology in comparison with the FDM method;
- The specimens fabricated using DLP 3D printing method have shown a huge tendency to stick-slip phenomenon especially those who were built in 45° and Z orientations. While the parts printed in 45° orientation by the FDM technique has reflected the lowest trend for this phenomenon;
- The products which are made employing FDM technology is more appropriate for the tribological applications since the DLP samples exhibited considerable wear and friction values during the tribology test due to the fast surface melting behaviour of its materials.

For further work, studying the effect of changing other parameters and print settings on the surface morphology and tribological properties is worthwhile. In addition, different 3D printing technologies (not those used in this work) could also be involved in manufacturing the suggested objects.

Acknowledgements
This work was supported by the Stipendium Hungaricum Programme and by the Mechanical Engineering Doctoral School, Szent István University, Gödöllő, Hungary.

References

[1] Berman B 2012 *3-D printing: The new industrial revolution* Bus. Horiz. 55 pp.155-162
[2] Bhushan B and Caspers M 2017 *An overview of additive manufacturing (3D printing) for microfabrication* Microsyst. Technol. 23 pp.1117-1124
[3] Ngo T D, Kashani A, Imbalzano G, Nguyen K T Q, Hui D 2018 *Additive manufacturing (3D printing): A review of materials, methods, applications and challenges* Compos. Part B Eng. 143 pp.172-96
[4] Wu P, Wang J, Wang X 2016 *A critical review of the use of 3-D printing in the construction industry* Autom. Constr. 68 21-31
[5] Turner B, Strong R, Gold S 2014 *A review of melt extrusion additive manufacturing processes: I. Process design and modeling Rapid Prototyp* J. 20 pp.192-204
[6] Parandoush P and Lin D 2017 *A review on additive manufacturing of polymer-fiber composites* Compos. Struct. 182 pp.36-53
[7] Sun Q, Rizvi G M, Bellehumeur C T, Gu P 2008 *Effect of processing conditions on the bonding quality of FDM polymer filaments Rapid Prototyp* J. 14 pp.72-80
[8] Torrado A R, Shemelya C M, English J D, Lin Y, Wicker R B, Roberson D A 2015 *Characterizing the effect of additives to ABS on the mechanical property anisotropy of specimens fabricated by material extrusion 3D printing* Addit. Manuf. 6 pp.16-29
[9] Tymrak B M, Kreiger M, Pearce J M 2014 *Mechanical properties of components fabricated with open-source 3-D printers under realistic environmental conditions* Mater. Des. 58 pp.242-246
[10] Senatov F S, Niaza K V, Zadorozhnyy M Y, Maksimkin A V, Kaloshkin S D, Estrin Y Z 2016 *Mechanical properties and shape memory effect of 3D-printed PLA-based porous scaffolds*

J. Mech. Behav. Biomed. Mater. 57 pp.139-48

[11] WIJK A V and WIJK I V 2015 *3D Printing with Biomaterials: Towards a Sustainable and Circular Economy* (Amsterdam: IOS Press)

[12] Redwood B, Schöffer F, Garret B 2017 *The 3D printing handbook : technologies, design and applications* (3D Hubs B.V.)

[13] Stansbury J W and Idacavage M J 2016 *3D printing with polymers: Challenges among expanding options and opportunities* Dent. Mater. 32 pp.54-64

[14] Kolan K C R, Leu M C, Hilmas G E, Brown R F, Velez M 2011 *Fabrication of 13-93 bioactive glass scaffolds for bone tissue engineering using indirect selective laser sintering* Biofabrication 3 025004

[15] Mironov V, Boland T, Trusk T, Forgacs G, Markwald R R 2003 *Organ printing: computer-aided jet-based 3D tissue engineering* Trends Biotechnol. 21 pp.157-161

[16] Lee J M, Zhang M, Yeong W Y 2016 *Characterization and evaluation of 3D printed microfluidic chip for cell processing* Microfluid Nanofluidics 20 5

[17] Invernizzi M, Natale G, Levi M, Turri S, Griffini G 2016 *UV-Assisted 3D Printing of Glass and Carbon Fiber-Reinforced Dual-Cure Polymer* Composites Materials (Basel) 9 583

[18] Zsidai L and Kalácska G 2014 *"Stick-slip" PA és PEEK kompozitok súrlódásánál henger/sík modell vizsgálati rendszerben* Műanyag és Gumi 51 pp.462-470

IOP Conf. Series: Materials Science and Engineering **749** (2020) 012016 doi:10.1088/1757-899X/749/1/012016

Some aspects of quality and risk management in natural gas measurement

P E Ilea[1a] and A Stoica[2]

[1] Transilvania University of Brasov, Department of Product Design, Mechatronics and Environment, Brasov, Romania
[2] Lucian Blaga University of Sibiu, Department of Engineering, Sibiu, Romania

E-mail: [a] paul.ilea@unitbv.ro

Abstract. The paper refers to some general aspects regarding the natural gas measurement. In this paper, we try to find methods for checking the quality and risks for gas meters. It is trying to find answers to the question of how we could design and operate a gas meter so that the measurement is as accurate as possible. A process has been proposed to identify, analyse and control the emerging risks. The desirable quality of the measuring instrument requires a rigorous analysis of all aspects that could adversely affect. It also shows the interdependence between quality and risk, one cannot exist without the other. Based on these principles, the following papers propose the approach of case studies on the innovation of a new type of natural gas meter, focusing on the quality of the new product and its ability to accurately measure and prevent unauthorized consumption.

1. Introduction

In this paper, we try to find methods for checking the quality and risks for gas meters.

It is trying to find answers to the question of how we could design and operate a gas meter so that the measurement is as accurate as possible.

Measurement technique is part of the science of applied metrology and has as its object the theoretical and practical aspects of measuring a certain size or a certain range of sizes or those of a particular field of activity. The essence of measurement technique is to create and develop measurement methods and tools that can take, process and transmit quantitative information as close to the truth as possible.

The science of measurement is based on the fact that the process of knowing the surrounding world is based on the experiment that is the qualitative or quantitative evaluation.

In the modern economy, measurement is a matter of great responsibility and importance, present in all activities. For example, scientific research is inconceivable without the use of precise measuring tools, and the design and development of new products requires them to be tested in exploitation conditions with measured performance [1].

Annual gas consumption increased in 2016 compared to 2015, reaching about 11.7 billion cubic meters, with an increase of about 2%, amid a slight increase in end-user consumption that the number recorded in 2016 compared to 2015, also an increase of about 116,000 customers.

In 2016, the total natural gas consumption at the Romanian market was 124.12TWh. Out of this consumption, the final customers' consumption amounted to 111.7TWh, of which 80TWh represented the consumption of non-household customers (71.65%) and 31.7TWh represented the consumption of household customers (28.35%).

The total number of clients registered at the end of 2016 was 3596574, of which 188,253 non-household customers (5.23%) and 340,8321 household customers (94.77%) [2].

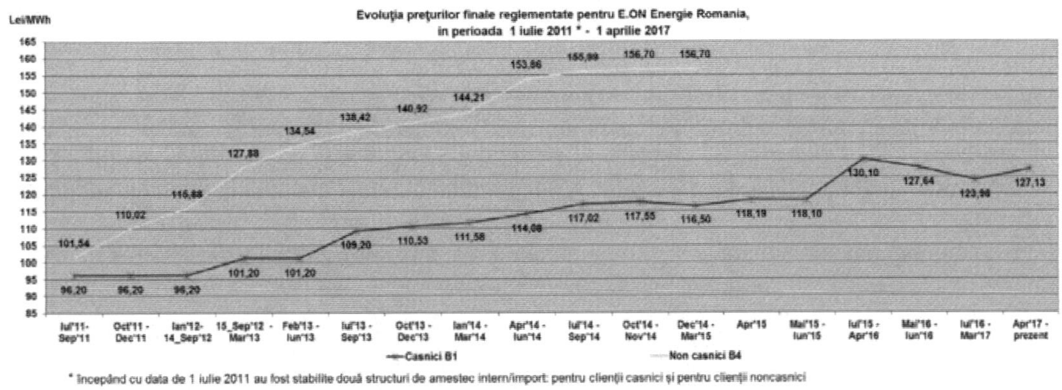

Figure 1. Natural gas price evolution, 2011-2017 [Source: ANRE, 2019]

Given the increase in gas consumption, the increase in the number of consumers, but also the increases in prices, consumers are tempted to intervene on measuring devices.

The desired quality of the metering device requires a risk analysis to identify existing or potential problems.

Risk management aims to manage threats that could have a negative impact on the company, so it is proposed to create a measuring device that highlights real consumption, and to alert the supplier quickly about the existence of interventions on measuring devices [3].

2. Risk and quality management

2.1. Quality management

The concept of quality can be defined as the degree to which a product, through the totality of technical, economic, aesthetic and exploitation characteristics, satisfies the needs for which it was conceived.

Quality management is a set of activities aimed at achieving objectives, through the optimal use of resources.

A good quality management system must have the following features:

- To be established in writing;
- Ensure that the customers's requirements are met;
- Ensure that the organization's requirements are met;
- To be applicable to all activities of the organization.

Advantages of implementing a quality management system:

- Increase revenue and reduce costs;
- Crediblity;
- Improving the company's image;
- Differentiation from competition;
- Control all processes in the company;
- Customer Satisfaction.

The basic principles of Crosby's quality management are:

- Ensuring compliance with requirements, quality means meeting consumer demands, but these requirements must be clearly defined and measurable;

- Quality assurance through prevention, quality must not be controlled, it must be done;
- Promoting the concept of "0 defects", Crosby believes that we can not operate with acceptable levels of quality;
- The quality measure is represented by the costs of the non-satisfaction of the requirements, in his opinion, the quality does not cost, the non-quality is the one that costs.

Crosby's approach is that the basic principle of quality is that of defect prevention described in the following programs:

- **To do everything right the first time and every time** - is based on the work done well for the first time, and there is no room for other levels or quality categories;
- **Zero defects and day zero defects** - the main objective of the quality improvement process is to obtain products / services without defects. 0 defects do not mean that the products are perfect, it means that every person has to perform his / her duties at the first moment and every time, and the failure to fulfill the attributions is unacceptable for their activity, and each activity is done according to the given specifications.

The quality improvement process begins with the four quality truths: quality is compliance with the requirements; the quality system means prevention; the performance standard achieved at organization level is zero defects; quality measure is the price of nonconformity.

Rinne believes that quality management has the following basic functions:

- Planning;
- Checking;
- Keeping under control [4].

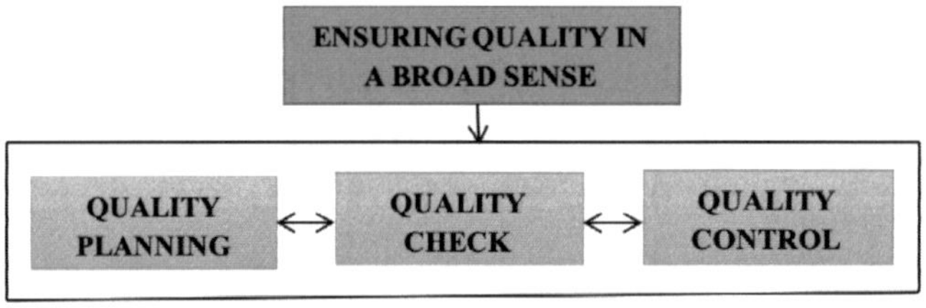

Figure 2. Functions of quality assrance in a broad sense
[Source: Olaru (1999), Quality management]

Total quality is a set of principles and methods organized in a global strategy to mobilize the entire company to achieve better customer satisfaction at a lower cost.

Total quality management can be highlighted as a new philosophy, a new culture model of the company, which orientates the client to all its activities and processes and optimizes them, so as to bring long-term benefits or can also be highlighted through the aspects technical and social, considering that motivation and training of staff is more important than technical aspects.

The principles of quality management are fundamental rules for the management and operation of an organization, which aims to improve continuously, focusing on customer satisfaction, but also on the needs of shareholders and employees:

- **Customer Orientation** - the success of any organization depends on the customer, therefore, all needs of customers, whether current or future, must be understood, satisfied. The organization must come with products beyond the customers' expectations;
- **Ledearship** - an organization is successful when the organization manages to mobilize staff to achieve the set goals. Employees will understand what the company's goals are and will be motivated to achieve these goals; the activities will be evaluated, coordinated and implemented in a unitary manner. Lack of communication between different hierarchical levels will be less;

- **Staff commitment** - success is mainly due to the staff of the organization. It is very important the total involvement of the staff regardless of the hierarchical position or the kind of work performed, in order to achieve the best performance and the continuous development of the organization. By applying this principle, employees are motivated and involved in the work of the organization, are rewarded according to their participation in the organization's success, and their creativity is freely expressed and can be used for the benefit of the company;
- **Process-based approach** - any activity that uses resources to convert inputs into outputs can be considered a process. For a good functioning, the organization has to conduct its activities and resources as processes, identify and manage processes and relationships between them. Managing and developing activities establishes a process leader. By this method the results are improved and predictable, the costs are reduced but also the efficiency of resource use increases;
- **Improving** - it must be a permanent goal of every organization. Improvement is done through employee training on how performance can be improved by regularly evaluating established criteria to see where they can be improved by transforming continuous improvement of products / services into a target for each employee;
- **Decision-making based on evidence** - all decisions taken by an organization must always be based on data analysis that can be demonstrated as true. By applying this principle, it is possible to re-analyse and change decisions and opinions, to make realistic substantiated decisions;
- **Relationship management** - this principle is based on the importance of the resources acquired from the suppliers, they must rise to the quality level expected by the organization. A mutually beneficial relationship develops both for suppliers that ensure a high level of quality and for the organization. This type of win-win relationship between the organization and vendors enhances the ability of both entities to create value.

In the concept of total quality, the client is sovereign, and the object of total quality is to exceed the expectations of the clients.

The overall quality management concepts are excellence, exceeding customer expectations and zero defects.

Reaching objectives is done by motivating and training employees, setting measurable indicators and using old and new methods and tools. Another very important aspect is the involvement of top managers, employees, and suppliers.

Total Quality Management translates into practice a new method of organizing quality activities, which integrates inspiration, statistical control, quality assurance and quality management [5].

2.2. Risk management

Risk management refers to translating a business aspect into all possible scenarios, so risk management becomes a process of identifying, analysing and responding to the potential risks of a firm.

Risk management involves 5 stages of work:

1) Risk planning;
2) Risk identification;
3) Risk analysis (qualitative/ quantitative);
4) Establishment of strategies for risk approach;
5) Risk monitoring and control.

First of all, risk management involves risk planning. Individuals who respond to each risk are established, regardless of hierarchical levels and risk categories. Errors are not accepted in risk planning because these errors can lead to disastrous consequences.

The next step in risk management involves identifying all risk sources that could affect your business in any way. Identifying sources of risk can be done through a Brainstorming session, through which multi-category specialists need to discover all the vulnerabilities of the project. The end of the

IOP Conf. Series: Materials Science and Engineering **749** (2020) 012016 doi:10.1088/1757-899X/749/1/012016

meeting is reflected in a list of possible issues that may arise along the way. The main purpose of this risk identification process is to avoid situations where the manager can get when he is surprised by an unwanted event. In practice it is quite difficult, almost impossible to determine absolutely all sources of risk. The next step is the risk analysis, which is a process of examining both qualitative and quantitative risk. There are aspects such as: if an unwanted event occurs, which repercussions it has on the firm. Depending on the probability and magnitude of the risk, the manager may or may not take into account this risk.

The next step in the risk management process is to develop strategies for addressing risks identified and analysed previously. The risk of accident can be transferred to an insurance company.

The main purpose of this risk analysis is to shelter the company from unpleasant events.

The last step is monitoring and control. This stage include the last phase of the management process, here operations are carried out which can again reach the first stage, the risk planning.

In practice, it is quite difficult to identify all the risks, so the main goal of the risk monitoring and control phase is to announce the emergence of new risk situations, preventing the surprise managers with unpleasant events.

The identification process aims at discovering all possible sources of risk in order to reduce or eliminate the effects they can produce.

Risk identification is achieved through:

- Questionnaires;
- Brainstorming;
- Journals;
- Behavioural Patterns;
- Charts;
- Flow charts;
- Periodic meetings with the staff involved.

The main advantage of a risk management program is economic efficiency: managers are aware of the risks to which the organization is exposed and administer properly so that they do not materialize.

The role of risk management in an organization is to help understand the risks to which an organization is exposed so it can be managed [6].

2.3. Risk-quality correlation

The quality-risk correlation represents the mutual relationship that is manifested between quality and risk, two fundamental notions in the production processes that cannot exist without one another.

The quality-risk correlation is the measure of the relationship between the two quality and risk concepts.

Figure 3. Risk-quality correlation [Source: Unitbv.ro]

The role of the two quality and risk management systems are to create the necessary conditions for the activity to unfold the first time, becoming at the same time:

- A way of continuing training of staff;

- A way of highlighting quality costs;
- A way of analyzing the activities;
- A way of correcting and adjusting activities;
- A way of measuring global risk;
- A modern and performing leading style.

Quality and risk theory applies to both products and services, with the idea being that it is easier to sell products than services, and an unsuccessful service from the start leads to 90% of the cases to the final loss of the customer.

For any organization, quality and risk integration should become key priorities.

Industrial excellence is the achievement of superior quality products/services offered in a shorter time and at a lower price.

The policy that governs industrial excellence is overtaking the expectations of its own employees, competitors and customers. Industrial excellence is trying to achieve at the same time three aspects: maximum quality; minimum time; maximum results.

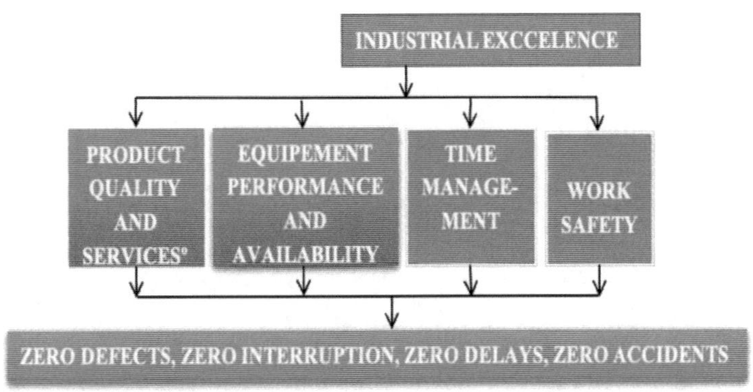

Figure 4. Industrial excellence [Source: Unitbv.ro]

3. Methods of quality verification

A very important activity in quality assurance is quality analysis. Quality analysis activity is done by analysing current non-conforming states in order to identify ways of intervening on the problem.

There are many ways to check quality, but are generally used: brainstorming, the cause-effect diagram, quality circle method, matrix methods, multi-voting methods, the structured problem-solving cycle, the Pareto diagram method [7].

The product is the one that ensures functionality, solves the problem and generates revenue. Product quality is the most important element because it depends on the quality of the whole project.

Part of product quality assurance refers to the compliance of standard procedures for doing it, but they do not have to be in the form of rigid rules followed with holiness on the grounds that they would be more important than the quality of the product.

Man is subjected to mistakes, and it is therefore necessary to introduce protection procedures to verify the product as it is being mastered and the management process as it is enforced. This measure is necessary to ensure that the product is executed correctly each time and with 0 defects.

There is a big difference between production and project activity in terms of quality assurance and control. Millions of products are manufactured in production, and quality assurance can be done on samples using statistical control techniques that determine if production deviates from the original specification.

In the case of a project, we can't do a destructive test because we only produce one product. If I misjudged him once, I wrong it wrong and wrong once is wrong every time. That's why it's important to make the right product from the beginning.

IOP Conf. Series: Materials Science and Engineering **749** (2020) 012016 doi:10.1088/1757-899X/749/1/012016

There are three simple steps to verify the quality:

- Let's check what we're doing;
- Compare what we accomplished with what was planned;
- Take action to recover any delays.

3.1. Pareto analysis

Pareto's analysis is based on the assumption that 80% of all quality events are the result of only 20% of all cases and that 20% of events are the result of 80% of the causes.

This analysis teaches us to discover those few and vital issues instead of identifying and solving the vast majority of the causes that do not result in only a small part of our problems.

As we eliminate vital causes one at a time, we can focus on the next vital cause, but that has a much diminish effect.

By this method you can quickly obtain high efficiency cost efficiencies.

3.2. Six Sigma

Six Sigma is a management methodology that aims to increase quality by determining and removing the causes of defects so as to ensure customer satisfaction.

Six Sigma is a management methodology based on 5 simple steps: define, measure, analyse, improve and control.

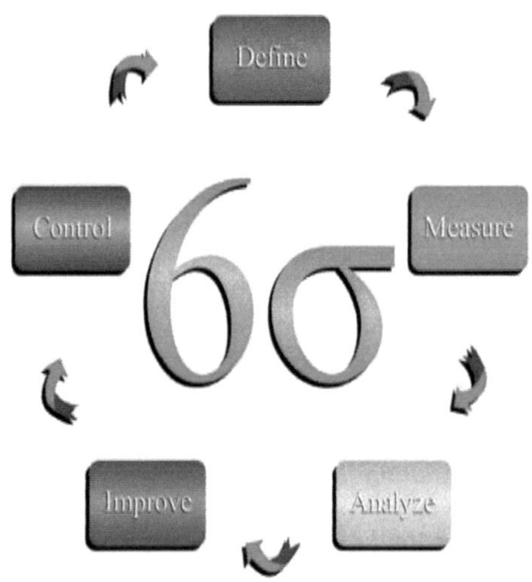

Figure 5. Six Sigma [Source: ttonline.ro]

This management methodology uses two statistical tools to improve quality: FMEA - Failure Mode and Effects and QFD - Quality Function Defection, together with modern management methods [8].

4. Conclusions

Based on the above, we can see that risk and quality management is a very important process in the manufacture of a product and that it is absolutely necessary to implement it within the company if we want to be successful.

We can see the interdependence between quality and risk, one can't exist without the other.

In the following papers, some case studies will be dealt with on the innovation of a new type of natural gas meter, focusing on the quality of the new product and on its ability to accurately measure and prevent unauthorized consumption.

IOP Conf. Series: Materials Science and Engineering **749** (2020) 012016 doi:10.1088/1757-899X/749/1/012016

References

[1] Stoica A and Feier D 2016 *Measuring and control devices in the natural gas industry*
[2] "National report 2016" accessed in Novermber 2018
[3] Ilea P, Savescu D, Stoica A 2018 *Some aspects of quality and risk management for natural gas meters* PRASIC 2018
[4] Olaru M 1995 *Quality management* Bucharest
[5] "*Quality Management course*" accessed in January 2019
[6] "Risk management course" accessed in December 2018
[7] "Methods of product and production quality analysis" accessed in February 2019
[8] "Quality management course" accessed in January 2019
[9] "Sig Sigma" accessed in January 2019

IOP Conf. Series: Materials Science and Engineering **749** (2020) 012017 doi:10.1088/1757-899X/749/1/012017

Analysis of temperature in different cooling methods

A Kári-Horváth[a] and T Pataki

Szent István University, Institute for Mechanical Engineering Technology, Mechanical Engineering Faculty, Gödöllő, Hungary

E-mail: [a] kari-horvath.attila@gek.szie.hu

Abstract. Cooling is one of the main issues in every machining process. By cooling, we can increase the tool life of our tool and maintain the quality of our surface. There are many options in the industry. A wide variety of cooling lubricants are available. There are general requirements for cooling and lubricating fluids used during machining, such as good thermal conductivity, proper lubrication, protection of workpiece and tool against corrosion, anti-foaming or stability during storage and use. These parameters are extremely important for the machining process, so you should be careful when selecting the coolant fluids. In our article we would like to introduce a measurement method, a measuring arrangement, with which we tested many coolant lubricants. Our main aim will to show that the MQL (Minimum Quantity Lubricant) technology is how better than the other cooling method. We used different types of cooling lubricant. Among our choices are mineral oils, synthetic oils, and biological degradable oils too. We have studied a number of cooling methods, including the Dry lubrication, where lubricant is ambient air, spontaneous emission, induced emission. Doppler cooling with emulsion flood lubrication and molecular feed lubrication, the emissions were spontaneous and induced. Doppler cooling with MQL-molecular oil, here the emissions were the same as by the lubricants. We compare and analyse the results in our article. We tell the measurement parameters and our experiences.

1. Introduction

The lubrication method (MQL) applied during cutting will meet the requirement of minimum quantity lubrication, when the amount of lubricant does not exceed 50 mg per hour [13], occasionally even less than that. The German term MMS for this lubrication method is derived from "Minimal Menge Schmierung", the English MQL from 'Minimum Quantity Lubricant' [1.]. MQL is often referred to as semi-dry lubrication [6], or micro lubrication [7]. MQL-lubrication, however, is not a precise term for the lubrication applied with minimum lubricant in everyday practice. Minimum lubrication can be applied for:

- Lubrication of machine and machine parts;
- Machining.

The literature around 2000 frequently mentioned the need to re-examine the multifunctional role of cooling lubricants [1], [2], [9], [11], [12], [13]. The debate was about whether the prevailing system should be developed according to certain functions or whether there should be a total change and the high-speed, high-temperature dry machining, yielding small surface-size chips, should be introduced. Neither the theoretical nor the practical, technological conditions for total change are present yet, but there exists a transitory stage between the two, as shown in figure 1 [3], [9]. Since the reduced amount of cooling lubricant stands between dry and wet working, attempts have been made to reduce the amount of cooling lubricant. These attempts were successful and the method spread widely within a short period of time (five years). The production technology develops, the new procedures appeared. [5], [8] Today it is called MQL cooling, lubricating. The benefit of the quick success is that less cooling lubricant, which is highly polluting, is being used; the drawback is that studies of the new method have not started. The researchers have been limiting their efforts to trying to find out which one of the cooling lubricants on the market was right for their purposes.

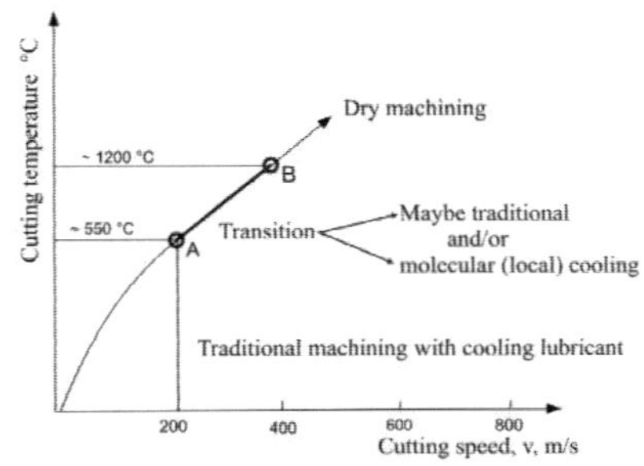

Figure 1. Cutting temperature change as a function of cutting speed in the case of steel and heavy metal tools

1.1. Experimental conditions

Experimental environment: normal ambience DIN50014 and ISO554-1970. Climate sign: normal ambience DIN 50014-20/65-1. The experimental protocol is shown in figure 2.

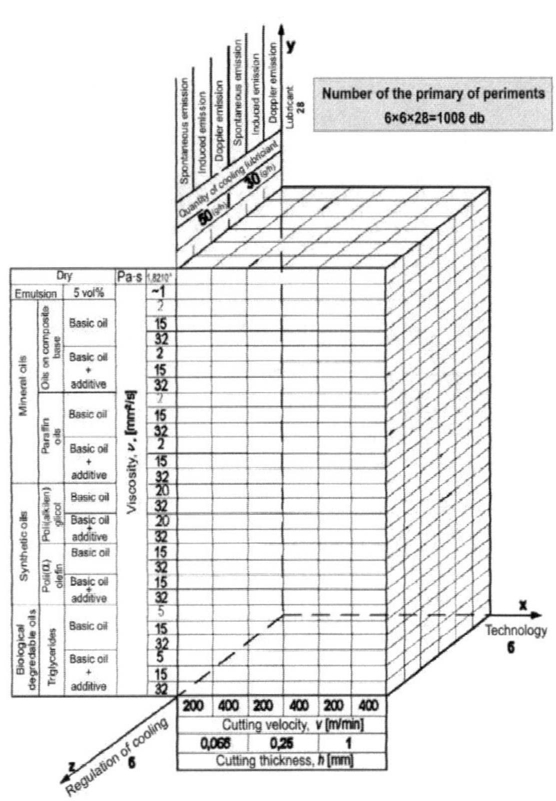

Figure 2. Experiment guideline

IOP Conf. Series: Materials Science and Engineering **749** (2020) 012017 doi:10.1088/1757-899X/749/1/012017

The lathe chisel used in the experiments was modified from a CMA brand one-line turning lathe, with the following dimensions:

- ✓ top length: 1500 mm
- ✓ top height: 250 mm
- ✓ revolution range: 16÷2000 min^{-1}
- ✓ grading factor 1.25
- ✓ morze: 5
- ✓ engine revolution: 1440 min^{-1}
- ✓ engine efficiency: 15 kW
- ✓ pump efficiency: 0.6 kW
- ✓ tool rigidity: 250 N/μm
- ✓ vibration: free

1.2. Lubricating processes and the quality requirement of the lubricants used

New terms had to be given to the cooling processes used in the experiments:

- Dry lubrication, where lubricant is ambient air:
  - spontaneous emission;
  - induced emission;
  - Doppler cooling;
- Emulsion flood lubrication;
- Molecular feed lubrication:
  - spontaneous emission;
  - induced emission;
  - Doppler cooling;
- MQL - molecular oil cooling:
  - spontaneous emission;
  - induced emission;
  - Doppler cooling.

1.3. Cooling methods and conditions

Figure 3 shows the measuring circuit. The following cooling methods were used in the measurements:

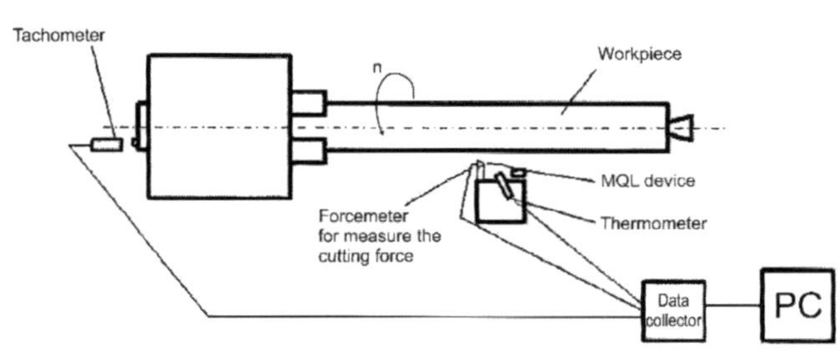

Figure 3. Theoretical design of measurements

Induced emission cooling: In this case a lamp was used to produce parallel beams of light. The lamp was attached to a stand on the lathe tool holder. This is shown in figure 5. During Doppler cooling a laser lamp was attached to the lathe tool holder, as shown in figure 4.

Doppler-cooling: energy distraction accomplished on an atomic level with lasers (refrigeration). The temperature rises steadily during the friction, with what the moving atoms amplitude is growing, this effect regulate the laser some that the laser brakes the oncoming moving atoms [4], [9], [10], [11].

IOP Conf. Series: Materials Science and Engineering **749** (2020) 012017 doi:10.1088/1757-899X/749/1/012017

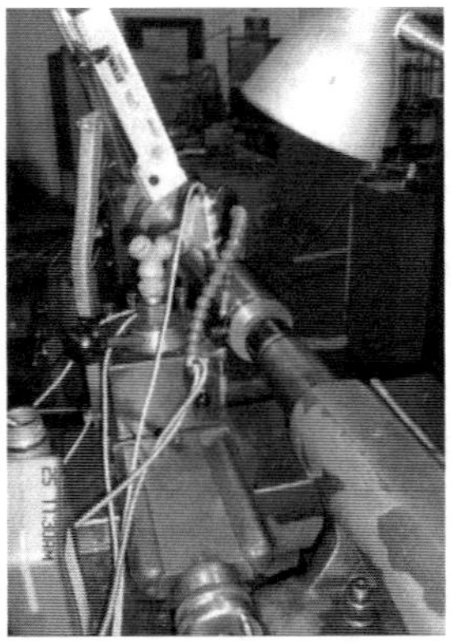

Figure 4. Use of Doppler cooling

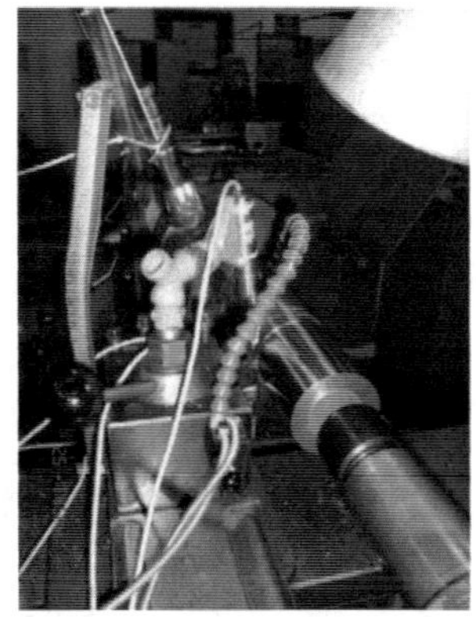

Figure 5. Induced emission cooling

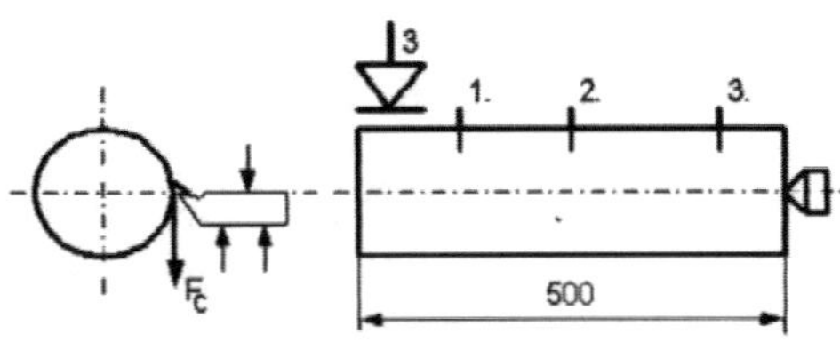

Figure 6. Measurement points

Experimental data	Technological data
Normal ambience DIN50014 and ISO554-1970	Feed: f(h) = 0.025 mm/rev.
feed: f(h) = 0.25 mm/rev.	Feed: f(h) = 0.067 mm/rev.
Lathe: HC/TiN, CNMG 1204 08 PF 4015 by DIN/ISO 513	Grip depth: a(b) = 1 mm
Quality of workpiece: 42CrMo4 (matter number: 1,7225)	Revolution: n = 1730 min^{-1}
Machine tool: C11A type lathe	
MQ-Lubrication applicator: Cobra 2000	
Measuring equipment	**Measured data**
Micrometer (range 50-75mm, 75-100mm	Measurement points 1, 2, 3
Regular slide gauge (1/20 – 150 mm)	D_1[mm], D_2[mm], D_3[mm],
Light microscope (for thin sections)	R_a[μm], R_z[μm], R_q[μm],
WA33 (TYP PRLTA13) scales (Precision: 0.001g)	R_y[μm]
Inductive signalling device for measuring revolution (special production)	n[min^{-1}], F_c[N], T[°C]
Data collector (Spider8 Control, Carman 4.5 program) GA 300 thermometer	
Mitutoyo SJ 201P diamond pointed roughness meter	
Photon-electron generating devices	
	Calculated data
	D_a[mm], V_c[m/min], s[m], V_f[mm^3/s], VB[mm], P[kW]

After gathering data in the experiments, the results were displayed in a table. These were then plotted in various graphs: R_a - V_f (surface roughness against the amount of chips removed during a period of time), VB - s (back wear against distance), VB - V_f (back wear against the amount of chips removed during a period of time).

Various effects were studied during MQL lubrication at 50 g/hour and 30 g/hour oil consumption. We investigated the viscosity, molecular structure, the effect of application in manipulated and non-manipulated molecular conditions. [9]

1.4. Dry cooling lubrication

Changes in temperature in the cutting zone for the three types of cooling are shown on a bar chart in figure 7.

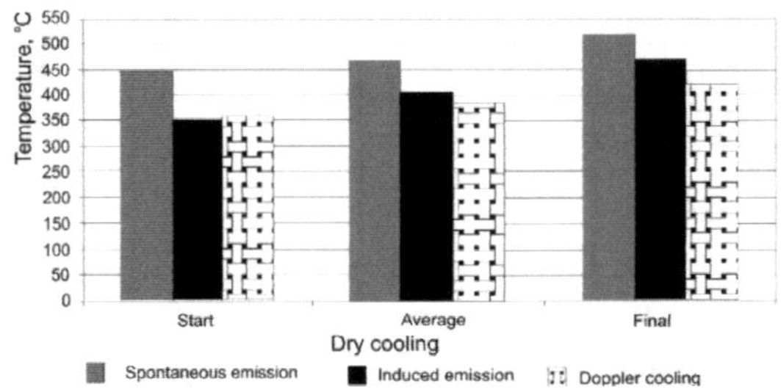

Figure 7. Cutting zone's temperature during dry cooling

The highest initial temperature for spontaneous emission is 450°C. The average temperature is 460°C, i.e. increasing. The final temperature is 515°C. In the lgh -lg kc coordinate systems the specific cutting energy decreases significantly by 10%. In the case of induced emission, the temperatures were significantly lower (initial: 350°C, average: 400°C, final: 470°C). On the whole, there was a roughly 5% decrease in specific cutting energy.

The most significant temperature decrease was in the case of Doppler cooling (initial: 360°C, average: 390°C, final: 420°C). Even so, the specific cutting energy decreased significantly by about 10%. This change can be explained by the fact that Doppler cooling is controlled, which makes it more intense at the chip base. This is more favourable due to its rigidity. The detachment areas of the material weakness: that is, the thermal plasticity increases, which is also favourable.

Spontaneous emission	increase from start	$\Sigma s \approx 740$ m
Induced emission	metal transfer	a_c=420m, $\Sigma s \approx 600$m
Doppler cooling	metal transfer	a_c=700m, $\Sigma s \approx 900$m
Spontaneous emission	metal transfer	a_c=1100m, $\Sigma \approx 1210$m

1.5. Flooding lubrication

The specific cutting energy change for spontaneous emulsion is shown in figure 8. The concentration of emulsion is 5% by volume. It can be seen in figure 8 that the initial temperature for spontaneous emission is 360°C with the average being 360°C. The final temperature rose to 418°C. The specific cutting energy was reduced significantly by 26%.

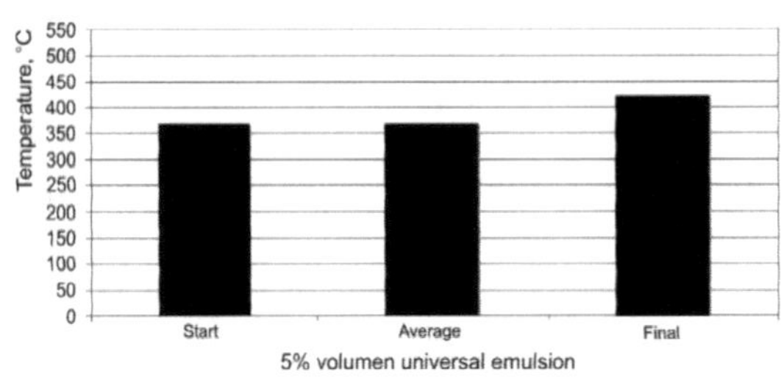

Figure 8. Cutting zone's temperature during 5% volume emulsive cooling

Tool wear:

Spontaneous emission	metal transfer	a_c=1100m, $\Sigma \approx 1210$m

The average surface coarseness (R_a) for spontaneous emission improves till Ac and deteriorates afterwards. The temperature of spontaneous emission used with dry cooling in the cutting zone is approximately 510°C greater than that of emulsive cooling.

1.6. Non MQL lubrication, non-soluble mineral oil based universal cutting oil

The temperature changes of the cutting zone - under different experimental conditions and cooling methods - are shown in figure 9.

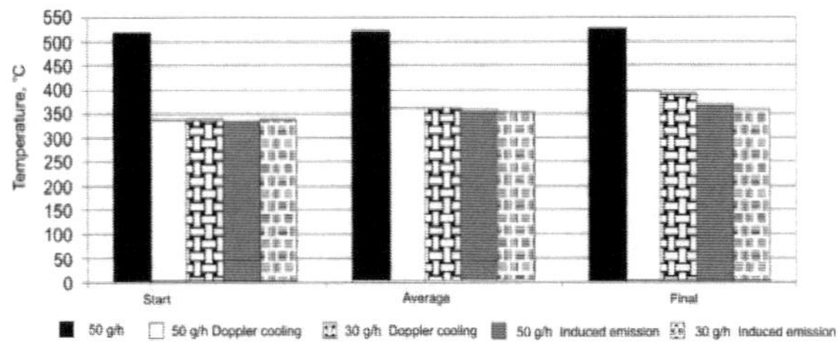

Figure 9. The temperature of the cutting zone in experiments using commercially available, non-MQL lubricant developed, complex cutting oil

Figure 9 show that the highest initial temperature of spontaneous emission (50g/h applied lubricant) is 510°C with the average being 522°C. The final temperature is 25°C. The specific cutting energy remains relatively unchanged.

In the case of induced emission (30g/h applied lubricant) the temperatures dropped significantly compared to spontaneous emission cooling (initial: 336°C, average: 353°C, final: 360°C). The specific cutting energy dropped by roughly 8%.

In the case of induced emission (50g/h applied lubricant) the temperatures dropped significantly compared to spontaneous emission cooling (initial: 331°C, average: 357°C, final: 367°C). The specific cutting energy dropped by roughly 10%.

In the case of Doppler cooling (30g/h applied lubricant) the temperatures dropped significantly compared to spontaneous emission cooling (initial: 335°C, average: 35 °C, final: 387°C). The specific cutting energy dropped by nearly 29%. (Figure 8/b)

In the case of Doppler cooling (50g/h applied lubricant) the temperatures dropped significantly compared to spontaneous emission cooling (initial: 335°C, average: 357°C, final: 396°C). The specific cutting energy increased by roughly 2%.

Tool wear:

Spontaneous emission (50g/h)	increase from start	$\Sigma s \approx 510m$
Induced emission (30g/h)	metal transfer	$a_c = 400m$, $\Sigma s \approx 655m$
Induced emission (50g/h)	increase from start	$\Sigma s \approx 530m$
Doppler cooling (30 g/h)	increase from start	$\Sigma s \approx 480m$
Doppler cooling (50 g/h)	metal transfer	$a_c = 660m$, $\Sigma s \approx 680m$

The average surface coarseness (R_a) shows steady improvement for spontaneous and induced cooling (for both 30g/h and 50g/h) and Doppler cooling (30g/h), while Doppler cooling (50g/h) improves till ac, then suddenly deteriorates. Tool wear is lowest with Doppler cooling (50g/h). On the basis of these results, it is safe to say that in contrast to spontaneous emission there was significant decrease in the cutting zone when induced emission or Doppler cooling was used.

Taking into consideration the amount of lubricant used in the experiments, it can be said that there is little to no temperature difference between the four molecular manipulation tests. It can also be seen that when induced emission cooling was used the average temperature showed only minor changes (approx.: 30°C).

1.7. MQL lubrication, non-aggregated cutting oil

When the amount of lubricant and the cooling method were changed, the specific cutting energy dropped by 8÷10% with the most significant drop being 30%. Figure 10 shows the results of experiments done with the D model liquid.

IOP Conf. Series: Materials Science and Engineering **749** (2020) 012017 doi:10.1088/1757-899X/749/1/012017

In the case of induced emission (30g/h applied lubricant) the temperatures rose compared to the common complex aggregate lubricant oil (initial: 392°C, average: 443°C, final: 477°C). The specific cutting energy dropped by roughly 29%.

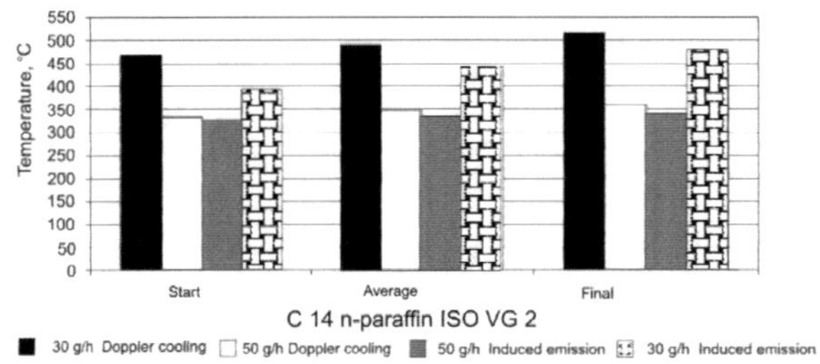

Figure 10. The temperature of the cutting zone using experimental MQL lubricant

In the case of induced emission (50g/h applied lubricant) the temperatures fell compared to the 30g/h applied lubricant experiments (initial: 325°C, average: 335°C, final: 343°C). The specific cutting energy dropped roughly 9%.

In the case of Doppler cooling (30g/h applied lubricant) the temperature values increased significantly (highest initial: 466°C, average: 493°C, final: 517°C). The specific cutting energy dropped by roughly 29%.

Tool wear:

Induced emission (30g/h)	metal transfer	a_c=660m, $\Sigma s \approx$900m
Induced emission (50g/h)	metal transfer	a_c=1100m, $\Sigma s \approx$1450m
Doppler cooling (30g/h)	increase from start	$\Sigma s \approx$760m
Doppler cooling (50g/h)	increase from start	$\Sigma s \approx$780m

The average surface coarseness (R_a) for induced emission (both 30g/h and 50g/h) and Doppler cooling (both 30g/h and 50g/h) shows signs of steady improvement. The specific cutting energy decreases if the amount of applied lubricant is decreased from 50g/h to 30g/h. The surface coarseness slightly improves by increasing cooling lubricant. The lifespan of a tool with induced cooling at 50g/h of applied lubricant proved the best.

1.8. MQL lubrication aggregate cutting oil

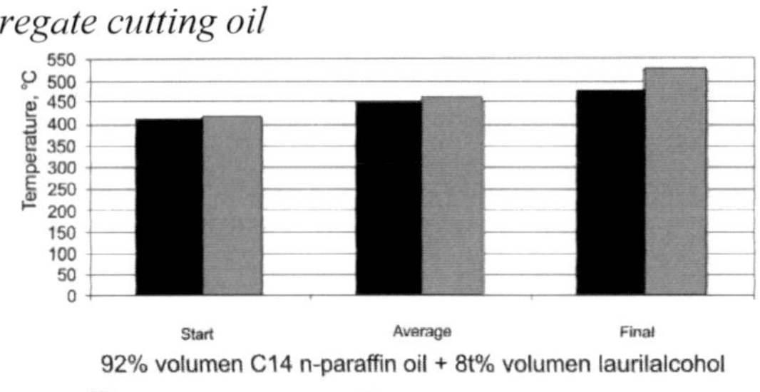

Figure 11. The temperature of the cutting zone using experimental MQL lubricant (M3-2)

The specific cutting energy compared to the D-lubricant was within the margin of error. The results for the series of experiments using E-model fluids are depicted in figure 11.

In the case of Doppler cooling (30g/h applied lubricant) the temperatures showed a slight increase (highest initial: 415°C, average: 454°C, final: 526°C). The specific cutting energy dropped by roughly 32%.

Tool wear:

Spontaneous emission (50g/h)	metal transfer	a_c=660m, $\Sigma s \approx$ 880m
Doppler cooling (30g/h)	metal transfer	a_c=650m, $\Sigma s \approx$ 870m

The average surface coarseness (R_a) for spontaneous emission (50g/h) and Doppler cooling (30g/h) improved till a_c and then deteriorated. Tool wear can be delayed by the model liquids and cooling procedures used in the experiments.

2. Conclusions

A modified lathe was used as the equipment for the experiment. The measuring equipment was secured. A parallel light emitting LED lamp was used, as was a small output (P<1mw, 630<λ<680nm) laser and a 'Cobra 2000' drip feeder. A theoretical method based on comparing the kc value of the chips' thickness was created for assessing the experimental results.

It has been proven with in the boundary of the experiment that local cooling can be achieved at a molecular level. Thermoplasticity (reduction of internal friction) and chip base rigidity can be increased by the use of molecules manipulated by induced emission and the Doppler Effect, which is how internal friction can be decreased. The internal stress created by the cutting energy is decreased by 3%. This way the crystallite changes in the layers near the surface become more favourable.

The action mechanism of MQL lubrication has been clarified, and the specific requirement for MQL lubricants has been determined. The surface oil retention ability of coated heavy metals has been examined as have the properties and amounts of the tool wear process.

More research still needs to be done, however. Experiments are being conducted, which are expected to provide more and clearer information on localized cooling. This is why γ-photon experiments have also been added to the programme. More MQL lubricant-specific experiments will be carried out as a result of the current test results.

The results summed up in the article and the results of tests that are currently underway supplement the presently existing knowledge of MQL lubrication, which is important not only for immediate use; it creates the basis for a paradigm shift in cutting technology.

References

[1] Avila R F, Abrao A M 2001 *The Effect of cutting fluids on the machining of hardened AISI 4340* Steel Journal of Materials. Processing Technology 119 pp.21-26

[2] Bartz W J und 70 Mitarbeiter 2000 *Kühlschmierstoffe und Zerspanung* Remingen Malmsheim Expert-Verlag

[3] Kári-Horváth A, Valasek I 2010 *Demand of Energy for Chip Detachment* Materials Science Testing and Informatics pp489-497 ISBN-13978-0-87849-235-0

[4] Kári-Horváth A, Valasek, I 2009 *Energiatakarékos forgácsolás molekula manipuláció alkalmazásával* Mezőgazdasági Technika I. évf. HU ISSN 0026 1890 pp.2-5

[5] Kátai L, Szabó I, Lágymányosi A, Lágymányosi P, Szakál Z 2018 *Additív gyártástechnológiában alkalmazott anyag szilárdsági jellemzőinek vizsgálata a 3D nyomtatás paramétereinek függvényében* GÉP 69 pp.45-48

[6] Klocke F, Eisenblatter G. 1997 Proceedings of the Opening Session of the Dry Cutting Annuals of the CIRP. Vol. 46/2 pp.519-526

[7] Ma Clure T F, Adams R, Gugger M D 2006 *Comparison of Flood vs. Micro-lubrication on Machining Performance* http://www.unitcom techsolve Html (accessed. Jan 23)

[8] Pataki T, Szakál Z 2014 *Polimerek erősítésére alkalmas szén nanoszerkezetek mechanikai teherbírásának vizsgálata* MŰANYAG ÉS GUMI 51 pp.112-114

[9] Valasek I, Kári-Horváth A 2011 *The action mechanism of minimum lubrication and the increase of its efficiency* TRIBIOLOGIE UND SCHMIERUNGSTECHNIK 58 Jahrgang 3/2011 ISBN 978-3-8169-2830-0 pp.34-47

[10] Valasek I 1996 *Tribológiai kézikönyv* Tribotechnik Kft. Budapest

[11] Zrínyi M 2004 2006 *A fizikai kémia alapjai I-II. kötet* Műszaki Könyvkiadó Budapest

[12] Walter A 2001 *Potenzial von Kühlschmierstoffen gezielt nutzen. Trocken oder nass -wohin geht die Metallbearbeitung?* VDI-Berichte 1635 VDI-Verlag Düsseldorf, S pp.33-47

[13] Weinert K 1999 *Trockenbearbeitung und Mineral Menge Kühlschmierung* Springer Berlin

IOP Conf. Series: Materials Science and Engineering **749** (2020) 012018 doi:10.1088/1757-899X/749/1/012018

Evaluation of the technological effects of application of the FMS elements

D Lukic[1a], A Antic[1], S Borojevic[2], M Jocanovic[1] and I Kuric[3]

[1] University of Novi Sad, Faculty of Technical Sciences, Novi Sad, Republic of Serbia
[2] University of Banja Luka, Faculty of Mechanical Engineering, Banja Luka, Bosnia and Herzegovina
[3] University of Zilina, Faculty of Mechanical Engineering, Zilina, Slovakia

E-mail: [a] lukicd@uns.ac.rs

Abstract. Flexible manufacturing systems (FMS) represent complex systems with great technological capabilities that are able to simultaneously achieve high level of flexibility, cost effectiveness and productivity. However, due to high investment cost and cost of exploitation of the FMS elements, the conditions for their rational use are required. Taking into account the increasingly complex conditions of production and launch of products, a systematic approach in the planning and implementation of FMSs is needed, thereby including the most significant technological and economical characteristics. This paper introduces two methodologies for evaluating the technological effects of application of the FMS elements at the conceptual level along with their verification for a single family of rotational parts.

1. Introduction

Flexible manufacturing systems (FMS) represent the basis for production systems enabling them to simultaneously achieve a high level of flexibility, cost effectiveness and productivity [1], [2]. The basic units of FMSs are CNC machine tools, especially at today's level when they are multifunctional and can integrate various machining processes (turning, milling, grinding) and various manufacturing technologies (cutting, heat treatment). By interconnecting machine tools with manipulation, measurement and transportation systems as well as their computer numerical control, FMSs of a different level of complexity, productivity and flexibility are created, figure 1 [3].

In modern industrial production there are several levels of complexity of flexible manufacturing systems: Flexible Manufacturing Modules (FMM), Flexible Manufacturing Cells (FMC), Flexible Manufacturing Groups (FMG), Flexible Manufacturing/Transfer Lines (FML) and as the most complex which include all the previous-Flexible Manufacturing Systems (FMSs) [4], [5].

FMSs are designed to manufacture one or more product families whose development and implementation are based on the principles of group technology [1], [4]. By applying the concept of group technology it is possible to set the necessary bases for defining and evaluating the effects of the application of the FMS elements that will be applied in manufacturing processes of an appropriate product range or product families. These bases can be set as a result of a required analysis of all parts from the observed production program, or perhaps as a result of an appropriate analysis of representatives from a product group [6].

This paper introduces two methodologies for evaluating the technological effects of application of the FMS elements for the example of a formed technological group of rotational-symmetric parts.

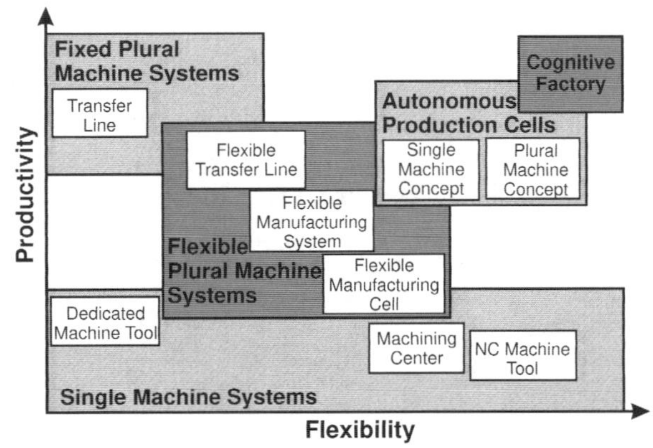

Figure 1. Productivity and flexibility of FMSs of different levels of complexity [3]

2. Theoretical bases of methodologies

The proposed methodologies for evaluating the effects of application of the FMSs elements consist of the following phases:

- Analysis of a production program, classification of parts and formation of part groups (families);
- Design or selection of complex parts (representatives of part families);
- Process planning for group of parts (GT process plans are designed);
- Precision of process plans for selected products and determination of manufacturing times;
- Evaluation of technological effects of the application of FMS elements.

Starting from the principle of product classification and grouping, a product range of a production system can be systematized according to the design and technological similarity among certain part families for which the concept of group technology is rationally applied. On the basis of group technology established by Mitrofanov and implemented by Burbidge, a group approach to planning of effective production and its structures has been developed.

GT Process planning enables easy and rapid precision of process plans for individual parts of a single group. This is achieved in the following way [6]:

1. Specification of process plans for all parts is done in the case of the detailed/final planning of FMSs, thus providing detailed techno economic data (time, costs, degree of utilization, etc.).
2. Specification of process plans for representatives of an appropriate product group is done in the case of the conceptual/preliminary planning of FMSs, thus providing approximate techno economic data, whose accuracy depends on the methodology that is being applied.

In order to achieve more rational and faster determination of manufacturing time of a single part group within the conceptual planning, it is necessary to specify manufacturing operations for an appropriate representative on the basis of the adopted method. These methods are used for determining processing time (t_k) and setup time (T_{pz}) of operations of the considered technological groups. They are used in practice as analytical or graphical methods. As far as this study is concerned, these two methods will be considered:

- Similarity method (graphical and/or analytical);
- Method based on part representatives of technological groups (analytical).

Both methods enable approximate time determination with good reliability which in production practice provides a good basis for determining the level of complexity, the number of elements of machining systems and the degree of their utilization. The following chapter shows the application of the observed methods used for evaluating the effects of application of the FMS elements for the case of turning process of the specified part group.

3. Evaluation of the effects of application of the FMS elements

3.1. Process planning for group of parts

By analyzing the production program of a single production system, the technological group of rotational parts was formed using the design-technological classification system. One of the observed groups is a group of axles whose main information is given in table 1.

Table 1. Basic information about the axle group

Part designation		Quantity [pcs/batch]	Mass [kg]	Price [€/pcs]
P1	708 033	320	0,126	15.000
P2	708 038	300	0,134	15.400
P3	708 041	280	0,109	14.200
P4	708 046	400	0,109	14.200
P5	708 047	350	0,109	14.200
P6	708 169	260	0,133	15.400
P7	708 170	340	0,133	15.400
P8	708 181	360	0,149	15.700

Figure 2. 3D model of the complex part

Table 2. Content of the GT Process plan

Process No.	Process designation	Machine type /workplace
10	Cut-off	Metalsaw
20/1	Turning left side	CNC Turning machine
20/2	Turning right side	
30	Control	Control place
40	Milling	CNC milling machine
50	Dressing	Work place
60	Annealing	Heat Treatment Furnace
70	Heat treatment control	Hardness testing device
80	Grinding central nests	CNC Grinding machine
90	Grinding holes	CNC Grinding machine
100	Grinding outside	CNC Grinding machine
110	Final control	Control place

Based on the geometric and technological characteristics of these parts, the imaginary complex part with all the corresponding features is designed, figure 2. Based on the designed 3D model/ 2D drawing of the complex part, production quantity, available resources and other necessary data, a GT process plan is planned. Content of the GT process plan is given in table 2.

3.2. Application methods

3.2.1. Similarity method. In order to determine processing time of the observed part group, it is necessary to point out the technologically simplest and the most complex part of this group as well as to specify the GT processes (or just manufacturing operations) for them. As a criterion for selecting the simplest and most complex part of the group, the degree of technological complexity is usually applied, that is, the number of operations in the observed process. The total processing time (t_k) for realization of machining process of the part group is determined on the basis of the following expression:

$$t_k = \sum_{i=1}^{k} (Q_i t_{k_i}) \quad (1)$$

in which: k- number of different parts in the group;
Q_i - quantity of individual parts of the group;
t_{ki} - processing time for certain parts of the group.

Processing times (t_k) for machining processes include machining (t_g) and auxiliary times (t_p), and they are determined from the similarity diagram. In this example, the diagram showing processing times of the turning process for the group of axles performed on the CNC turning machine is shown in figure 3. The diagram shows dependency between the processing time and the number of operations where the simplest part have 23 operations and processing time of 3.1 minutes per piece, while the most complex part has 30 operations and processing time of 4.5 minutes per piece. Processing times of turning process for the specific parts of the technological group can be determined graphically using the diagram in figure 3 or analytically with the predetermined path coefficient which describes the change in processing time (t_k) depending on the number of operations (z).

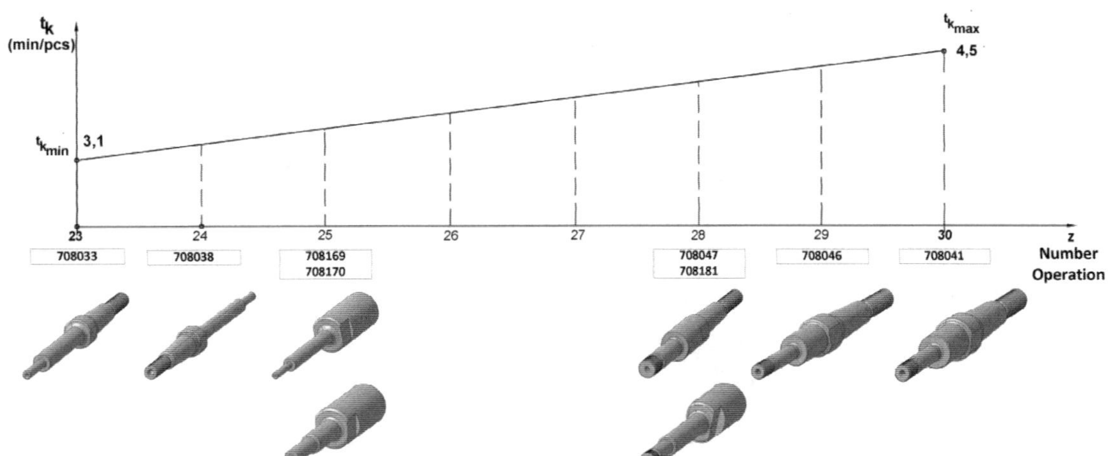

Figure 3. Dependency between the processing time and the number of operations for the turning process

According to figure 3, the path coefficient (k) is determined in the following way:

$$k = \frac{t_{kmax} - t_{kmin}}{z_{max} - z_{min}} = \frac{4.5 - 3.1}{30 - 23} = 0.2 \quad (2)$$

As an example, we show the procedure for determining processing time for the part P4 with designation "708 046" which in this process (no. 20) has 29 operations:

$$t_{k4} = t_{k_1} + k \cdot (z_4 - z_1) = 3.1 + 0.2 \cdot (29 - 23) = 4.3 \quad (3)$$

3.2.2. Method based on representatives of part groups. The application of this method begins with selecting a part representative based on the ABC analysis which determines the dependency between the type of part and the corresponding quantities, masses, values, profit etc. Then, it specifies machining processes based on the appropriate GT process plans, including the determination of processing times of these representatives (t_{kp}). Based on table 3, a quantitative, mass and value ABC analysis is performed on the basis of whom the part P8 (708181) is selected as a product representative of the group of axles. As an example, figure 4 illustrates the diagram of value ABC analysis.

Table 3. Data for ABC analysis

Part	Q_i [pcs/yr]	m_i [kg/yr]	V_i [€*103/yr]	Q [%]	m [%]	V [%]
708 033	3200	403.2	48000	12.26	12.36	12.33
708 038	3000	402	46200	11.49	12.32	11.87
708 041	2800	305.2	39760	10.73	9.36	10.21
708 046	4000	436	56800	15.33	13.36	14.59
708 047	3500	381.5	49700	13.41	11.69	12.76
708 169	2600	345.8	40040	9.96	10.60	10.28
708 170	3400	452.2	52360	13.03	13.86	13.45
708 181	3600	536.4	56520	13.79	16.44	14.52
Σ	2610	3262.3	389380	100%	100%	100%

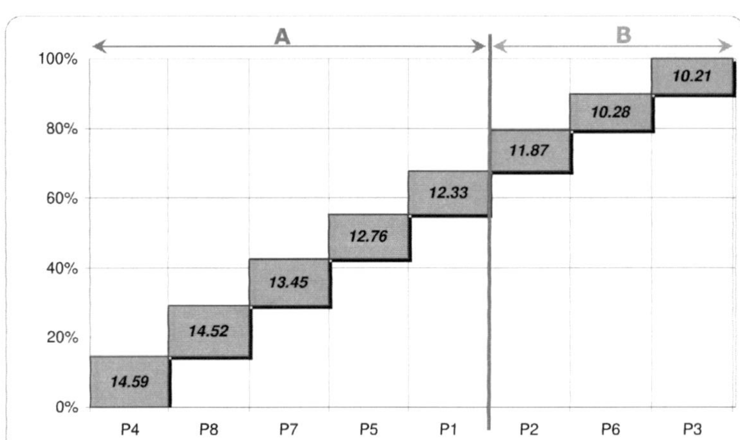

Figure 4. Value-based ABC analysis

In order to determine the processing time of the observed technological group for the process no.20, a reduced quantities of the group (Q_r) are determined, according to the expression (4):

$$Q_{r_i} = Q_i \cdot r_i \tag{4}$$

in which: Q_i - quantity of individual parts of the group;
r_i - degree of reduction of the i-th part.

The degree of reduction of a specific part of the observed group includes reduction for mass (r_m) and complexity (r_s), which can be expressed by (5):

$$r_i = r_{m_i} \cdot r_{s_i} \tag{5}$$

The degree of reduction for mass and technological complexity is determined respectively according to this expression (6):

$$r_{mi} = \frac{m_i}{m_p} \quad r_{si} = \frac{s_i}{s_p} \tag{6}$$

in which: m_i - mass of individual parts;
s_i - degree of technological complexity of parts;
- m_p - mass of the representative of the part group;
- s_p - degree of complexity of the representative.

The degree of technological complexity of specific parts of the group is determined by the ratio between the number of operations of individual parts and the number of operations of the corresponding representative. The reduced quantity of all parts of the group is determined by the expression (7) and given in table 4:

$$Q_r = \sum_{i=1}^{k} Q_i r_i = \sum_{i=1}^{k} Q_{ri} \tag{7}$$

Table 4. Reduced quantity of the axle group

No	Q_i [pcs/yr]	m_i [kg/pcs]	s_i	r_m	r_s	r_i	Q_{ri}
P1	3200	0.126	23	0.846	0.821	0.695	2223
P2	3000	0.134	24	0.899	0.857	0.771	2313
P3	2800	0.109	30	0.732	1.071	0.784	2195
P4	4000	0.109	29	0.732	1.036	0.758	3031
P5	3500	0.109	28	0.732	1.000	0.732	2560
P6	2600	0.133	26	0.893	0.929	0.829	2155
P7	3400	0.133	26	0.893	0.929	0.829	2818
P8	3600	0.149	28	1.000	1.000	1.000	3600
REDUCED QUANTITY				Qr=20894			

The total processing time of the observed group when machining on CNC turning machine is:

$$T_c = Q_r \cdot t_{kp} = 20894 \cdot 4.1 = 85665.4 \text{ [min/yr]} \tag{8}$$

IOP Conf. Series: Materials Science and Engineering **749** (2020) 012018 doi:10.1088/1757-899X/749/1/012018

If we assume that the planned production quantity of some product is Q_i, then the total workload of machine/workplace in a specific machining process of creating this product in the specified time period is determined by the expression (9):

$$T_i = Q_i \cdot t_{ki} + Q_i \cdot \frac{T_{pzi}}{z_s} = Q_i \cdot t_{ki} + n_{si} \cdot T_{pzi} \text{ [min/yr]} \quad (9)$$

where: T_i - total processing time;
T_{pz} - setup time for batch of parts;
z_s - number of parts within the batch;
n_s - number of batch in time period.

If it is assumed that the observed facility will work for $m_e = 250$ days per year in $s_e = 2$ shifts per day with $n_e = 7.5$ hours per shift and efficiency of $\eta_e = 0.8$, what follows is that the effective work capacity of machine/workplace equals $K_e = 180000$ minutes per year.

Based on the previously calculated processing time, assuming that $n_s = 10$ batch per year, the required number and degree of machine utilization for the observed turning process is:

$$T = 85665.4 + 10 \cdot 30 = 85965.4 \text{ [min/yr]} \quad (10)$$

$$N = \frac{T}{K_e} = \frac{85965.4}{180000} = 0.4776 \quad (11)$$

$$\eta = \frac{0.4776}{1} = 47.76\% \quad (12)$$

If it is assumed that operators work for $m_e = 250$ days per year in $s_e = 1$ shift per day with $n_e = 7.5$ hours per shift and efficiency of $\eta_e = 0.8$, it follows that K_e equals 90000 minutes per year:

$$N = \frac{T}{K_e} = \frac{85965.4}{90000} = 0.9552 \quad (13)$$

$$\eta = \frac{0.9552}{1} = 95.52\% \quad (14)$$

4. Conclusions

The main objective of this paper was to present two rational methodologies for evaluating the technological effects of application of the FMS elements at the conceptual level.

The paper shows that by classifying and grouping of parts, and then by designing the content of GT process plans and corresponding group operations on the basis of complex parts, good bases for selecting types and levels of complexity of the FMS elements for individual group operations may be created. It was shown that by specifying GT process plans for the selected products–part representatives, quality technological bases for evaluating the effects of application of the FMS at the conceptual level can be generated. It can be achieved through the calculation of the number and degree of utilization of the elements of the FMS.

Acknowledgements
This paper is part of a research on projects TR 35025, supported by the Ministry of Education, Science and Technological Development of Republic of Serbia.

References

[1] Radharamanan R 1994 *Group technology concepts as applied to flexible manufacturing systems* International Journal of Production Economics Vol. 33 ISSN 0925-5273 pp.133-142

[2] Zäh M F, Beetz M et al. 2009 *The cognitive factory* H.A. ElMaraghy (ed.) Changeable and Reconfigurable Manufacturinf Systems ISBN 978-1-84882-067-8 pp.355-371
[3] Gania I P, Stachowiak A, Oleśków-Szłapka J 2017 Flexible *Manufacturing Systems: Industry 4.0 Solution* Proceedings of 24th International Conference on Production Research (ICPR 2017) Poznan (30.7.-3.8.2017) pp.57-62
[4] Shivanand H K, Benal M M, Koti V 2006 *Flexible Manufacturing System*3 ISBN 978-8122418705 pp.22-4
[5] Satya S C, Douglas NH 2008 *The evolution of manufacturing cells. An action research study* European Journal of Operational Research Vol. 188 ISSN 0377-2217 pp.153-168
[6] Todić V, Zeljković M, Tepić J, Milošević M, Lukić D 2012 *Techno-Economic Method for Evaluation and Selection of Flexible Manufacturing Systems* Metalurgija Vol. 51 No. 3 ISSN 0543-5846 pp.349-353

IOP Conf. Series: Materials Science and Engineering **749** (2020) 012019 doi:10.1088/1757-899X/749/1/012019

Digital Transformation of Manufacturing Towards Industry 4.0 Concept

M Milošević[1a], D Lukić[1], M Đurđev[2] and J Vukman[1]

[1] University of Novi Sad, Faculty of Technical Sciences, Department of Production Engineering, Novi Sad, Serbia
[2] University of Novi Sad, Technical Faculty "Mihajlo Pupin", Zrenjanin, Serbia

E-mail: [a] mido@uns.ac.rs

Abstract. Markets are nowadays oriented towards high quality products that are produced in small series production with fast delivery dates. Therefore, manufacturing companies are forced to adapt their production processes and make them more flexible and efficient. In that sense, this paper presents digital transformation of manufacturing as a concept of product innovations in which conventional methods and models are replaced by digital ones through all stages of product lifecycle. Digital transformation or digitalization of manufacturing refers to the interaction between physical and digital models within the production process. That is the crucial step towards the implementation of the Industry 4.0 concept which is fully focused on the integration of manufacturing with information technologies and digital services.

1. Introduction

Reduction of product lifecycle and changes resulting from frequent modifications in production program, increase requirements for much faster and more efficient transfer of information between engineers as well as other participants involved in planning and manufacturing processes. Application of the concept of digital factory, in other words, digital manufacturing, facilitates transfer of information which also leads to faster reduction of new product development time. Complete digitalization of manufacturing represents the framework for the implementation of novel technologies such as Smart Factory and Industry 4.0 concept [1, 3].

In recent years, implementation of information and internet technologies has accelerated the integration of engineering activities, equipment and all other components involved in planning and manufacturing. Digitalization has brought many innovations in the fields of product development and production preparation.

2. Digital engineering

Digital engineering accelerates the product development cycle, from design to its use, thereby increasing efficiency of engineering processes through the integration of CAx and PLM systems. Efficient exchange of digital information within the engineering processes is achieved in this way, figure 1.

Modern software tools enable the cooperation among project teams based on the exchange of digital information which ensures greater innovation and better quality, while reducing the new product cost and time to market. Frequent product changes result in the occurrence of multiple product

IOP Conf. Series: Materials Science and Engineering **749** (2020) 012019 doi:10.1088/1757-899X/749/1/012019

variants and digital engineering brings new tools and methods for efficient work administration during all stages of product lifecycle for various product variants and customer needs, figure 2.

By applying engineering information in a digital form, an analysis and evaluation of products, its functions and manufacturing conditions can be performed. In doing so, starting parameters include targeting cost, expected sale, possible prices as well as required resources for the planned product. In this way, planned costs are defined in the earliest stage of production planning

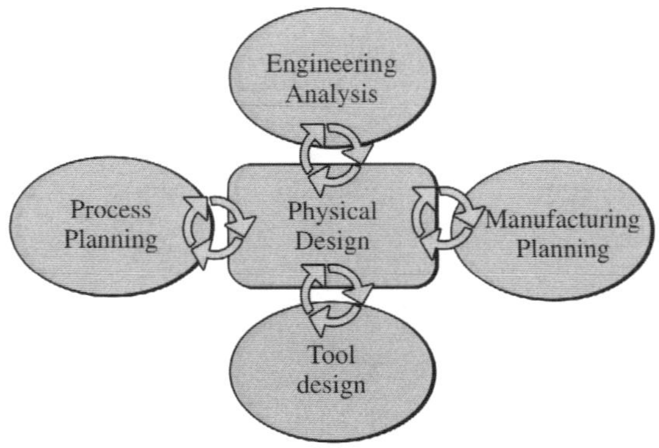

Figure 1. Exchange of digital information in engineering processes

Stages that follow are product design, including various engineering analyses of future product's characteristics, then definition of process plans and planning of direct manufacturing processes. Management of digital information also involves sales analysis, maintenance as well as recycling of end-to-life products. Efficient analysis of all stages with product lifecycle also includes continuous collaborative work among all subjects involved in digital engineering processes.

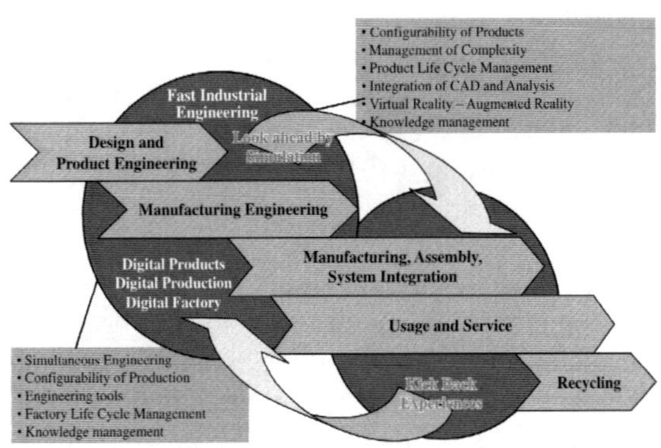

Figure 2. Digital engineering - Knowledge-based engineering

The digital engineering framework contributes to the efficient solving of engineering tasks and provides:

- Reduction of planning errors,
- Increase in productivity,
- Reduction of time losses,
- Reliability of products,
- Introduction of innovations etc.

Digital manufacturing is a universal concept for the entire network of digital models, methods and tools, including simulation and 3D visualization, integrated with a continuous data management system. Digital manufacturing represents the integration of virtual and real manufacturing, figure 3.

IOP Conf. Series: Materials Science and Engineering **749** (2020) 012019 doi:10.1088/1757-899X/749/1/012019

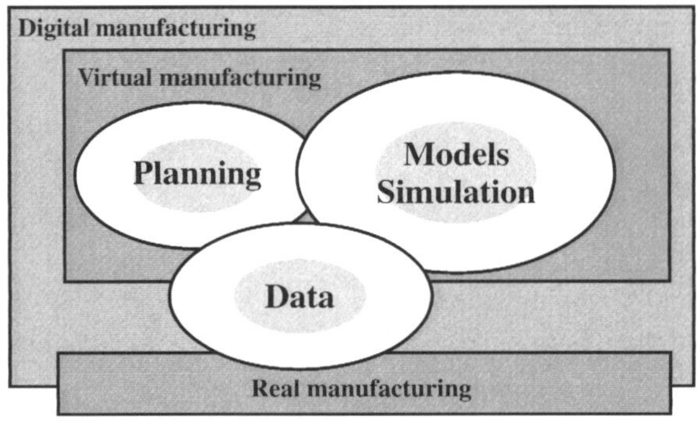

Figure 3. Digital manufacturing: integration of virtual and real manufacturing

Therefore, the concept of digital manufacturing is considered as an integrated concept aimed at improvement of products and engineering activities with production process and simulation. Different forms of simulation can be applied in virtual models on different levels of planning in order to improve products and manufacturing processes. Thus, digital manufacturing represents the simulation of all activities in the field of research and development as well as in the field of production processes.

Vision of the concept of digital manufacturing focuses on the integration of available methods and tools for design and control of products at different levels of manufacturing and operational control with the company. Consequently, the concept of digital manufacturing integrates the following elements [1, 5]:

- Product development, testing and optimization,
- Development and optimization of process plans,
- Production system planning and improvement and
- Operational production planning and control.

Digital transformation of manufacturing can also be viewed as the information strategy of a company for improvement and collaboration of production processes in global environment [2, 7]. The concept of digital manufacturing includes modelling of products, processes and resources on the basis of real data. Planned products as well their processes are intensively verified and enhanced using virtual models until they are fully developed [6]. When all potential errors are eliminated, then the models can be applied in real manufacturing.

3. Goals of digital transformation of manufacturing

Digital manufacturing is a technology or a discipline that provides a strategic approach for development, implementation and optimization of all elements of a production process. The concept of digital manufacturing covers a network of digital models and methods used for defining all aspects of a production process in companies of the future [4]. Digital manufacturing environment represents the combination of a digital product and digital processes and resources. That means the framework that integrates a virtual manufacturing model with real physical model.

Digital transformation of manufacturing includes the following aspects:

- Process definition,
- Process planning,
- Modelling of production system,
- Visualization,
- Simulation of operations,
- Analysis of ergonomics,
- Study of manufacturing time,
- Study of productivity,
- Balancing of production lines,
- Analysis of alternative resources,

- Automation of manufacturing and
- Programming of control structures.

The concept of digital manufacturing starts with defining and planning of manufacturing processes, then managing information on manufacturing processes which serves as a condition for efficient collaboration, figure 4.

Figure 4. Framework for manufacturing process control using digital manufacturing

Simultaneous product design and process planning is the principle applied in concurrent engineering. In this way, process planning is carried out not only to define product shape and its functional criteria, but to simultaneously determine manufacturing and technological criteria as well. Different simulations and engineering analyses result in the optimal product design and optimal process plan and after validation, the adopted solution is applied in a real manufacturing system. Results of manufacturing operations are further analysed and they represent useful information within the digital reengineering of products and manufacturing operations. Also, digital manufacturing includes integration with equipment for direct manufacturing and control, systems for enterprise resource planning, supply chains and other relevant factors that have an impact on the production plan, Figure 5.

Figure 5. Digital manufacturing-synthesis of process planning and production execution

IOP Conf. Series: Materials Science and Engineering **749** (2020) 012019 doi:10.1088/1757-899X/749/1/012019

Digital transformation of manufacturing is an integrated solution package for planning of production processes and visualization using tools for virtual simulation which enables engineers to optimize production processes.

Therefore, this concept integrates various tools for:

- Product design,
- Engineering analysis,
- Process planning,
- Control of machining systems,
- Time management and business applications,
- Planning the allocation of production processes,
- Ergonomics,
- Robot work simulation,
- Production planning,
- Product lifecycle management,
- Simulation of production processes,

4. Digital manufacturing and Industry 4.0

Digital manufacturing plays a crucial role in the development of enterprises on the principles of Industry 4.0 concept, as it supports innovations and improvements of product development [3]. This technology provides shorter time to market and reduces development and manufacturing costs.

Application of digital manufacturing opens the possibility of efficient monitoring and improvement of a production process by using and controlling data related to the development, planning and validation of production resources. The goal is to integrate data from different areas in the domain of product design and production, Figure 6.

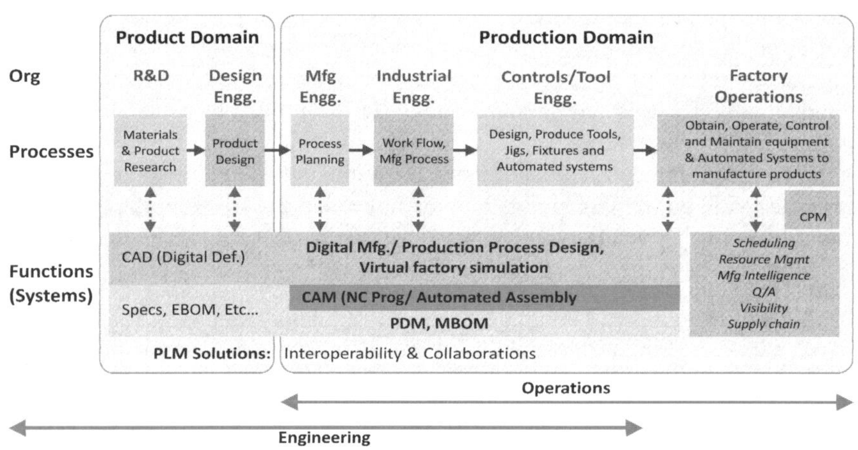

Figure 6. Integration of processes from the domain of product design and the domain of production

The starting point in the application of Industry 4.0 is the application of the concept of digital manufacturing as it provides the following [5]:

- Definition of logistical standards that can be applied to different products,
- Time saving for correction of possible errors in process and production planning,
- Optimization of logistics planning,
- Standardization of assembly processes,
- Optimization of planned time norms in all stages of production and
- Efficient collaborative data management in the domain of product design and production planning.

Information exchange and engineering collaborative processes are crucial in digital manufacturing and the implementation of Industry 4.0 technologies [3,7], Figure 7.

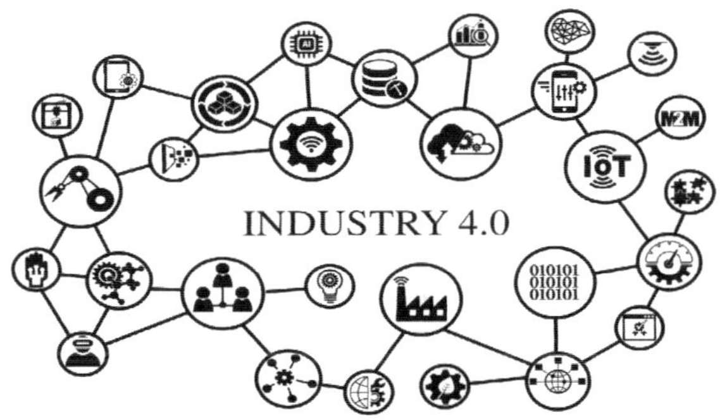

Figure 7. Industry 4.0 technologies and tools based on the digital transformation of manufacturing

5. Conclusions

Digitalization of manufacturing is a critical step towards the application of novel technologies that are presented within the Industry 4.0 concept. It is necessary to digitize not only physical models of complex products, but also production processes, systems, resources and other elements that are a part of product lifecycle. That is the only way for manufacturing companies to respond to future technological challenges and improve their efficiency in modern business. One of the important reasons for digital transformation of manufacturing is efficient communication between the experts responsible for solving common problems and successful achievement of common goals. That primarily relates to the optimization of products and processes which leads to significant reduction in development and manufacturing costs.

Acknowledgements
Some results in this paper are a part of the project: Contemporary approaches to the development of special solutions related to bearing supports in mechanical engineering and medical prosthetics - TR 35025, carried out by the Faculty of Technical Sciences, University of Novi Sad, Serbia, are presented. Ministry of Education, Science and Technological Development of the Republic of Serbia support the project.

References

[1] Elias Hans Dener Ribeiro da Silva, Jannis Angelis, Edson Pinheiro de Lima 2015 *In pursuit of Digital Manufacturing* Procedia Manufacturing vol. 28 ISSN 2351-9789 pp.63-69
[2] Gregory Vial 2019 *Understanding digital transformation: A review and a research agenda, The Journal of Strategic Information Systems* In press, ISSN 0963-8687 pp.1-27
[3] Jian Qin, Ying Liu, Roger Grosvenor 2016 *A Categorical Framework of* Manufacturing *for Industry 4.0 and Beyond* Procedia CIRP vol. 52 ISSN 2212-8271 pp.173-178
[4] Lihui Wang 2018 *Combining digital twin and prognosis in factories of the future* Proceedings of the 13th International Scientific Conference MMA 2018 Flexible Technologies keynote paper ISBN 978-866022-094-5 Novi Sad, Serbia
[5] Lihui Wang, Andrew Yeh Ching Nee 2009 *Collaborative Design and Planning for Digital Manufacturing* Springer-Verlag London Ltd. ISBN 978-1-84882-286-3
[6] Phani Kumari Paritala, Shalini Manchikatla, Prasad K.D.V. Yarlagadda 2017 *Digital Manufacturing- Applications Past, Current, and Future Trends* Procedia Engineering vol. 174 ISSN 1877-7058 pp.982-991
[7] Theodor Borangiu, Damien Trentesaux, André Thomas, Paulo Leitão, Jose Barata 2019 *Digital transformation of manufacturing through cloud services and resource virtualization. Computers in Industry* vol. 108 ISSN 0166-3615, pp.150-162

IOP Conf. Series: Materials Science and Engineering **749** (2020) 012020 doi:10.1088/1757-899X/749/1/012020

Analyses of abrasive wear behaviour in Pin-on-Plate Tribo-System for several materials

H Muhandes[1a], A Kalacska[2] and G Kalacska[1]

[1] Szent Istvan University, Institute of Mechanical Engineering Technology, Godollo, Hungary
[2] Ghent University, Department of Electrical Energy, Metals, Mechanical Construction and Systems, Ghent, Belgium

E-mail: [a] hasanmuhandes@gmail.com

Abstract. To improve the performance of the Agricultural machinery, several composite materials were suggested as replacements of critical fast-wearing steel parts. Three possible replacements were selected (PA6E, PA6G, and PA66GF30) and examined by a pin-on-plate system with sliding abrasive cloths. The results were compared according to the pin-on-plate system with a combined property (H/E) of these materials. Then to see how the friction affected these specimens, 3D microscopic photos of the worn surfaces were taken. As results, an inverse relationship between the wear rate and H/E property for these materials, an increasing of H/E factor leads to a decrease in the wear rate. While for the 3D microscopic PA6G was the most outstanding worn surface.

1. Introduction

Nowadays, abrasive wear of engineering and agricultural machine components caused by the abrasive particles is a major industrial problem. Therefore, a full understanding of the effects of all system variables on the abrasive wear rates is necessary to undertake appropriate steps in the design of the machinery and the choice of materials to reduce/control wear [6].

La Mantia et al. mentioned that the recycling of composite is a complicated process and very limited, for example, when we have two different components in the composite, it will be hard to separate and recycle them, and often they use the incineration which considered as an unsatisfactory way [4].

Lee et al. introduced a simple model of composite materials abrasive wear. This model is based on the mechanics and mechanisms associated with sliding wear in soft matrix composites, which contain solid reinforcement particles. The model supposes that any removed part of the reinforcement as wear debris participate to the wear resistance of the matrix material [5].

Walczak et al. presented an analysis of the tribological properties of a ball on a disc as sliding elements in agricultural machinery. They conducted dry friction tests for polymer-metal pairs to obtaining the corrosion rate and friction coefficient of the samples used. Three pairs of Iglidur samples were used [10].

Patnaik et al. reviewed the solid particle erosion behaviour of fibre and particulate filled polymer composites. This review focused on polymer matrix composites and showed the problem related to the processes and modes during erosion. During the experimental studies, several aspects have been affirmed. Moreover, they discussed the use of the statistical techniques to analyse that erosion behaviour [8].

Ando et al. selected cast polyamide 6 material for semi-finished structural materials because it has excellent mechanical and chemical properties and easy for composite production, and they selected magnesium catalytic cast polyamide 6 natural polymer matrix as basic grade to have some new composites according to the need of the agricultural and they decided to develop a new composite version of it to improve tribology grade having abrasion resistance, antistatic composite version and improve fire-resistant version [1].

Kalacska studied improving the tribology properties by adding several materials like graphite, silicon dioxide, polytetrafluoroethylene, polyethylene, lead, oils, calcium silicate, waxes and silicone [3].

By using a scanning electrochemical microscopy (SECM) high-resolution three-dimensional images were generated by measuring the variation in the tip current due to perturbations in the diffusion layer [2].

A study of Surface Roughness of Thin Silicon Films Deposited on SiO_2 was done by using an Atomic Force Microscope. Several Silicon Films were used like amorphous silicon and Low-pressure, chemically-vapour-deposited silicon films on silicon dioxide [7].

Also, the Atomic Force Microscope and Profilometer were used to calculate the Surface Roughness in Ceramics specimens where the surfaced had different Finishing Techniques. Three ceramic materials were used: Vitadur Alpha, IPS Empress 2 and AllCeram. Five surfaces finished systems were used. The stylus profilometer and an atomic force microscope (AFM) are the two used roughness measuring instruments [9].

2. Materials and Methods

At that point, as per the above brief survey for some written literature, two types of composite materials which are (PA6G, PA6E, and PA66GF30) materials were recommended to supplant these steel parts.

Table 1 shows the main properties of the used materials.

Table 1. The main properties of the PA6G, PA66GF30 and PA6E

Properties	Method	PA6G	PA66GF30	PA6E
Density [g/cm3]	DIN EN ISO 1183	1.15	1.34	1.14
Yield stress [MPa]	DIN EN ISO 527-2	80	91	78
Modulus of elasticity [MPa]	DIN EN ISO 527-2	3500	5500	3300
Elongation at break [%]	DIN EN ISO 527-2	130	13	130
Flexural strength [MPa]	DIN EN ISO 178	109	135	100
Impact strength, Charpy [kJ/m2]	DIN EN ISO 179-1eU	No break	87	No break
Tensile strength [MPa]	DIN EN ISO 527-2	83	91	79
Compressive strength [MPa]	EN ISO 604 1% / 2%	19/ 36	25/ 46	24/ 41
Notched impact strength, Charpy [kJ/m2]	DIN EN ISO 179-1eA	4		7
Ball indentation hardness [MPa]	ISO 2039-1	170	216	155
Shore D hardness [-]	DIN EN ISO 868	83	86	75

To test these materials, a pin-on-plate test with a sliding abrasive cloth. Pin-on-plate test made by using the following device:

Figure 1. Pin-on-plate wear device

This device can give us a nonstop wear interface and furthermore a possibility to control several parameters:

1. Opportunity to control the speed of the interface wear;
2. Possibility to use several wear interfaces;
3. Opportunity to add a load over the specimens:
    - Two types of wear interfaces were used (P60 and P150);
    - Two speeds are used which are (20%=0.032 m/s and 40%=0.056 m/s);
    - Moreover, we used three loads (9.81 N, 29.43 N and 49.05 N).

Wear of different structural plastic materials was analysed by measuring relevant features during abrasion tests (time, force, abrasion rate). Experimental equipment has been utilised to trace the wear of plastic materials under the same conditions. Abrasion rate has been recorded, together with pressure (force) applied to the sample. Length (duration) of the experiments have been standardised by taking identical experimental runtime. Simple apparatus assembled (figure 2) and sensors were connected to a computer data collecting interface. The interface was converting the sensor signals into digital values and provided the data to a computer that was recording the values.

Figure 2. The Specimen holder and the connected sensor

Several strain gages are connected to a PC by an estimation device (Spider 8) to take these parameters as results. The interface was converting the sensor signals into digital values and provided the data to a computer that was recording the values.

IOP Conf. Series: Materials Science and Engineering **749** (2020) 012020 doi:10.1088/1757-899X/749/1/012020

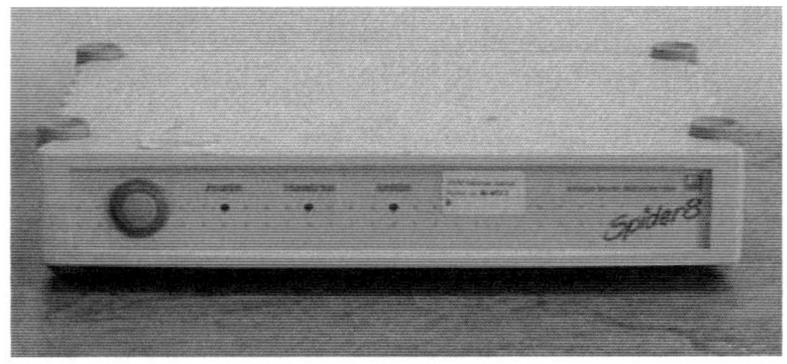

Figure 3. Spider 8, a strain gage measurement device

The used specimens have an exact dimension with 8mm diameter and 20 mm length. We tested each sample for 10 minutes; some samples did not bear this time due to its specifications, enormous load, and substantial sliding speed. These cases, we can see it amid the results.

After testing these samples, we took them to do the 3D microscopic check to calculate some parameters and to get a 3D profile of the worn surfaces. To do this process, we used (Taylor Hobson CCI HD) device. The Measurement type is Non-contact 3D optical profilometry.

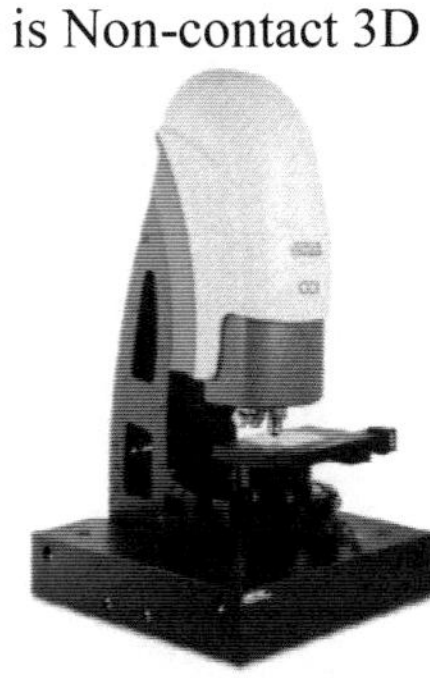

Figure 4. Taylor Hobson CCI HD device

This device has several features like:

- Reduced inaccuracy and noise compared to 2D;
- High resolution, precise results;
- Data extraction from whole 3D surface.

Moreover, during the measurement, the used settings are:

- 50X lens;
- Scanning speed: x1, 4m mode, filter 2, rough surface.

3. Result

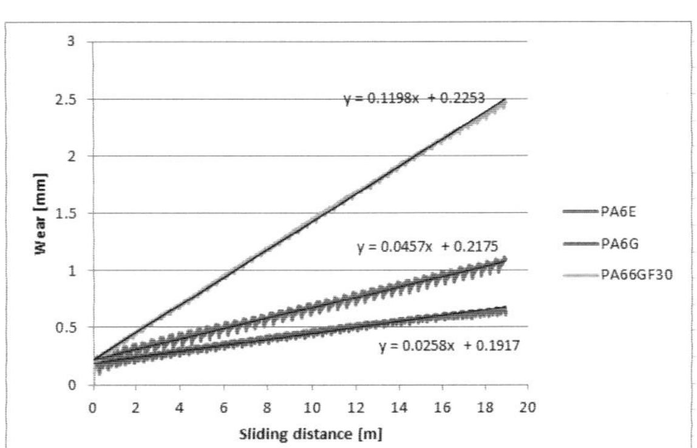

Figure 5. The relation between the wear and sliding distance for several polymers types (0.031m/s, 9.81N)

IOP Conf. Series: Materials Science and Engineering **749** (2020) 012020 doi:10.1088/1757-899X/749/1/012020

Figure 5 demonstrates a comparison between the wear and sliding distance for several polymers types; the load is 9.81N, the sliding speed is 0.0315m/s, and the wear interface is P60.

We can compare the different types of polymers by using the slope value of the wear lines.

Table 2. Wear line equation and the slope value of the different polymers types

Type	Wear line equation	Slope value
PA66GF30	y=0.1198·x+0.2253	0.1198
PA6E	y=0.0457·x+0.2175	0.0457
PA6G	y=0.0258·x+0.1917	0.0258

From figure 5 and table 2, we can see that PA66GF30 was subjected to the highest wear rate and had the maximum slope value, while PA6G had the lowest wear rate and had the minimum slope value. Moreover, in other words, we can state that the high slope value implies high wear occurred in the specimen.

Moreover, to study the effect of the speed we compared these three specimens with a new three specimens were the speed was 0.056 m/s with the same load (9.81N) and same wear interface P60.

From figure 5 and figure 6, we can see that by increasing the sliding speed that will lead to increase the wear values.

Figure 6 shows the result of this comparison.

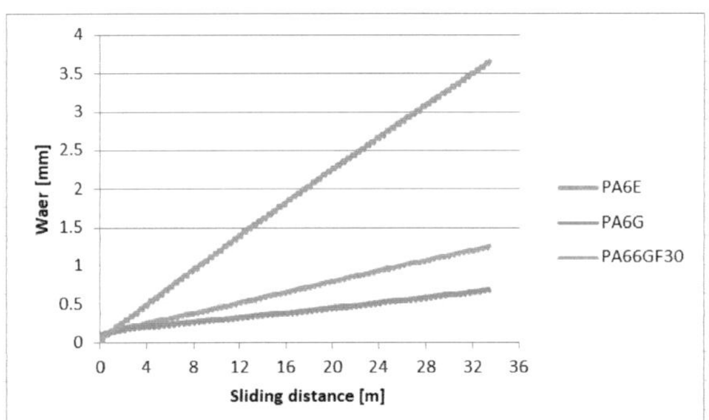

Figure 6. The relation between the wear and sliding distance for several polymers types (0.056m/s, 9.81N)

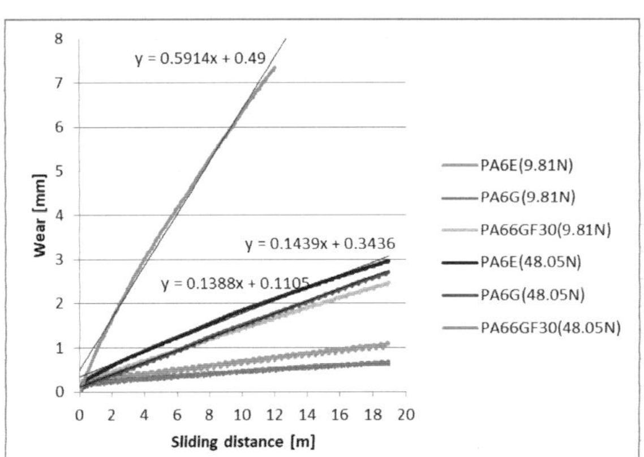

Figure 7. The relation between the wear and sliding distance for several polymers types

IOP Conf. Series: Materials Science and Engineering **749** (2020) 012020 doi:10.1088/1757-899X/749/1/012020

While to study the effect of changing the load, we compared these three specimens with a new three specimens were the speed was 0.031 m/s with same wear interface P60 and the load was (48.05N). Figure 7 shows the result of this comparison.

The wear line equations for the new three specimens are shown in the following table:

Table 3. Wear line equation and the slope value of the different polymers types

Type	Wear line equation	Slope value
PA66GF30	y=0.5914·x+0.4900	0.5914
PA6E	y=0.1439·x+0.3436	0.1439
PA6G	y=0.1338·x+0.1105	0.1138

By using figure 5 and 7 with table 2 and 3, we can see that by adding a higher load will lead to increase the values of the slope values of the wear lines, and of course, an increase in the wear values.

The obtained results were compared with two main properties of these materials, which are Modulus of elasticity (E) [MPa] and Shore D hardness, and were used as a ratio of (H/E). However, we had to convert the value of the hardness to [MPa] as a unit firstly by using the following equation:

$$\text{Hardness}\left[\text{MPa}\right]=\text{EXP}\left(\left(\text{Shore D}+50\right)\cdot 0.0235-0.6403\right) \tag{1}$$

According to a table 1, we have the Shore D hardness values as the following:

Properties	Method	PA6G	PA66GF30	PA6E
Shore D hardness [-]	DIN EN ISO 868	83	86	75

And by using the previous equation, we got the hardness values in MPa as the following:

Table 4. Hardness values in MPa for the used materials

Properties	PA6G	PA66GF30	PA6E
Hardness [MPa]	12.003	12.88	9.946

Then we calculated the ratio (H/E) and got:

Table 5. H/E ratio for the used materials

Properties	PA6G	PA66GF30	PA6E
$\frac{\boldsymbol{H}}{\boldsymbol{E}}$	$\frac{12.003}{3500}=0.0034$	$\frac{12.88}{5500}=0.0023$	$\frac{9.946}{3300}=0.00301$

As an order of these numbers, we can see that PA6G had the highest value as (0.0034), then PA6E as (0.00301), then PA66GF30 as (0.0023).

By comparing this order with figure (5) and table (2), where:

- PA6G had the lowest wear rate and had the highest (H/E) value;
- PA6E had the mid wear rate and had the mid (H/E) value;
- PA66GF30 had the highest wear rate and had the lowest (H/E) value.

We can see that there is an inverse relationship between H/E and the wear rate.

Then these samples were tested by the 3D microscopic to have a further idea about the worn surfaces after the pin-on-plate process, and the results were as the following figure:

IOP Conf. Series: Materials Science and Engineering **749** (2020) 012020 doi:10.1088/1757-899X/749/1/012020

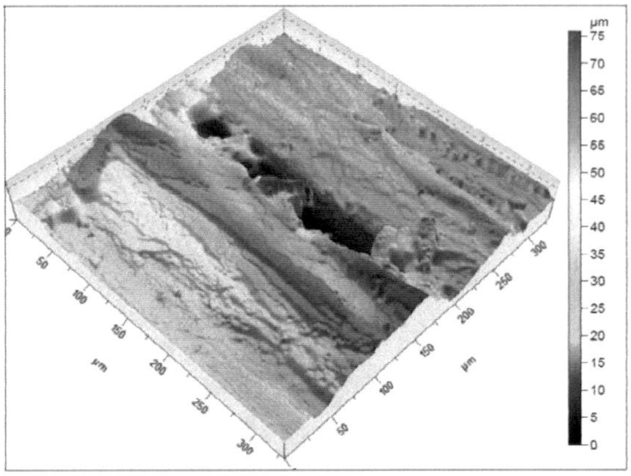

PA6E

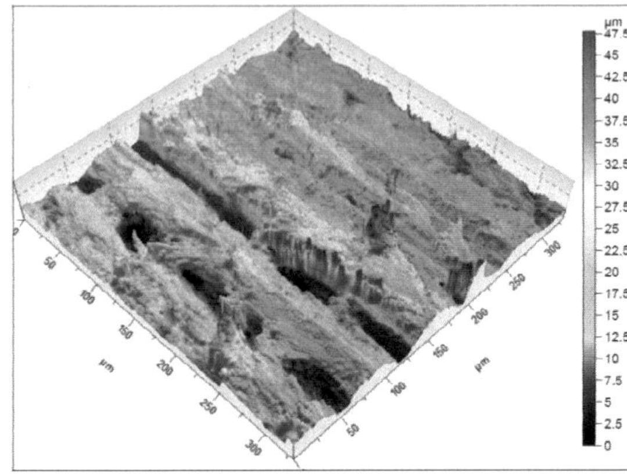

PA66GF30

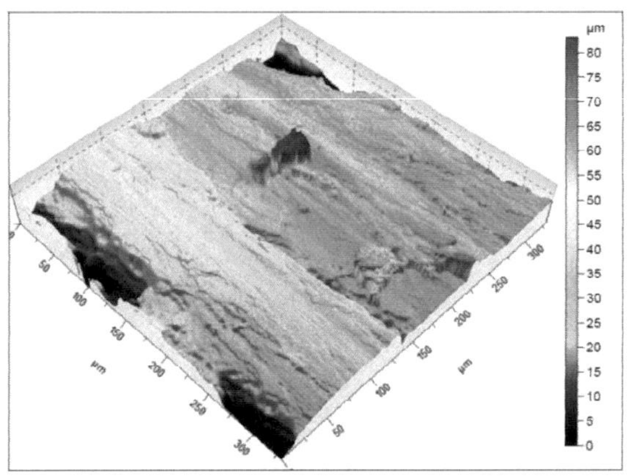

PA6G

And as results for these photos, we got the following table:

Table 6. The Height Parameters of the worn surfaces according to ISO 25178

	PA6E	PA6G	PA66GF30
S_q (µm)	15.03766	12.46001	6.67540
S_{sk}	-0.33826	-1.15040	0.54640
S_{ku}	2.91990	5.04750	3.77496
S_p (µm)	34.30528	41.84994	29.88581
S_v (µm)	41.67346	41.30695	17.96119
S_z (µm)	75.97874	83.15689	47.84700
S_a (µm)	12.17908	9.32774	5.21821

From the previous table we can recognise that:

- The height distribution is skewed below the mean plane for PA66GF30 since S_{sk} value is bigger than zero, while for PA6E and PA6G the height distribution is skewed above the mean plane since S_{sk} is less than zero;
- The sharpness of the roughness profile of the PA6G is the most spiked then PA66GF30 since S_{ku} values are bigger than 3, while PA6E have a normal distribution since the value is almost 3;

- PA6G had the highest peak within the defined area as 41.84994μm, while PA66GF30 had the lowest as 29.88581μm;
- PA6E and also had the largest pit height as 41.67349μm. Also, PA33GF30 had the lowest one as 17.96119μm;
- The Maximum height where for PA6G as 93.15689μm, which is the sum of largest peak height value and the largest pit depth value, but the minimum height was for PA66GF30 as 47.84700μm;
- PA6E had the highest arithmetic mean of the absolute value of the height from the mean plane of the surface as 12.17908μm, while PA66GF30 had the lowest as 5.21821μm.

As a result of the previous points, we can see that the PA6G was the most outstanding worn surface, and that is obviously also according to the surface photo.

4. Conclusions

In this work, several kinds of Polyamide polymers were used to study their wear performance with various changing parameters such as the speed of friction and the applied loads. As a result, PA6G was the best choice of these used polymers because it had the lowest wear rate.

Also, we got that by increasing the load and increasing the speed, the wear values increased. By comparing the wear rate result with the combined properties (H/E) ratio, we found that there is an inverse relation between (H/E) ratio and the wear rate for these tested polymers.

While for the 3D microscopic photos, we had a further understanding of the worn surfaces. The PA6G was the most outstanding worn surface compared to other materials according to the surface photos, where the sharpness of the roughness profile was the most spiked, with the highest peak and largest pit height.

References

[1] Andó M, Kalácska G, Czigány T 2008 *Cast polyamide 6 polymer composites for agricultural machine application* Hungarian Agricultural Engineering Gödöllő pp.67-69

[2] Ellis KA, Pritzker, M D, Fahidy, T Z 1995 *Modeling the Degradation of Scanning Electrochemical Microscope Images Due to Surface Roughness* Analytical Chemistry Vol. 67 pp.4500-4507

[3] Kalácska G 2007 *Műszaki polimerek es kompozitok a gépészmérnöki gyakorlatban* Gödöllő 3C-Gragika Kft

[4] La Mantia F P, Morreale M 2011 *Green composites: A brief review* Composites: Part A 42 pp.579–588

[5] Lee G Y, Dharan C K H, Ritchie R O 2002 *A physically-based abrasive wear model for composite materials* Wear 252 pp.322–331

[6] Mishra V 2014 *Physical, mechanical and abrasive wear behaviour of jute fiber reinforced polymer composites* Doctoral dissertation National Institute of Technology India

[7] Nasrullah J, Tyler G L, Nishi Y 2005 *An Atomic Force Microscope Study of Surface Roughness of Thin Silicon Films Deposited on* SiO_2 IEEE Transactions on Nanotechnology Vol. 4 pp.303-311

[8] Patnaik A, Satapathy A, Chand N, Barkoulad N M, Biswasb S 2010 *Solid particle erosion wear characteristics of fiber and particulate filled polymer composites: A review* Wear 268 pp.249–263

[9] Tholt B, Miranda-Júnior W G, Prioli R, Thompson J, Oda M 2006 *Surface Roughness in Ceramics with Different Finishing Techniques Using Atomic Force Microscope and Profilometer* Operative Dentistry Vol. 31 pp.442-449

[10] Walczaka M, Cabanb J, Marczukb A 2017 *Evaluation of tribological properties of polymer materials used for sliding bearings in agricultural machinery* Agricultural Engineering 21 pp.95-103

IOP Conf. Series: Materials Science and Engineering **749** (2020) 012021 doi:10.1088/1757-899X/749/1/012021

Milling convex surfaces with toroidal cutting edge

A R Osan, M Cosma[a] and V Nasui

Technical University of Cluj-Napoca, Faculty of Engineering, Department of Engineering and Technology Management, Baia Mare, Romania

E-mail: [a] marius.cosma@cunbm.utcluj.ro

Abstract. The toroidal cutting edge, also called corner cutter, is a combination of the cylindrical-frontal milling cutter and the spherical head milling. The paper follows at generating the optimal trajectory for making the finished product with the toroidal cutting edge without having to use the flat end mill and the ball nose end mill. This trajectory is accomplished using a CAM (Computer Aided Manufacturing) program that is generated on a 3D model. The proposed method was experimentally verified on a 3-axis numerical control centre, making a roughness comparison according to the climb milling or conventional milling.

1. Introduction

Complex and convex surfaces are encountered in many situations, such as aerospace, cars, molds, and the field of pneumatic robots. The surfaces are 3D generated using a CAD program, helping to provide a geometric description for generating the trajectory of the milling tool.

Computer-Aided-Manufacturing (CAM) is defined as the use of software for controlling machine tools in processing processes representing computer assistance in processing processes with the primary purpose of creating a faster production process.

According to Xun Xu [1], computer-assisted manufacturing can be interpreted both from a general point of view and from a particular point of view. In a general sense, computer assisted manufacturing refers to any application used in manufacturing processes such as: machine tool control; monitoring machine tools; simulation of the manufacturing process; communication on the production platform as well as mechanical testing. In a particular sense, computer-assisted manufacturing is computer-based machining and addresses machine tools with numerical control [1].

Computer-assisted production covers a wide range of activities. At the lowest level, the automation of individual processes or a group of processes can be accomplished by microprocessors, programmable logic commands and microcontrollers [2].

Computers or microprocessors can control production equipment such as machine tools, welding machines, assembly machines, and so on. The type of control used in machine tools is called computerized numerical control (CNC). [2] The concept of numerical command was made at the Massachusetts Institute of Technology in 1951.

The term "numeric" means the inputs of command numbers that take the form of numbers. They are presented in binary code and can be processed directly by the equipment. The entered figures are intended to describe the geometry of the workpiece, the data related to tools and work speeds, in numerical form. This description by numbers is characterized by a literary prefix. These command formats are indicated in DIN 66 025 of 1983 respectively ISO 6983 in 2009 and ISO 646 since 1991 and are called blocks.

2. Numerically controlled machine tool axis analysis

In the numerical machine tool sector, the notion of axle has been initiated to define each fixed or circular motion direction. In order to recognize the machine axes, some internationally agreed symbolism and meanings of movement are assigned by ISO R-841-1968, and nationally by STAS 8902-1971. The axial displacement axes of the numerically controlled machine constitute, by association, a geometric coordinate system. According to STAS 8902-1971, the normal coordinate system is a straight-line clamp, which refers to a piece fixed to the machine and having the edges (s) parallel to its main guides.

3. Effects of climb and conventional milling

Milling is the machining of flat, cylindrical or profiled surfaces by means of a multi-pronged tool, called milling machine. [3]

I activating in the field of machining as a CNC programmer engineer, I encounter different cases and situations where different factors have to be taken into account, which can influence the quality of the workings positively or negatively, so I have learned that when it comes to chip processing the difference makes the details, however small they are.

Machining of surfaces by climb milling is the most frequently encountered milling method, whether it is planning a surface or profiling. As the name implies, the milling teeth pull the material toward itself trying to climb through the material following the trajectory of movement. In this movement, the milling teeth first contact the outer surface of the cutting where the chip thickness is greater, continuing to taper as the milling tooth passes through a complete movement. [4]

This feature has the advantage that the chips will be removed and stored behind the cutting without contact with the milling cutter so that a cleaner surface will be produced and will increase the life of the tool. This cutting method requires less force to achieve cutting, and the tool movement generates a force on the part that helps stabilize it, especially in surface planning and finishing. Conventional milling has a major advantage namely that the milling tooth is required as it penetrates into the material on the surface previously machined by the tooth thus increasing the friction between the material and the milling tool generating also a considerable amount of heat. As a result, when conventional milling, the tooth enters the cutting at the scissor's zero thickness and reaches the cutting edge to the maximum thickness, so there are no shocks. [3]

4. The experimental part

The study of climb milling and conventional milling with classical tools has been done by many authors [3], [4] but a concrete part of what involves the toroidal milling is not yet what obliges me to do it in order to know the optimal principle to apply in future experiments.

4.1. The experimental-theoretical part

The CAM program was generated using Powermill. According to [5] Powermill is a 3D Cam program that generates the tool trajectory for CNC milling machines from 2 to 5 axes developed by Delcam.

Figure 1a. Climb milling

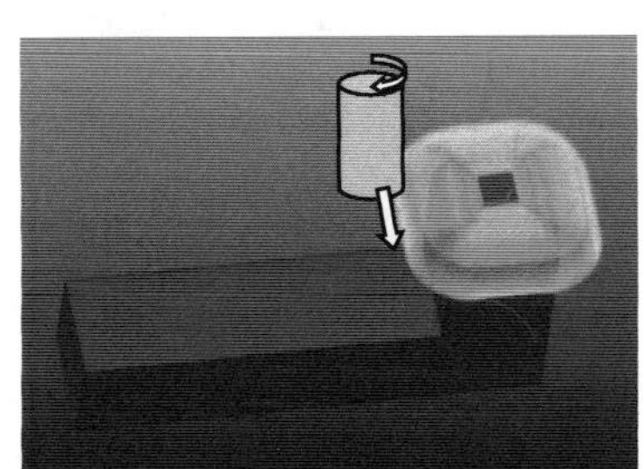

Figure 1b. Conventional milling

IOP Conf. Series: Materials Science and Engineering **749** (2020) 012021 doi:10.1088/1757-899X/749/1/012021

The program generated in the first phase is shown in figure 1.a. indicates the movement of the cutter in the direction of the feed and motion against the advance, figure 1.b.

4.2. Practical experimental part
For the practical part a MCV 1016 3 axis machining centre, 2015 manufacturing year of S.C. Ramira S.A.

Figure 2. Vertical machining centre in 3 axis MCV 1016

The cutting tool used was a toroidal cutter JHP780160E2R400.0Z4-M64 with a carbide coating of Ø16 with a number of 4 teeth and a radius of R4.

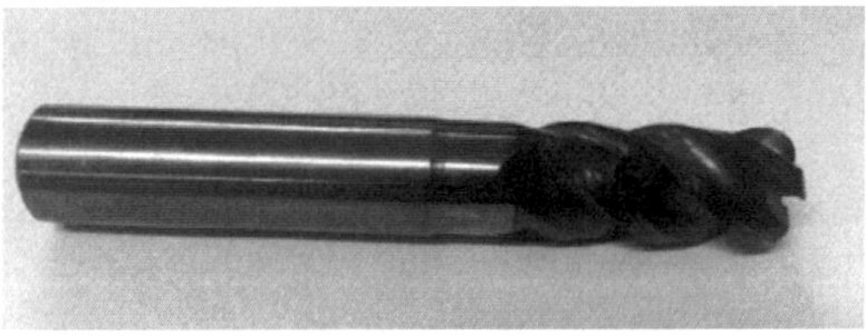

Figure 3. The toroidal milling cutter

4.3. The material and the shape of the piece
The material used in the experiment is C45 (1.0503) with the following characteristics: 0.42 ... 0.50% C, 0.5 ... 0.80% Mn, 0.17 ... 0.37% Si, maximum 0.040% P [6] etc. The piece has the shape of a square being machined only one of the ends representing the active part of the piece.

Figure 4. Form of the workpiece

Figure 5 shows a visual comparison of the two types of machining and it can be seen that the climb milling has a visual advantage due to the surface gloss compared to the conventional milling where some optical scratches can be observed.

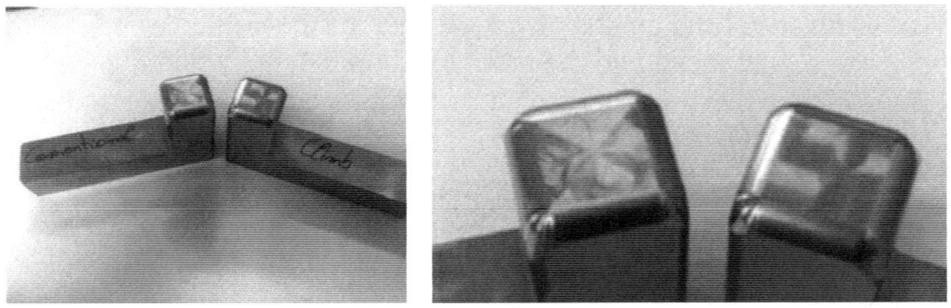

Figure 5. Processing of surfaces with climb milling and conventional milling

4.4. Comparing the roughness of the two pieces processed with toroidal milling

The device used to check the roughness is the TIME TR 200, this portable absolute measuring device has the sensitivity required to measure the very fine roughness of the tens and microns.

Following the two roughness determinations, a series of measurements were performed, following 6 parameters in terms of surface quality. A number of 6 measurements were made on each surface, as shown in Table 1.

Table 1. Measured roughness values

Surface roughness processed climb milling [μm]							Surface roughness processed conventional milling [μm]						
	Ra	Rq	Rz	Rt	Rp	Rv		Ra	Rq	Rz	Rt	Rp	Rv
1.	0.289	0.472	2.900	4.539	0.759	2.14	**1.**	0.446	0.399	2.956	5.019	0.551	2.404
2.	0.621	0.897	5.032	7.619	1.888	3.144	**2.**	0.621	0.838	4.127	5.400	1.572	2.555
3.	0.431	0.905	4.780	9.140	1.383	3.396	**3.**	0.647	0.817	3.896	5.480	1.516	2.38
4.	0.681	0.962	4.932	8.819	2.023	2.907	**4.**	0.531	0.473	2.483	3.920	0.868	1.616
5.	0.603	0.905	4.288	8.039	2.036	2.251	**5.**	0.437	0.383	2.500	4.039	0.615	1.883
6.	0.447	1.028	5.119	6.199	1.896	3.223	**6.**	0.594	0.774	3.831	5.780	1.636	2.196
Average	0.512	0.861	4.508	7.392	1.664	2.843	**Average**	0.546	0.614	3.298	4.939	1.126	2.172

The probable true value and the precision indices of a measured magnitude are calculated with the formula:

$$M_x = \frac{\sum_{i=1}^{r} m_i X_i}{\sum_{i-1}^{r} m_i} \tag{1}$$

in which: r is number of value ranges of measured values,

mi - the number of values appearing in any one i, in other words, the frequency of occurrence of Xi in the i range.

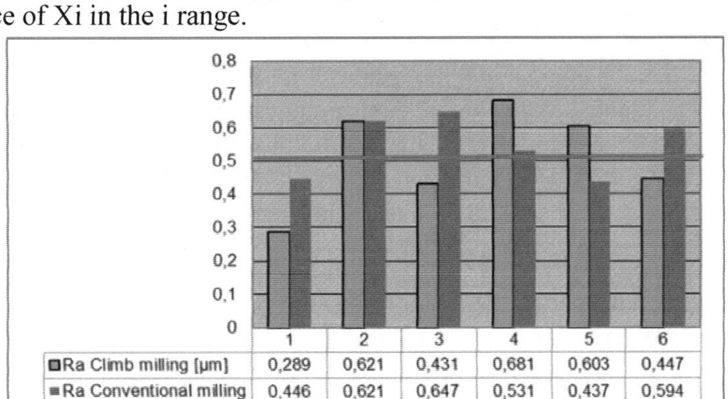

Figure 6. The average of roughness measurements

IOP Conf. Series: Materials Science and Engineering **749** (2020) 012021 doi:10.1088/1757-899X/749/1/012021

The measurement accuracy of the studied sizes can be evaluated by several indices, the most used being the deviation of the square mean and the probable error.

The deviation of the square mean (σ) is determined with the following relation:

$$\sigma = \sqrt{\frac{\sum_{i=1}^{n} m_i X_i^2}{\sum_{i=1}^{n} m_i - 1}} \qquad (2)$$

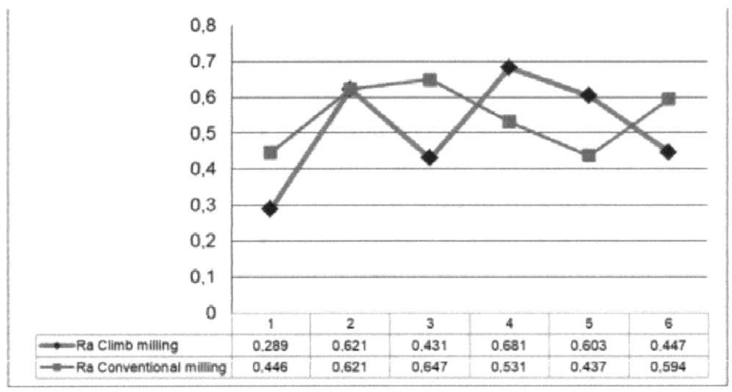

Figure 7. Diagram of roughness of the two machining

To verify the accuracy of the data presented in the above diagrams, I decided to use the Table Curve 2D program following the diagram of Figure 8.

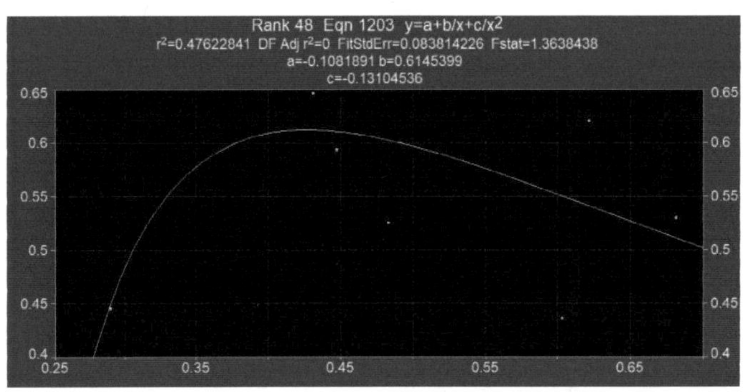

Figure 8. Diagram generated by the Table Curve 2D program

5. Experimental results

According to the experiments above, the comparison between the two types of processing, climb milling and conventional milling with the toroidal cutting edge the data centralized in Table 2 was reached.

Table 2. Data centralization

	Climb milling	Conventional milling
Cutting edge used	Toroidal end mill Ø16 - R4	Toroidal end mill Ø16 - R4
Material used	C45 (1.0503)	C45 (1.0503)
Cutting speed [m/min]	55 m/min	55
Cutting feed [mm/min]	328	328
Cooling	emulsion	emulsion
Roughness Ra [μm]	0.512	0.546

Large differences between the roughnesses of the treated surfaces do not exist, but small scratches can be observed visually on the surface treated conventional milling due to the shank pushed to what is exposed and to the measured values being a roughness higher by 0.034μm, these two values being compared in Figure 9.

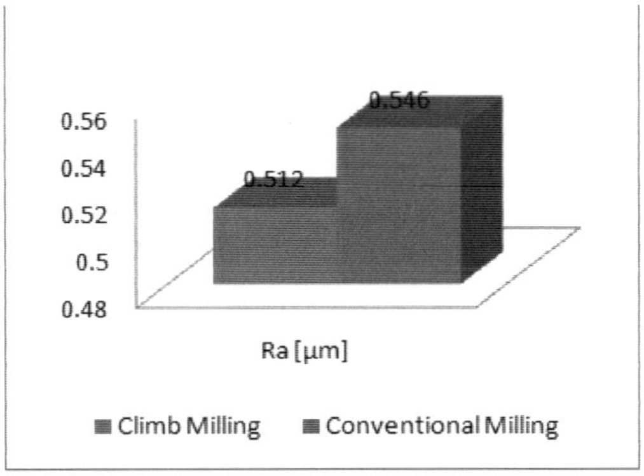

Figure 9. Comparison of roughness for the two types of milling

6. Conclusions

The use of the toroidal cutter in complex and coveted surfaces can be successfully applied. This can take the place of the spherical milling cutter and the cylindrical-frontal cutter in such situations.

Generating a CAM program according to these surfaces in accordance with optimal regimes and parameters leads to the best surface processing.

After comparison both visually and in terms of surface roughness, we found that when milling convex surfaces with the toroidal milling, processing in the advance direction is more efficient than processing conventional milling.

7. Acknowledgments

The experimental research carried out with the support of the processing department of S.C. Ramira S.A., Baia Mare.

References

[1] Xu Xun 2009 *Integrating Advanced Computer-Aided Design, Manufacturing, and Numerical Control: Principles and Implementations* Information Science Reference, Hershey New York ISBN 978-1-59904-714-0

[2] Radhakrishnan P 2014 *Computer Numerical Control Machines and Computer Aided Manufacture* New Academic Science Ltd ISBN 9781781830154

[3] Nasui V 2002 *Principiile mașinilor-unelte* Cluj-Napoca Editura Riscoprint ISBN 973-656-299-9

[4] Picos C, Pruteanu O, Bohosievici C, Coman Gh, Braha V, Paraschiv D, Slăteanu L and Grămescu T 1992 *Proiectarea tehnologiilor de prelucrare mecanică prin așchiere Manual de proietare I* Editura Universitas Chișinău

[5] ***http://www.delcam.ro/software/powermill.php

[6] ***https://www.matbase.com/material-categories/metals/ferrous-metals/high-grade steel/material-properties-of-high-grade-steel-c45.html#environment

Ontological Approach for the Conceptual Development of Products

A Petrovan[1], M Banica[2,a]
[1]Technical University of Cluj-Napoca, Faculty of Engineering, Department of Electrics, Electronics and Computers Engineering, Baia Mare, Romania
[2]Technical University of Cluj-Napoca, Faculty of Engineering, Department of Engineering and Technology Management, Baia Mare, Romania

E-mail: [a] mihai.banica@cunbm.utcluj.ro

Abstract. In order to improve the structuring process of a product a company often refer to adopt different solutions for correlating the structure of the newly product and its components with other complementary factors. In essence, these factors target the market, suppliers, solutions of manufacturing, terms for delivery, price, quality, etc. All these factors must be taken into considerations in order to make an optimal decision during the conceptual or detailed design process. These factors are collected preliminary and introduced into the proposed ontology. Despite other approaches, the method proposed into this paper uses also some limitations to determine the right decision-making act even for unqualified people and some other knowledge's incorporated into the ontology.

1. Introduction

Todays product development process is highly focused on users requirements. This means that companies spend a lot of time to capture users needs, organize and find solutions to these needs and finally to implement them into the new product. One of the most important aspect of these activities is related to the process of data management that requires a holistic approach around application systems, data processes, techniques and people skills. Techniques of management of product life cycle through specific tools allows the inclusion of all elements needed to ensure its traceability, such as modelling, document management, numerical analysis, the capitalization of know-how etc. [1, 2]

All activities involved in the product development such as drafting, designing, technology and production planning, manufacturing, requires processes that are based on knowledge reasoning. An appreciable part of the cost of product development is the result of the activities carried out since its design phase [3]. This is heavily influenced by the quantity and quality of knowledge involved in conducted activities. Considering this, manufacturers take a special risk if the design phase is not well coordinated and supported by the knowledge base in various fields. Sample risks can be: frequent changes of design solutions, budget alterations, furniture delivery delays, faulty staging etc.

Based on these considerations, in this paper the authors try to propose an ontology to support the process of product development from the beginning stages of their lifecycle.

2. On the conceptual product development

Conceptual design is the phase of the product development cycle, where finding a basic solution is determined by developing a search principle for the solution [4]. This involves the formulation of abstract ideas with approximate concrete representations. The early or conceptual stage of the design process is dominated by the generation of ideas, which are then evaluated in contrast to the initial requirements.

There is a process by which additional data is incorporated into the partial solutions, allowing a decision to be made between the competing alternatives. The literature describes the conceptual design activity as having the object of defining a product that has a particular functionality and which is in accordance with certain specifications. According to NF L00-007 the concept of a product is "a creative activity which, based on the needs clearly expressed, the existing means and the technological possibilities, leads to the definition of a product that satisfies these needs and which is industrially achievable". AQAP (Apllied Quality Assurance Publication), states that the design process is a task-definition approach to problem-solving tasks in order to create a product or service designed to meet the expressed needs. Certain industrial design situations, especially when the product is already known, can provide preconfigured procedures, rules, and methodologies applied to similar projects to design teams. Also, Bonnardel and Marmche [5] consider that this activity is a goal-oriented, purely mental activity focusing on pertinent aspects of the product, which leads to an understanding of the need to progressively construct a precise representation of the purpose of the concept. If the design team does not have any procedure that can be used to reach the goal, it will be required to adopt a strategy to develop a response.

The concept of a new product also involves describing the environment of the future product (highlighting user requirements for the product), describing its characteristics (highlighting the functional and structural aspects of the product) and product development planning (project management). Another component, project management, involves both the planning of activities to ensure the most favorable environment for the project to be carried out in optimal conditions as well as the project control by analyzing the differences between what was foreseen and what was achieved as well as the measures that will be taken to deal with the disruptive elements that inevitably occur during the course of the project. The competitiveness of an organization depends greatly on how it manages the overall process of conception. In general, effective use of a design method will help ensure the commercial success of the future product. At present there are over 100 methods and tools that can be used in product design. For this reason, when an organization wishes to develop a new product, it is in a difficult situation to choose the method that best suits its field of work, skills and, last but not least, the adopted strategy.

Often, the concept of product design is synonymous with the development of a product, defined as the set of information creation and communication activities that transform market data and technical opportunities into production information - models, specifications, prototypes, drawings, programs, necessary machine tools, tools, etc.

An approach of conception can be structured around three logics of action [6]:

- divergence - the action aimed at broadening the boundaries of the conceptual situation in order to widen the research space of the solutions;
- transformation - the action of building a structure, a model, a solution, starting from the results of the logic of divergence; is a stage of discovery and creativity;
- convergence - the action of progressively reducing uncertainty due to the multitude of possible solutions in order to select the most satisfactory solution.

IOP Conf. Series: Materials Science and Engineering **749** (2020) 012022 doi:10.1088/1757-899X/749/1/012022

3. On the ontology aspects of product development

The concept of ontology has been used in recent years, especially in many areas with applicability in knowledge management and computer assisted cooperation. By its peculiarities, ontology is used to share existing information in an area of interest. It is conducive to the representation and use of data relationships and is also effective in knowledge-based reasoning.

A broadly accepted definition for the notion of ontology is that given by Gruber [7], namely: Ontology is a formal specification of conceptualization. Gruber in [7] states that the term conceptualization is an abstract notion, a simplified view of the world we want to represent for a particular purpose. Regardless of the community type which t will change and share knowledge through an ontology, whether human actors or agents (in the framework of artificial intelligence), ontologies establish a common terminology, for this reason conceptualization requires a proper formalization. As defined by Gruber [7], conceptualization refers to objects, concepts and other entities that are supposed to exist in a particular area of interest, as well as the relationships that hold them together. Regardless of the field, an ontology consists of several elements, the most important being: concepts, relationships, attributes, instances, and axioms. In other words, a conceptualization is an abstract and simplified vision of the world we want to represent for a certain purpose. Formal specification involves the existence of a vocabulary of representation in which the objects of the domain and the relations between them can be formally represented.

An important role in product ontology has the "feature" of the product. The concept of "feature" first appeared in the field of product engineering, where it was given the following definition: "a representation of the shape aspects of a product that can be made in a generic form and that is functionally significant for the phase of the life of some products" [8]. Initially, it was closely related to the geometry of the product. However, because product development does not only include engineering activities, product information is not limited to geometry, it has a much richer and more complex semantic content (functional, structural, behavioral, technological, etc.). In order to capture this semantic content, the meaning of "feature" is extended to have a relevant definition depending on the context in which it is used, reducing the gap between geometry and other product information.

Considering the large amount of information associated with the product development process and noting that the formal specification is an important part of it, some authors have tried to get semantics from shapes [9]. Several approaches to developing ontologies for products and services are known in several areas: biology [10, 11, 12], medicine [13, 14], design [15] and geosciences [16, 17]. Regarding special products like machines or industrial equipment, problems were treated relatively briefly in [18, 19].

As an application in the technical domain, Lobontiu and Petrovan [20] have proposed a definition of ontology in product development: The ontology represents that set of concepts, hierarchically structured which describes a certain knowledge domain which can be used to create a knowledge base. The ontology of a technical domain contains concepts, a structure, a subordinated hierarchy, arbitrary relation-ships between concepts, axioms, functions and other constraints.

4. Approach

In order to build an ontology that covers the conceptual design process we have fol-lowed the methodology presented by Uschold and Gruninger [21] that combines a top-down and bottom-up practices for ontology engineering. The ontological approach for conceptual design is fostering the reuse, modularity, extensibility and maintainability as follows:

- **Reuse**. Our approach is based on existing ontology named Product Development Ontology (PDO) [20]. From a top-down perspective, the developed ontology was enriched with specific concepts linked to the conceptual product development and from the bottom-up perspective,

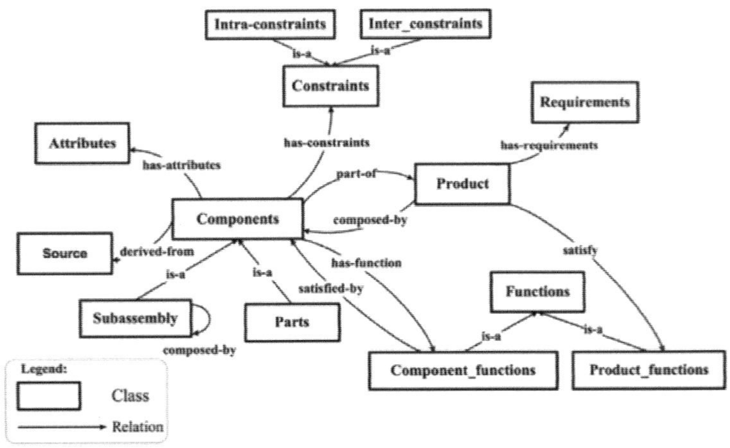

Figure 1. The generic ontology of product structuring

a very large amount of work was done to populate the ontology with objects and relations between them which are used to illustrate the usefulness of ontology.

- **Modularity**. This allow a separation of different parts of the ontology in order to be more flexible in the development of specific concepts. Also, this approach allows the involved team in product development to be more efficient and collaborative.
- **Extensibility**.This facilitate the further improvement of the ontology for the purpose of other product development applications. For example, the PDO that describes main aspects of product structuring was extended with new features related to conceptual product development.
- **Maintainability**. This characteristic simplifies the process of identifying and re-viewing deficiencies of the ontology, to implement new requirements in order to satisfy the changes that could occur in ontology.

Further, the generic model inspired by previous modeling work [22, 23], has been complemented with results from the analysis of the most used models of product structuring. In the first phase, we have identified the relevant terms and concepts within the domain of products structuring such as Components, Subassembly, Parts, Product, Functions, Attributes, Requirements, Constraints and Source. The result of this step is shown in Figure 1.

In order to extend this ontology to serve the entire conceptual product development activity, we consider the realization of necessary and sufficient concepts (see Figure 2) covering the whole field under consideration, such as:

- Person, which represents type of individuals that provide or request activities. In this class a few subclasses specify activities performed: Designer, Project-Manager, Procurement.
- Team, which represents groups of individuals from class Person involved in product development process. This organization is necessary to be able to closely follow the occupancy degree of personnel within the Company.
- Project contains relevant information about the resources involved in the conceptual product development activity. Through ontology relationships, we can aggregate information about product structure, number and type of component parts, their origin, delivery dates, product quality information, warranties, etc. It can also keep information about the team involved in the project or the person responsible for it.

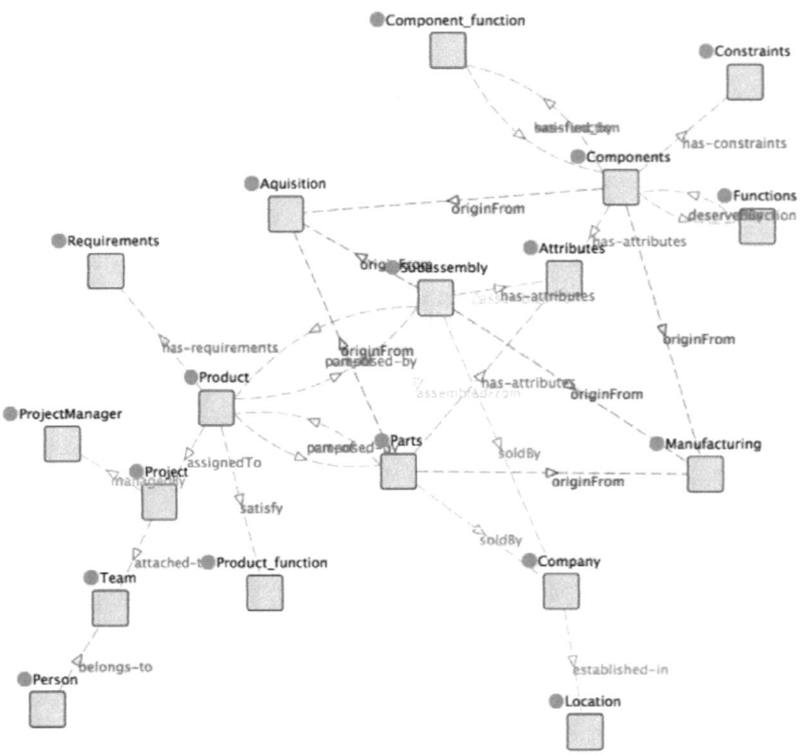

Figure 2. The generic ontology of product structuring

- The Company class includes information about the partners involved in developing a product. These include the origin of the component parts of a product, possibly specifying service chains of the future product.
- Location, which represents the geographical area or countries and cities used to mark where a Company is located. This information is relevant for scheduling, supplying and integration of the various Components within a Project.

5. Conclusions

In this paper the authors have tried to bring up front the conceptual product development ontology. Through this ontology, relevant information can be gathered on: decision-making processes, selection of component parts or sub-assemblies, payment and phasing of the entire process, organization of working teams, integration of sup-pliers and distribution chains, etc. Also, the use of ontology could be used to sup-port innovation on each component of a product and to support the collaboration of designers and other staff within the company.

This approach provides for the development of a network of ontologies based on a generic ontology, which explicitly describes the main concepts adopted in the product development process. This ontology can be further extended by creating new ontologies either directly or for the specific purpose of individual applications.

References

[1] Zina S, Lombard M, Lossent L and Henriot C 2006 Generic modeling and configuration management in product lifecycle management *The Proceedings of the Multiconference on" Computational Engineering in Systems Applications"* vol 2 (IEEE) pp 1252–1258
[2] Petrovan A and Lobontiu M 2015 Ontology based innovative support in industrial product development (ToKnowPress) pp 955–963
[3] Sullivan W, Bontadelli J and Wicks E 2000 *Engineering economy (11th ed.)* (Upper Saddle River, N.J.: Prentice Hall)
[4] Pahl G and Beitz W 2013 *Engineering design: a systematic approach* (Springer Science & Business Media)
[5] Bonnardel N and Marmèche E 2004 Evocation processes by novice and expert designers: Towards stimulating analogical thinking *Creativity and Innovation Management* **13** 176–186
[6] Asimov M 1982 *Introduction to Design* (Prentice Hall)
[7] Gruber T R 1993 A translation approach to portable ontology specifications *Knowledge acquisition* **5** 199–220
[8] Bronsvoort W F, van den Berg E, Bidarra R and Noort A 2001 Essential developments in feature modelling *Proceedings of the 7th International Conference on Computer-Aided Design and Computer Graphics, International Academic Publishers, Beijing, China* pp 6–15
[9] Brunetti G and Grimm S 2005 Feature ontologies for the explicit representation of shape semantics *International Journal of Computer Applications in Technology* **23** 192–202
[10] Consortium G O 2012 Gene ontology annotations and resources *Nucleic acids research* **41** D530–D535
[11] Lewis S E 2005 Gene ontology: looking backwards and forwards *Genome biology* **6** 103
[12] Myhre S, Tveit H, Mollestad T and Lægreid A 2006 Additional gene ontology structure for improved biological reasoning *Bioinformatics* **22** 2020–2027
[13] Bickmore T W, Schulman D and Sidner C L 2011 A reusable framework for health counseling dialogue systems based on a behavioral medicine ontology *Journal of biomedical informatics* **44** 183–197
[14] Matei O 2008 Defining an ontology for the radiograph images segmentation *9th International Conference on Development and Application Systems*
[15] Diana C and Alina P 2013 Defining spatial relations in a specific ontology for automated scene creation *Carpathian Journal of Electronic and Computer Engineering* **6** 50
[16] Hu Z, Tang G and Lu G 2014 A new geographical language: A perspective of gis *Journal of Geographical Sciences* **24** 560–576
[17] Jung C T, Sun C H and Yuan M 2013 An ontology-enabled framework for a geospatial problem-solving environment *Computers, Environment and Urban Systems* **38** 45–57
[18] Barbau R, Krima S, Rachuri S, Narayanan A, Fiorentini X, Foufou S and Sriram R D 2012 Ontostep: Enriching product model data using ontologies *Computer-Aided Design* **44** 575–590
[19] Panetto H, Dassisti M and Tursi A 2012 Onto-pdm: Product-driven ontology for product data management interoperability within manufacturing process environment *Advanced Engineering Informatics* **26** 334–348
[20] Lobontiu M and Petrovan A 2012 Product development ontology (1). information integration concepts *Revista de Management si Inginerie Economica* 43–56
[21] Uschold M and Gruninger M 1996 Ontologies: Principles, methods and applications *The knowledge engineering review* **11** 93–136
[22] Petrovan A, Lobonţiu G and Ravai Nagy S 2013 Broadening the use of product development ontology for one-off products *Applied Mechanics and Materials* vol 371 (Trans Tech Publ) pp 878–882
[23] Petrovan A, Lobonţiu M, Lobontiu G and Ravai Nagy S 2014 Overview on equipment development ontology *Applied Mechanics and Materials* vol 657 (Trans Tech Publ) pp 1066–1070

IOP Conf. Series: Materials Science and Engineering **749** (2020) 012023 doi:10.1088/1757-899X/749/1/012023

Experimental research methodology for physical distribution in bakery industry

A D Pop, G Rus and R F Drenta[a]

Technical University of Cluj-Napoca, Faculty of Engineering, Department of Engineering and Technology Management, Baia Mare, Romania

E-mail: [a] raul.drenta@cunbm.utcluj.ro

Abstract. The research done in the bakery industry worldwide and in special in the Romanian market had their goals on the production and technological process, focusing on the end result products. Bakery industry focused on profits changed and developed their products according to the market need and preferences and very few traditional businesses know how to make the profits continuing the distribution routes and using the old bread recipes. During this study we aim to go through an experimental process in order to observe the influence that different cost factors have on the end result such as performance and profits. A questionnaire was designed to answer questions regarding the existent technology in the bakery industry, transportation vehicles, clients, turnover, invoice overdue, and the structure of the production, costs with workers, transport, maintenance, marketing and raw materials. The questionnaire will be applied only in the North West area of Romania to all the bakery companies that have over 5 employees and deliver bread products to business customers using predefine routs. In a future stage, it would be of a great interest to apply a similar questionnaire in different parts of Europe for further studies in this industry, where management has a deficiency in scientific research and managers have to take decision based most of the time only on their own experience.

1. Management tool for bakery performance

The most complex business model in bakery industry is the sale of the bakery products through the distribution channel by own means on predefined distribution routes. For this business model we continued the elaboration of the algorithms describing the distribution of the bakery products, following that through software based on these algorithms and an intuitive interface, to provide a simple and useful tool to support the improvement of the business performance of bakery enterprises.

Through this study, we aim to improve managerial performance (expressed at company level through profit and profitability) by highlighting factors or groups of factors that contribute the most to performance. To this end, by gathering information from bakery companies, we can validate a generally valid level of costs and expenses that they must fit for each stage of the distribution process, depending on the main factors of influence.

The term performance is used to highlight both individual, group, process, or system capabilities. Due to these aspects, the existence of an all-encompassing definition seems to be almost impossible. However, in the scientific context, for the performance in the managerial area, we came to the definition formulated following a thorough research by Dr. Iulian Vişoiu, in his doctoral thesis. It defined performance through its component elements and proposed a graphical representation of

IOP Conf. Series: Materials Science and Engineering **749** (2020) 012023 doi:10.1088/1757-899X/749/1/012023

performance as a vector located in a three-dimensional space. According to the author, performance is defined by three elements: effectiveness, efficiency and adaptability.

Further on, we will present the structure and main characteristics of the software we are developing and how, using our questionnaire we can achieve the answers that will help the bakery managers to optimize their business.

1.1. Software structure for end users in bakery management

The computer program is conceived as a tool available to bakery managers. It is made up of three interactive segments to diagnose the situation and provide solutions to improve the company's distribution process by pointing out missing factors or those that exceed the values identified at the industry level as defining margins for profitable firms.

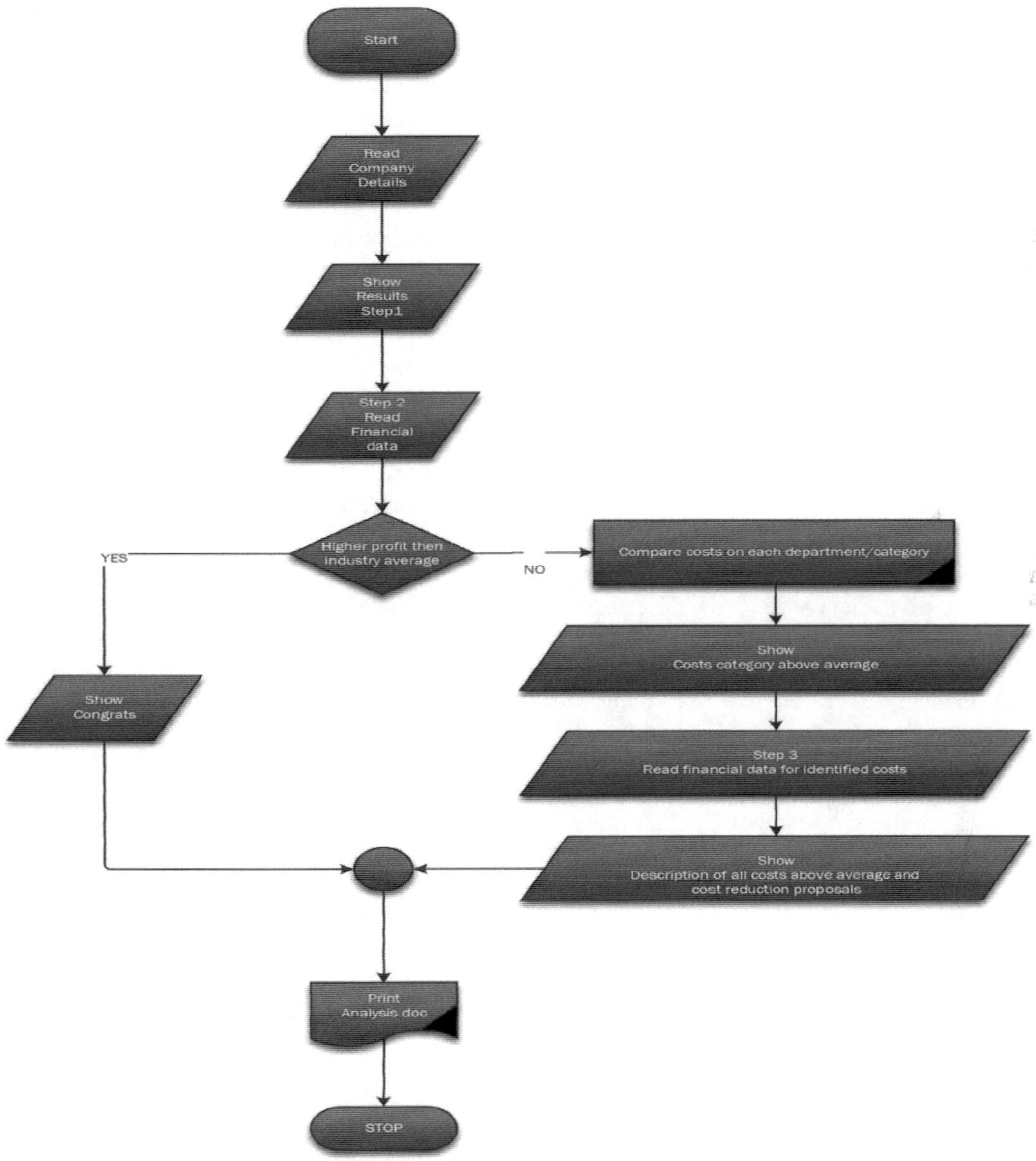

Figure 1.1. Software algorithm

The architecture of the software algorithm to be used is described in figure 1.1 "Software algorithm" and describes the steps to be taken by the end user (manager of the bakery production unit). In a first step, the user of the software program will introduce its own variables in the same way as all the respondents did in the questionnaire and the return returned to the first stage will be a relative one, positioning the company described by the user compared to the results obtained from the questionnaire reflects the influence factors and their share in activity for the companies that have profited in the last years of activity.

1.1.1. Step one. On the first step the information provided by the software application will include graphics and graphical user friendly expression as outlined in figure 3.2. Stage 1 information display Staying in step 1 of the questionnaire and software program the same set of questions, the information collected from both sources, highlighted in parallel, can provide the user with an overview of the strategic location of his business in the area which can provide a profit

The application provides the same perspective and concrete information for managers already involved in a business as well as those who are about to start and provide business plan information.

1.1.2. Step two. On the second step, the software application will ask the user to provide new information in order to continue to evaluate the business and to provide information on the improvements it can make to the production and distribution process in order to optimize them.

Starting from the information obtained by applying the questionnaire, at the level of the bakery industry, for each influence factor and for the overall activity, the average values of the industry will be determined, taking into account only the performances of the enterprises that obtained profit during the analysed period.

Step two will begin by comparing the user's profit with the average of the bakery industry. If it is found that the user is currently earning a profit equal to or above the average of the industry, the application will rewrite a greeting message that will mention this and the program will stop.

If by this comparative method, the profit position below the average of the industry is noticed, additional information will be required that will go through a mathematical model that will determine the activities within the company to be tracked, giving the user a set of performance indicators Incorporates profitable businesses from the bakery industry.

A mathematical model based on the described experience can be represented as follows:

Starting from the break-even formula where: Fixed Costs (Value-Cost Variable) = 1 or Amount of All Expenses + Profit = Income from Sales of Products,

=> 40% Gasoline x Price + 1% Marketing x Price + 1% Interest x Price + Z% Raw Material = 100% Price

=> Price (40% + 5% + 3% + 2.5% + 7% + 1% + 1%) + Z% Raw material = 100%

=> 59.5% Ch + Z% Raw materials + profit = 100% Price

=> Raw materials = 40.5% Price

We can observe that the approximate method is currently being applied to reach the profitability threshold, whereby the cost of raw materials cannot exceed one third of the sales price of the products.

Starting from this concrete example where we can see an overview of the cost and profit of societies, we aim to develop an application to support bakery managers to improve their production and distribution performance by identifying and addressing punctual the factors that need to be improved so that the end result is to maximize economic performance.

1.1.3. Step three. On the third step of the software application, the information provided by the user will be compared to those obtained from the analysis of the information provided by the participants in the study conducted by us and on the basis of the publicly available financial information.

Factors tracked in stage three: Staff expenditure, Expense Car Fuel, Car park maintenance costs, Maintenance costs, Gas/ power production costs/ wood/ fuels, Marketing costs, Cash Flow/ Investment (Interest) Costs.

2. Methodology and questionnaire

The research will be carried out by applying a questionnaire developed on the basis of the research studies analysed as well as relevant in the first chapters on the theoretical study of the distribution of finished products in the bakery industry studies that each of them analysed the influence of the identified factors on the efficiency, and generating profit.

Out of these studies, relevant factors have been extracted, capable of generating concrete and conclusive data for the present study, factors analysed by formulating and adapting some of the questions identified in the questionnaires identified, thus materializing in a questionnaire which will include questions about the business model, the structure of expenditures and those of the obtained economic results (profit), a questionnaire that will be applied online through specialized media platforms.

Thus, the qualitative and quantitative data extracted will be analysed with the statistical data analysis programs SPSS 20 and EXCEL 2010, thus generating statistical and mathematical testing, concluding that the hypotheses launched or their invalidity are supported, on the existence of the situation at the level of the Northwest Development Region of Romania

2.1. *Description and evaluation of control variables*

The following factors were selected:

The sector of activity comprises the area of activity of the selected target companies, in order to analyse and highlight the activity of production and distribution of finished bakery products.

Regarding the sector of activity, it was appreciated the introduction as a control variable the companies having the main field of activity and CAEN code 1071 Manufacture of bread and bakery products.

At North Western Region level, according to the information obtained from the Trade Register, CAEN 1071 is authorized for 700 active companies with at least one employee.

The present study aims to analyse and provide a management tool for bakery company managers but the CAEN 1071 activity is also defining the production of cakes and confectionery products or frozen bakery products. By determining the production structure in total revenues we aim to identify firms based on a predominant production in bread and pastries fresh baked products.

The size of the firm will refer to the average number of employees of the sectarian industrial firms in the chosen research area, with the selection of companies from the four categories: micro, small, medium and large micro enterprises.

Regarding the size of the company, it was appreciated the introduction as a control variable the size of the company, because at the level of the NW region of Romania, the activity of the organizations from the smallest to the multinational is starting, for which, by questioning a representative number within them, will lead to shaping a present situation of those categories of organizations that sell their products through distribution based on pre-established routes and their own fleet.

In order to highlight this aspect, only companies with a minimum number of 5 employees will belong to the target group, so that the number of companies that cumulatively meet these criteria is reduced to 331 firms. In order to highlight this aspect, only companies with a minimum number of 5 employees will belong to the target group, so that the number of companies that cumulatively meet these criteria is reduced to 331 firms.

The evaluation of all the above mentioned factors will be done through the SPSS 20 and EXCEL 2010 statistical program, which will generate statistical data based on the chosen mathematical model, to show the positive / negative correlations between these factors in order to validate / invalidate the hypotheses launched.

Considering that this questionnaire will be sent to all 311 companies in the chosen field of activity having at least 5 employees and operating in the North-West region of Romania, regarding the representativeness of the group, according to the Taro Yamane formula (Israel 2003), a minimum of 171 questionnaires were needed for representativeness.

A probability of 90% and a maximum error of +/- 5% were taken into consideration, for which the following calculation formula was used: (Israel 2003)

n = N / (1 + N (e2))

n - sample size;

N - the size of the total population;

e - the level of accuracy (maximum permissible error);

n = 311 / (1 + 311 (0.1 * 0.1)) = 311 / (1 + 311 (0.01)) = 311 / 4.11 = 76 questionnaires

Collecting the data when completing the questionnaire will be done through the Google Forms electronic application, an application that will generate a record of the number of completed questionnaires and statistics on respondents' answers.
The questionnaire is available online on http://www.UniversitatiRomania.ro/bakeryStudy

2.2. *The structure of the questionnaire*

The structure of the questionnaire is based on the structure described for the software as long as the main purpose of this interaction with companies in the bakery industry is to identify alghoritms that can provide guidance for new bakery managers to plan a profitable business plan or for other bakeries to restructure and increase the profitability of their activity.

The information we aim to identify and structure is related to cost factors that must be kept well under control and the positive experience of bakery companies with higher profits ill be able to offer valuable information and guidance to other bakery managers.

First questions are designed to identify and position the company in the market, offering information about their location in urban or rural area, a factor that can influence the strategy of the company and the profitability in different areas.

Very important for our study is to identify the marketing strategy and branding importance for our survey respondents.

Supply chain and production information are valuable and are present in our questionnaire right from beginning, trying to identify the supply strategy, negotiation capacity, cash flow issues due to suppliers rigidity, technology used in production and manipulation of the raw materials, last investment in technology and the capacity of the production and total use of the capacity.

The portfolio of the company and structure of production for each type of products will help us identify the new bakery companies that are not aiming to produce traditional bread and bakery products and are not part of our research.

Packaging is a complex activity for each company and it is reflected in our study but so far we don’t find it very relevant for final profitability compared with investment costs needed in the beginning of a new business.

Distribution channels used, the amount of different routes, sale strategy, transportation vehicles structured on capacity and volume for each type, age of the transportation fleet, mileage are a big concert for our study. Our studies indicate so far that an important role in costs control is played by these factors.

Distribution reflection is carried on with question related to the relation with the retail customer and end consumer.

By defining an average distance route used in the industry, we can provide valuable information for new bakery managers.

Cash flow is a strategic decision and can determine the success or failure in a new business but also can give the new companies a competitive advantage on the market.

Perishable products are a common problem for fresh products and we will study different approach in the bakery industry in the observed area.

Employees are the most valuable asset and an efficient structure of personal that is common on bakery industry level is the kind of information that each manager would like to possess.

Profitability questions in our questionnaire have two roles. To test the correctness of our survey companies because we can check the information from the available financial information we have access too. The second role is to find the profits companies estimate in different product.

Last questions relate to the most important aspect of our study. Percentage and intervals of cost estimated for each department and activity. We are working on five option grill structure, the middle answer being the one identified in our preliminary research on focus groups.

Acknowledgements
Being one of the first students to approach the topic of Business Management and Engineering on Distribution of bakery products in Romania, I am keen to highlight the step by step support, guidance and time offered by professor Nicolae Ungureanu from academic perspective and Andana Pan Baia Mare Bakery for their full cooperation in designing and testing the current research methodology.

In perspective we are looking forward for other researchers with similar interest in order to apply the questionnaire on bakery companies in different regions around the world as long as we can talk about fresh baked products and traditional bakery product, not yet merged into a general approach of food logistics. Feel free to contact us for a full version of the document.

References

[1] Cioca L I, Moraru R I, Băbuţ G B and Ungureanu N S 2015 *Integrating risk analysis with safety dyagnostic in complex industrial systems: modeling hazard* ACTA Universitatis Cibiniensis 66 (1) pp.17-22

[2] Isachi S E 2015 *The competitive advantage of microeconomical level in the context of globalization. Economy and Management*

[3] Isaksson L 2010 *Corporate Social Responsibility: A Study of Strategic Management and Performance in Swedish Firms* Australia Bond University Gold Coast

[4] Israel G D 2003 *Determing Sample Size* University of Florida Ifas Extension

[5] Ungureanu N, Duval P, Mocan M L and Taucean I M 2010 *Logistica Activitatilor de Mentenanta* Editura Universitatii de Nord Baia Mare

[6] Ungureanu N, Lung C and Cotetiu R I 2015 *E-maintenance-a new trend in industrial maintenance* Mechanical Engineering Letters HU ISSN 2060-3789 12 (1) pp.70-77

[7] Ungureanu M, Pop N and Ungureanu N 2016 *Innovation and technology transfer for business development* Procedia Engineering 149 pp.495-500

IOP Conf. Series: Materials Science and Engineering **749** (2020) 012024 doi:10.1088/1757-899X/749/1/012024

Comparison of S 235 steel and DLC coated steel using Pin-on-disc method in dry sliding conditions against PA 6 countersurface

Á Sarankó[a], G Kalácska and R Keresztes

Szent István University, Institute for Mechanical Engineering Technology, Department of Mechanical Engineering, Gödöllő, Hungary

E-mail: [a] saranko.adam@gmail.com

Abstract. This study presents tribological tests of S 235 steel and DLC coated steel using the pin-on-disc test method, where these materials were the pins to compare it with each other. The counter surface was Polyamide 6 (PA 6) as the disc. Two kinds of loads were used in three levels: static weight loads and static weight loads supplemented with continuously changing load. Friction forces were measured in two directions, and the temperature was also measured near the contact zone. Two dimensions diagrams were used to present the results. The same parameters were set for the two tested materials for maximum comparability. The coefficient of friction and temperature were compared. After evaluating the test results; we concluded that in the case of DLC coated steel the coefficient of friction was lower. With the added vibration-load tests this difference was significantly higher at lower load levels. Regarding the temperature, with lower loads, the temperature values were the same. When higher loads were used, the temperature values were significantly lower using DLC coated steel.

1. Introduction

Nowadays coating has an increasing role. High-performance coating materials are available, and it is a frequently used technology to improve the performance of equipment, prolong the service life of critical parts and reduce the energy consumption of systems.

For instance, in case of cutting tools, or tool inserts the friction, wear and the life of the tool is a crucial factor. With high-performance coating materials, these factors also can be easily improved.

Physical vapor deposition (PVD) technology is an environmentally friendly surface treatment. Three commonly used coatings are also produced by this process which is the following: diamond-like carbon (DLC), graphite-like carbon (GNC) and chromium nitride (CrN) [6].

Diamond is a high-performance tool material with excellent tool life. The properties of the diamond are similar to the properties of the DLC [1]. DLC provides high hardness, low friction, and excellent chemical stability [5]. Diamond-like carbon films can be divided into two major categories according to their hydrogen content. These categories have similarities in tribological performance, but the films also behave differently in different tribological conditions [3].

Due to their good wear resistance and low friction coefficient DLC coatings have been studied intensively for many years and a great deal of research has been carried out in the field of tribology [4].

In the literature we found, researchers made tribological tests against metals and even against DLC [2], but we did not see any data against polymer countersurface.

In this article, we studied the sliding properties of DLC coated steel against Polyamide 6 counter surface using the pin-on-disc test method.

2. Method

2.1. Materials

The pin-on-disc test method is a frequently used tribological test method. It is suitable for compare materials in continuously sliding systems. We chose this method because we wanted to compare two kinds of materials in a well controllable, continuously sliding system.

Figure 1. S 235 steel specimen

Two kinds of special shaped specimens were used as pins. The first type was a DLC coated AISI52100 ball bearing steel; the second was S 235 steel (Figure 1). The DLC coatings were produced by Oerlikon Balzers, in Hungary. The S 235 steel specimens were produced with the same geometry in our workshop. The height of the specimen was 5,3 mm; the contact surface was nearly a flat surface with diameter 21,5 mm, which is 363,05 mm^2. The surface roughness of the DLC coating was Ra 0,04. The sliding surface of the S 235 steel specimens was polished to reach the same value of surface roughness. A special specimen holder was made for the specimens. The material of the holder was Polyvinyl Chloride (PVC). PVC is a common and cheap polymer material. Usually, polymers have good thermal insulation ability, so the holder was suitable for measuring temperature during the tests. The holder is compatible with our pin-on-disc type tribological test device, which can be found in our institute's workshop in Szent István University, Gödöllő.

The material of the disc was Polyamide 6 (PA 6). It was 100 mm in diameter with 12 mm thickness. In this setup, the disc was the softer material. After every test, the sliding surface was damaged. To get better results, the damaged surface was removed after every test, using a conventional lathe machine. Therefore, the surface roughness of the disc was always different. In sliding systems, the surface roughness is a significant factor. Surface roughness values were measured on a Mitutoyo SJ-201 device. Table 1. shows the surface roughness values for each test. The parameters of the tests are described in detail in the next chapter.

Table 1. Average surface roughness (R_a) values of the PA 6 disc

Test 1	0,75 μm	**Test 7**	1,08 μm
Test 2	0,76 μm	**Test 8**	0,92 μm
Test 3	0,84 μm	**Test 9**	0,92 μm
Test 4	0,78 μm	**Test 10**	0,83 μm
Test 5	1,03 μm	**Test 11**	0,83 μm
Test 6	1,24 μm	**Test 12**	1,05 μm

2.2. Parameters of the tests

To reach better comparability, the same parameters were set to each test on both types of specimens. Table 2. shows the parameters of all tests.

Table 2. Parameters of the tests

	Test 1	Test 2	Test 3	Test 4	Test 5	Test 6	Test 7	Test 8	Test 9	Test 10	Test 11	Test 12
Material of specimen*	DLC	DLC	S 235	S 235	DLC	DLC	S 235	S 235	DLC	DLC	S 235	S 235
Diameter of sliding path [mm]	60											
Sliding velocity [m/s]	0,5											
Static Load [MPa]	0,055	0,055	0,055	0,055	0,11	0,11	0,11	0,11	0,165	0,165	0,165	0,165
Vibration load [MPa]**	-	±0,04	-	±0,04	-	±0,04	-	±0,04	-	±0,04	-	±0,04
Sliding distance [m]	1000	1000	1000	1000	1000	1000	1000	1000	1000	600	1000	1000
Ambient temperature [°C]	26											

* 'DLC' means DLC coated DIN 1.3505 steel, and 'S 235' means S 235 steel

** 'Vibration load' is exlpained in detail below

We can separate three different load levels regarding the 'Static Load' values in table 2.

In the case of Test 10, the sliding distance was only 600 m. With those parameters, the temperature went too high and raised continuously. The test was stopped to prevent damaging the holder and the measuring system.

The vibration load is provided by a so-called vibration unit device. It is suitable for produce continuously changing loads to the contact surface only in a normal direction. In case of this experiments, once the unit was turned off, or it worked at 28 Hz, added ±0,04 MPa to the static, normal direction load. Special dead weights were made to providing the static loads (figure 2.)

IOP Conf. Series: Materials Science and Engineering **749** (2020) 012024 doi:10.1088/1757-899X/749/1/012024

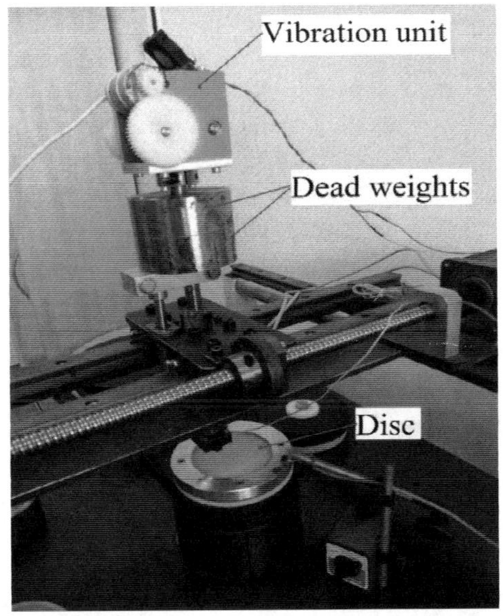

Figure 2. The pin-on-disc tribological test device setup

Three loads were used. First without vibration load, then with the continuously changing vibration load. The vibration unit and the system provided the first type of normal direction force, which was 20N. With one of the dead weights, the force increased to 40N, and after applying the second dead weight, the force was 60N. Regarding the sliding surface of the specimens, the loads were calculated.

The measurements always run at dry sliding conditions.

2.3. Method

Figure 2. shows the pin-on-disc type tribological test device.

In Fig. 3. we can see the special holder. There is a magnet in the bottom of the holder to keep the specimen on the place when fixing the holder to the testing device. After the holder was fixed to the device, we must adjust the screw to prevent any movement of the specimen during the measurements.

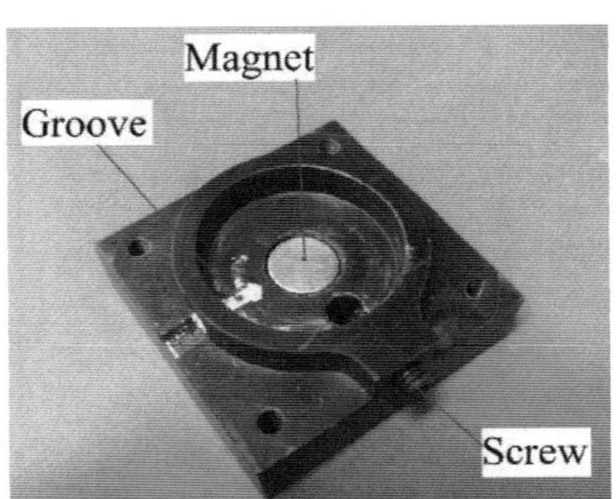

Figure 3. The specimen holder

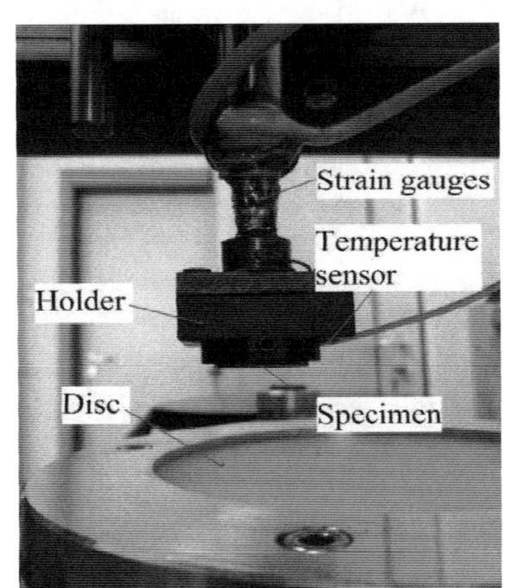

Figure 4. The specimen and its environment

We can also see a groove at the bottom of the holder. In this groove the temperature sensor takes place. We cannot measure the temperature directly in the contact zone. The temperature was measured on the opposite side from the contact zone of the specimens. Thermal conductive paste was used for more precise results. This method is suitable to compare the investigated materials.

In figure 4 we can see the disc, the holder and the specimen after a test.

After the correct input parameters applied to the system, the contact surfaces were cleaned with methylated spirit. In the following, we reset the force measuring sensors. Before commencing the test, we initialized parameters on the computer. The strain gauges found on the force measuring sensor and the type Spider 8 measuring amplifier are connected by wire converting the analog signals to digital, which are then transmitted to the computer. The chart-like visualization of the digital signals was achieved with software called Catman. The measuring frequency was 5 Hz in all cases.

During the tests, all data and the prepared diagrams were monitored and displayed on the screen in real time. After completing the test, all data were saved in the appropriate file for later processing, using a spreadsheet.

The contact surfaces of the disc were re-machined after every single run.

3. Results

During the entire tests, the computer registered the friction forces, the rotation speed of the disc, and the temperature of the specimens from the opposite side in the function of sliding distance. Comparative diagrams were made in Microsoft Excel to illustrate the results. In figure 5. we can see the results of Test 1, 2, 3 and 4 in one diagram. In these four cases, the static load was the same (Table 2.). With this method, the differences between the two examined material, and the effect of the vibration load also can be seen.

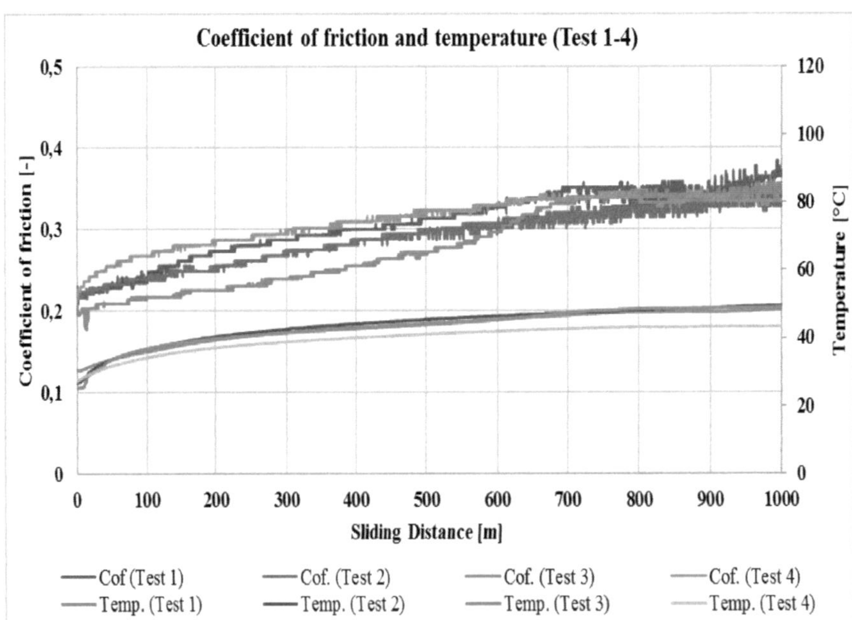

Figure 5. The coefficient of friction and temperature of Test 1-4

In figure 5. we can see the vibration load has a more significant effect on S 235 steel specimens. The coefficient of friction (Cof.) values were the highest and lowest in the case of S 235 steel. With these parameters, vibration has a negative effect on the S 235 steel specimen regarding the Cof. value. However, after the running in period (700 m) the Cof. values reached the steady-state in all four cases. In the case of temperature curves, there were no significant differences between the four tests. The Cof. curves are not as smooth as the temperature curves. In a continuously sliding system, this is a common phenomenon, so-called stick-slip. The materials are stick together in the contact zone, then suddenly move again

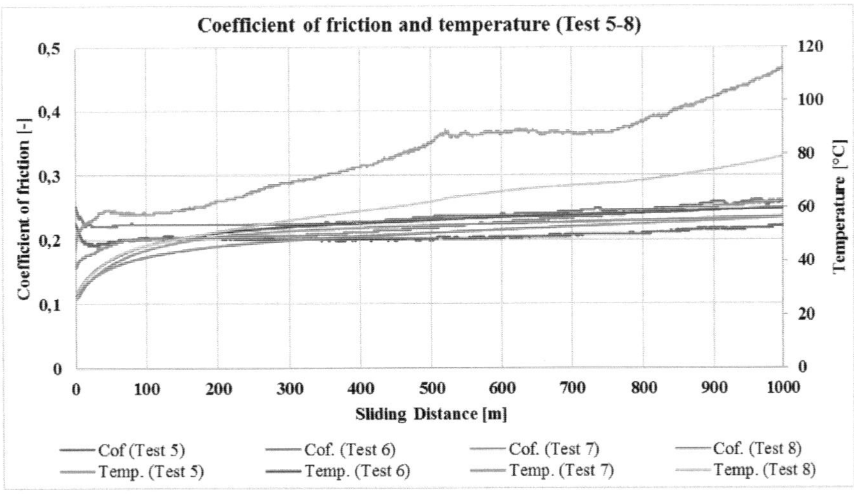

Figure 6. The coefficient of friction and temperature of Test 5-8

Figure 6 shows the results of Test 5, 6, 7 and 8. In this diagram different scale was used regarding the Cof. value to get a better visualisation. In case of this load level, we can see the vibration strongly influenced the Cof. and temperature curves of S 235 steel specimens. Without vibration load, the Cof. curve is nearly steady-state during the measurement. With the supplemented vibration load, the Cof. values are much higher. After the running in period there is a short steady-state part of the curve. After this phase the Cof. highly increases. The temperature curves run together except in case of S 235 steel specimen with vibration load.

We can see the DLC coated specimens are not influenced strongly by the vibration in this setup. There is a non-significant difference between the Cof. curves. Both curves are nearly steady-state during the tests

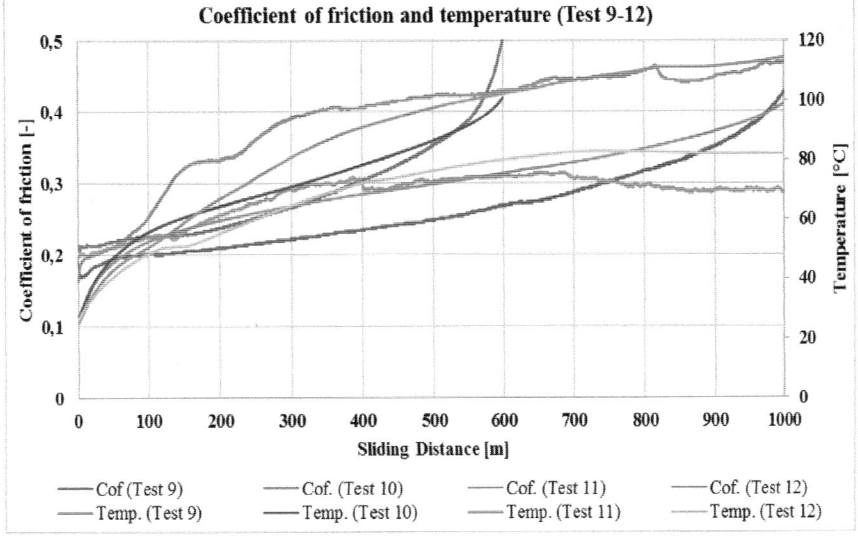

Figure 7. The coefficient of friction and temperature of Test 9-12

Figure 7 shows the third comparative diagram with the highest loads in this experiment. Without the vibration load, the measured values were lower with the DLC coated specimen. However, both curves were progressive, while the curves of the S 235 steel were regressive. With vibration load, the shapes of the curves were the same regarding the two examined materials, but the Cof. and temperature values of S 235 steel were lower during the test. In this load level with the added vibration load, the temperature went too high in case of the DLC coated material. The measurement was stopped to prevent damaging the measuring system. The vibration load was strongly and positively influenced the S 235 steel curves.

4. Conclusions

In this experiment two kind of specimens were used to compare them with each other in dry, continuously sliding conditions. The counter surface was PA 6. Two types of loads were applied on three levels. The specimens were tested with the same parameters. From the comparative diagrams, we can see the differences.

At the first level of loads, there are no significant differences. Both the Cof. and temperature curves reach the steady-state after about 700 m. With the second level of loads, all values are increasing during the whole test. In this level, the S 235 steel has the highest values of Cof. and temperature with the supplemented vibration load. The sliding properties of the DLC coated steel material were not influenced significantly by the vibration load.

At the third level of load, without the supplemented vibration load the DLC coated steel produced more favourable values in the examined interval. With vibration, the temperature curve of the DLC coated material increased to an extremely high value, while with the same parameters the S 235 steel produced favourable values. After about 700 m the curve reached the steady-state.

As a further result, we found that this test method is suitable for compare materials under sliding conditions.

References

[1] Erdemir A and Martin J M 2018 *Superior wear resistance of diamond and DLC coatings. Current Opinion* Solid State & Materials Science vol. 22 pp.243-244

[2] Humphrey E, Morris N J, Rahmani R and Rahnejat H 2018 *Multiscale boundary frictional performance of diamond like carbon coatings* Tribology International

[3] Ronkainen H, Varjus S and Holmberg K 1998 *Friction and wear properties in dry, water-and oil-lubricated DLC against alumina and DLC against steel contacts* Wear vol. 222 pp.120-121

[4] Sedlaček M, Podgornik B and Vižintin J 2006 *Tribological properties of DLC coatings and comparison with test results: Development of a database* Materials Characterization vol. 59 pp.151

[5] Lu Y, Huang G and Xi L 2019 *Tribological and mechanical properties of the multi-layer DLC film on the soft copper substrate* Diamond & Related Materials vol. 94 pp.21

[6] Li Z, Guan X, Wang Y, Li J, Cheng X, Lu X, Wang L and Xue Q 2017 *Comparative study on the load carrying capacities of DLC, GLC and CrN coatings under sliding-friction condition in different environments* Surface&Coatings Technology vol. 321 pp.350-351

IOP Conf. Series: Materials Science and Engineering **749** (2020) 012025 doi:10.1088/1757-899X/749/1/012025

Study and model on university-industry relations in cluster management within the North-West Region of Romania

D Stoicovici and M Banica[a]

Technical University of Cluj-Napoca, Faculty of Engineering, Department of Engineering and Technology Management, Baia Mare, Romania

E-mail: [a] mihai.banica@cunbm.utcluj.ro

Abstract. In previous research on the N-W Region of Romania, it appeared that there are conditions conducive to the emergence of early-type clusters in this region. These conditions are manifested first and foremost from the bottom up, in the sense of the existence of material equipment with fixed means and human strength, both as professionals and as specialists. These conditions are also top-down in the sense of the authorities' preoccupation with developing strategies and a legislative framework to support and encourage the formation of these clusters and their further development. It has also been highlighted that in parallel with cluster structures, there are networks of innovation of IVth and Vth generation structures, including universities and research and design institutes, which are somehow superimposed on the structure clusters and which, by their interaction with all these elements of the clusters, have the effect of stimulating the technological transfer, for example, among others. In order to maximize the existence of these two networks, a model is proposed to be able to simulate different scenarios of evolution and interaction between all the components of the two structures, in order to highlight ways to deepen the integration of collaboration between these networks. The novelty of the model would be to introduce a negative feedback on actual and incidental external actions on the system and to introduce a particular property that takes into account the actions of the individuals involved, actions related to the free will of individuals, which cannot be restricted and cannot be assumed a priori that they are exercised only in the sense of amplifying some actions, even if these actions are positive.

1. Introduction

Starting with Romania's accession to the EC, a complex of studies and researches were carried out by accessing European funds, regarding the determination of natural evolving cluster conditions emergence in the north-western area of Romania and the economic areas of their emergence. On the other hand, studies have been carried out on the impact and importance of technological transfer on the development of economic activities in the area as well as on its importance in integrating the economy into the European economic circuit [1].

During these investigations, there was a downstream development system through governmental and legal governmental rules to allow clusters to develop in the geographic area of existence, depending on the results of their economic operation system. At the same time, there was a bottom-up development system that allows clusters to increase business competitiveness and efficiency by concentrating occupational human resources and specialists from university and research and design

IOP Conf. Series: Materials Science and Engineering **749** (2020) 012025 doi:10.1088/1757-899X/749/1/012025

institutes. This facilitates all participating companies through a so-called "triple helix" [2] relationship to help increase the level of industrial development of their region.

This type of relationship integrates into the economic tendency of the countries of the European Union where relations already exist that lead to interactions between the economic agents within the cluster groups and the IV and V innovation networks, including the cadres of the universities and research institutes and industrial engineering.

On the other hand, in parallel with all these economic phenomena, there are the innovation systems of processes networks that in [3] was suggested that overlap the cluster networks, interfere and influences with each other.

Since, at present, networks of institutions involved in technological innovation and technology transfer, due to the complex and effective links between them and industrial companies, can be assimilated as belonging to the IV and Vth generations of innovation structures of processes, regional development must be tackled mainly by increasing the level of innovation and technological transfer.

This logical conclusion has been reached due to the similarities between the structures and modes of operation of cluster precursors and process innovation systems, which facilitate the expansion and complexity of links between universities, enterprises and research institutes, and through which it is possible to increase the efficiency of the development of the incipient clusters in the region.

Thus, it is desirable to study the links between the university structures, the specialists within them and those belonging to the industrial clusters in the incipient clusters of the N-W of Romania, based on the overlapping of the networks of the two structures, in order to construct a model simulating the interaction of the structure of the clusters with the structure of the process innovation system.

2. Methodologies

2.1. Preconditions *and success factors for university/company relations in innovative clusters*

It is obvious that intense correlations between a company efficient absorptive capacity of new developments in scientific and technological knowledge and its innovation outputs will increase the probability of generating a strong competitive advantage on the market [4].

The transfer of knowledge depends on a variety of conditions. For example, it depends on knowledge type, that is, explicit knowledge can be transferred easily by learning processes, while implicit knowledge depends on the individual work experience and his capability of giving or receiving information.

Also, the efficiency of the whole phenomenon of technology transfer depends on how people apply and engage themselves in the acquisition, processing, adaptation and application of new technologies or new products. These are actually the steps to implement an "*autonomous management of strategic absorption capacity (such as: i-acquiring knowledge, ii-assimilation, iii-transformation, iv-application of new knowledge to a new product)*" [4].

In order to achieve an efficient technology transfer process, companies can group into incipient clusters. The success of such a process lies also in intensifying collaborations with universities and research institutes to gain access and to implement cutting-edge technologies.

These collaborations involve the transfer of technologies, of know-how, and of specialists, specialized knowledge exchanges, technical consultancy and management of transfers of all kinds, mainly through joint projects in the field. Thus, the legislative and organizational work of the government to fund joint research projects and to encourage research programs in the field of development and the progress of both universities and companies becomes decisive.

In all these aspects, direct personal ties between academics and industry members are essential as they ensure an optimal deployment of technology transfer and knowledge transfer. This strengthens the possibility of solving technical problems, the emergence of innovative ideas and the emergence and development of new businesses.

This is all the more important as there are some differences between the way in which academic and industry communities behave, as the assigned time for solving problems, differences between the

IOP Conf. Series: Materials Science and Engineering **749** (2020) 012025 doi:10.1088/1757-899X/749/1/012025

corporate and academic social and technical characteristics, and also additional cultural national issues involves in cross-border collaborations.

These considerations lead us to the idea that there are a multitude of factors, favourable or disturbing, which positively or negatively influence the assimilation of new technologies, university-industry collaboration and cluster development.

But economic processes have to be analysed in their real aspect and taken into account by all events that may occur incidentally and/or accidentally, including the decisions taken by the participants as stand-alone units whose actions are related to their will and whose free will cannot be restricted, anticipated or directed only to a small extent.

2.2. Approaches to cluster/innovation systems of processes modelling

In order to develop development plans for the North-West region of Romania, it can be considered that economic activities take place according to those economic models in which interactions between clusters and process innovation systems take place.

They have been theorized that there are five generations of innovation process models. In the EU area, the last two generations are of interest at the moment. In Figure 1 a generic synthesis model for these two generations of innovation processes is proposed.

Functional Integrated Model - IVth Generation basic components are:

- Marketing, research and development and distribution departments of innovation-producing companies;
- Similar departments of other supply companies;
- Internal departments in the innovation process;
- Developing networks of suppliers, customers and partners;

Functional Integrated Model - IVth Generation basic processes are:

- Functional integration of different activities and the simultaneous development of all levels of product innovation, instead of a sequential evolution;
- Complex business processes, feedback and reciprocal relationships.

Integrated System Model - Vth Generation basic components are: a deepening of integration into complex networks in the design of systems and organizations of innovative networks, customer units, universities, suppliers and other industrial companies and local communities, including by addressing state-of-the-art technologies dedicated to complex issues within these networks team of system specialists interconnected in R&D-design, CAD/CAM software, models for simulations, rapid-expert prototypes.

Integrated System Model - Vth Generation basic processes are: information flows and actions occur simultaneously and closely with the activities and interests of the communities in which businesses operate.

The types of clusters that may occur in areas where universities and research and design institutes are involved in regional development are [4]:

i. Regional partnerships: interconnection of a group of companies in a region which in industrial production and services to obtain beneficial economic benefits, following the specific conditions of the region;
ii. Innovative clusters: cooperation between companies focused on the development of innovative companies' activities;
iii. University clusters: organization of co-operation between industrial enterprise groups, scientific and university institutions, concentrated especially in university research parks.

According to their level of development, there may be the following clusters:

a) Strong clusters: with high internal integration in cluster activities;
b) Stable clusters: with positive dynamics in interactions between their elements, but without consolidation;
c) Potential clusters: with poor development;

d) Latent clusters: organizations and companies with beneficial work together, but without an integrated structure.

On the other hand, there are three stages of cluster-based development:

a) The initial stage: a number of economic agents, having the necessary specialists and know-haw, initiate an industrial activity;
b) The quantity expansion stage: the best innovative techniques are implemented from leading, innovative companies;
c) The quality improvement stage: trends to increase the performance of companies through technology transfer and the introduction of innovations.

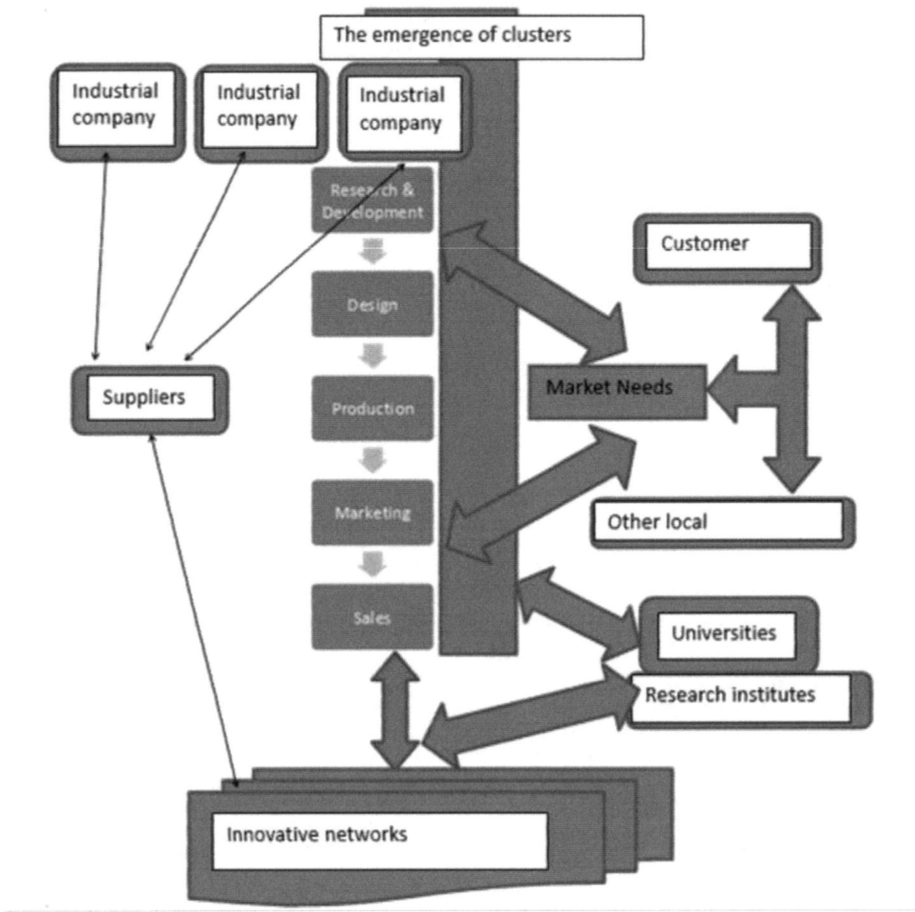

Figure 1. A generic minimal structural model proposed for generation IVth and Vth of innovation systems of processes

As can be seen from the above-mentioned classifications, the work carried out in the two networks - those of clusters and innovation process models - is based on component entities and in stages of evolution.

Components evolve from incipient forms to complex organizational forms in order to increase efficiency. This translates from products of low quality and complexity to high quality products that embody intelligence and innovation, as their technical capacity increases.

The efficiency of clusters also depends on inputs and outputs of material and financial resources, as well as access to specialized human resources.

Access to human resources also depends as well, on the regional concentration of institutions with innovative potential and resources in communication and research and design, through which to build the organizational structures characteristic of IVth and Vth generations of the innovation process models in the area.

At the same time, the way in which clusters evolve depends also on the pace of their occurrence and their life span, in order to have the necessary time to cluster them. Also, the percentage in this number of innovative cluster enterprises should be considered [5].

All these possible structures, processes and evolutions justify us to consider an archetype of evolution in stages, involving: i) transformations, ii) integrations, iii) inputs and outputs of human, financial, and material resources.

This must lead, for example, to the integration of a company into a cluster: i) in the first stage, to the structuring of these companies towards a technology, ii) intensive structure in a second stage, iii) and finally to companies that launch new technically evolved products on the market, in a third stage.

These stages involve also other entities that appear, interact and evolve from one stage to the next one.

3. Results

The elements and their interaction must be the fundaments for all attempts to build an evolution model of cluster-innovation systems of processes networks simulation.

For example, scientific research requires the development of technological innovations and this influence can be considered as a two-way flow of information and influences, an interaction that intervenes decisively between the elements of innovation processes and cluster networks. This influence can be considered as a negative feed-back, in the sense that as the pressure of innovative ideas increases in quantity and quality, the cluster enterprises and innovation networks must restrict their material and human resources to deal with more opportunities.

Most clusters include:

- Companies produce products or services of interest;
- External suppliers of raw materials, production materials and energy;
- Financial institutions and service companies.

There are also government organizations, and often other consumer organizations, in collaborative networks.

The main collaborative structures within the innovation networks are universities and research structures, whose economic role to support clusters are [1]: a) consolidating the knowledge base of clusters; b) creating workforce through education and training; c) supporting scientific and technological infrastructure; d) testing to meet the cluster's technological needs to provide opportunities for new industrial opportunities [4].

Therefore, for a model that can be used in the case of considering also incidental events, the main components are: a) industrial productive companies which are grouped in regional clusters; b) external suppliers of raw materials, production materials and energy; c) specialists from universities and design and research institutes; d) new created workforce through education and training; e) financial institutions which provides financial resources, services; f) companies in horizontal industries which provides by-products, external suppliers of raw materials, production materials and energy;
g) universities, and research structures; h) consumers organizations.

In such a model, the main actions to consider are: a) the transformation from a company with low quality products to a company with the production of high-quality products; b) inputs and outputs of staff employed; c) inputs and outputs of financial resources, materials, by-products etc.

Also, the model should contain a network component which inputs and outputs are knowledge, intellectual resources, and information. By this component the staff of universities for example can work and influence the structure of clusters and they can really superpose the innovation systems of processes networks over the clusters networks.

IOP Conf. Series: Materials Science and Engineering **749** (2020) 012025 doi:10.1088/1757-899X/749/1/012025

4. Conclusions

The analysis suggests that superposition of innovation systems of processes networks over cluster formation is not only probable but even mandatory given since technological progress is vital to maintaining the competitiveness of cluster companies.

Also, scientific and technological research is itself a force that produces the premises for the development of local communities present in geographic clustering areas.

From this point of view, the innovation systems of processes networks of IVth and V^{th} generations are mostly complementary, which ensures an increase in the speed of assimilation of technological progress, which in turn ensures the safety and stability of the companies present in the clusters.

On the other hand, the literature on the development of regional clusters, results in a three to four cluster development stages. So the model structure must respect this type of evolution [6].

The components of such a model are those involved in the creation of clusters and innovation systems of processes networks, the actions being taken are the inputs and outputs of products, personnel and financial resources, and the coordination between all these elements is made by negative feedback in order to provide the technological transfer.

5. Implications

One must take into account as implications for future research works the following considerations about the construction of the economic model:

- There are a multitude of factors, favourable or disturbing, which positively or negatively influence the assimilation of new technologies, university-industry collaboration and cluster development;
- The events according to the real, incidental aspect must be considered, as the presence of hazard on the markets regarding a multitude of aspects such as the prices of materials and raw materials, and bank and financial fluctuations.
- The personal decisions and actions of the people involved cannot be restricted and assumed a priori that they are exercised only in the sense of amplifying some actions, even if these actions are positive.

References

[1] Stoicovici D, Banica M, Ungureanu M and Stoicovici M 2017 *The Impact of Collaborations Between Universities and Private Organizations on Cluster Development and Competitiveness in Romania* Innovative Ideas 2017 IOP Conference Series: Materials Science and Engineering Vol. 200/2017 http://iopscience.iop.org/issue/1757-899X/200/1

[2] Etzkowitz H 2008 *The triple helix: university-industry-government innovation in action* Routledge New York; London

[3] Stoicovici D I, Banica M and Dascalescu A 2014 *Potential Links between Competitiveness of the Cluster-Type Economic Agglomerations and Systems of Innovation Process* Cers 2014: 5th Central European Conference In Regional Science International Conference Proceedings Kosice SK 2015 ISBN: 978-80-553-2015-1 pp.961-967 WOS000379207200093

[4] Klimova N, Kozyrev O and Babkin E 2016 *Innovation in Clusters. Understanding Universities, Special Economic Zones, and Modeling, In Springer*, ISBN 978-3-319-21108-4.D

[5] Laursen K and Salter A 2004 *Searching high and low: what types of firms use universities as a source of innovation?* ELSEVIER Research Policy Bind 33 Hæfte 8 Side 1201-1215 DOI:10.1016/j.respol.2004.07.004

[6] Yuzhuo C 2014 *Implementing the triple helix model in a non-western context: an institutional logic's perspective* Triple Helix 2014 1:1 Springer

IOP Conf. Series: Materials Science and Engineering **749** (2020) 012026 doi:10.1088/1757-899X/749/1/012026

Aspects regarding the capitalization of remanufactured scrap in the automotive industry using a Paternoster system

C Torcatoru[a] and D Savescu

Transilvania University of Brasov, Department of Product Design, Mechatronics and Environment, Braşov, Romania

E-mail: [a] constantin.torcatoru@unitbv.ro

Abstract. Among the current challenges of the global automotive industry include changes in global markets that lead to a variety of products, regulation that drives pressure on new technologies and competition from new competitors. This paper attempts to find answers to the question, how can we capitalize the semi-finished parts of the automotive industry that have emerged from the parameters demanded/imposed by the final customer or which have some visual/ design defects, but which retain their functional role. In almost all cases, information and communication technology plays a major role in future strategies to meet the challenges mentioned above, which can stimulate the production process more efficient by using intelligent manufacturing approaches. Thus, this paper proposes to automate and maximize the capitalization of these parts, a Paternoster system for eliminating the logistics/storage space, reducing the handling time, increasing the efficiency and traceability of the scrap

1. Introduction

In this paper, we try to find answers to the question of how we can maximize the use of remnants of the automotive industry in this era of advanced technologies, for trying to rebuild the parts with certain visual or design defects but retaining the role functional is a very old subject, even in the period immediately following the start of production. This problem has arisen even in traditional craftsmanship when our trees have realized that some of these products have defects, some of which are major, and can't be used for the purpose for which they were created.

With the passing of the years, the manufacture of products has evolved from handicraft to mechanical, automated.

Vehicles represent the pursuit of freedom of each individual. The relevance of the car industry to economic prosperity and its impact on jobs can be seen in all major areas of the market and countries. Interestingly, both developed and developing countries see this industry as the backbone for their own economy.

The automotive industry has wide-ranging effects across the economy, supporting a vast supply chain and generating a range of business services. In the European Union, 19.6 million automobile units are produced annually, the turnover generated by this sector accounting for 6.8% of EU GDP. [4] This sector employs around 13.3 million people or 6.1% of the workforce in the EU. Approximately 3.4 million jobs in the automotive industry account for 11.3% of EU employment. [4] The automotive industry is the largest private investor in R & D in Europe, with nearly €54 billion invested annually.

In 2017, over 8400 patents were granted to the automotive sector by the EPO. Worldwide, 98.9 million vehicles were produced in 2017.

A paternoster or paternoster elevator is an elevator for passengers consisting of a chain of open compartments, which moves slowly and without interruption in a loop up and down, inside a building. The same method is used for the storage cabinets that store large amounts of paper documents or small parts. [8].

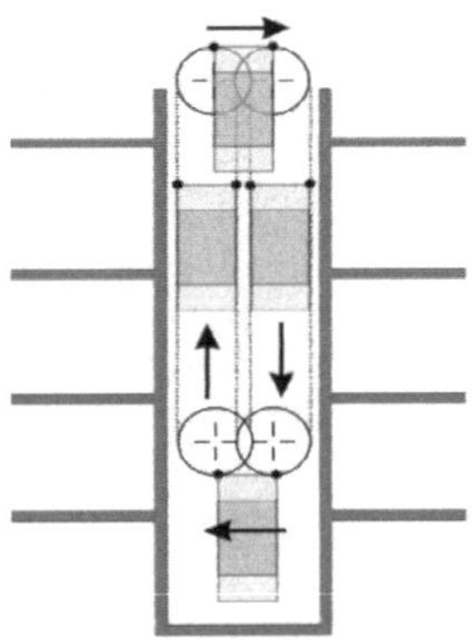

Figure 1. The 2D scheme of a Paternoster [Source Own work, author Helfmann, 01.10.2009]

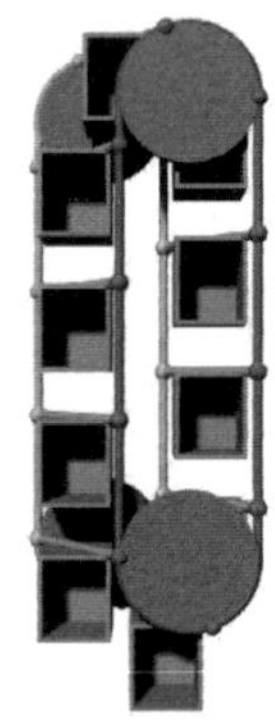

Figure 2. The operating principle of paternoster [Source: Own work, author RokerHRO, 20.01.2019]

Paternoster name ("Our Father", the first two words of the Lord's prayer in Latin) was first applied to this device because the lift is in the form of a loop and is similar to rosary beads used as an aid in reciting prayers. [9]

Peter Ellis installed the first lifts that could be described as a paternoster lift in Oriel Chambers in Liverpool in 1868 [6] In 1877, British engineer Peter Hart has obtained a patent for the first paternoster [1]. In 1884, in Dartford, England, the engineering firm of J & E Hall installed its first "Cyclic Lift", using Hart's patent, in a London office building. [5]

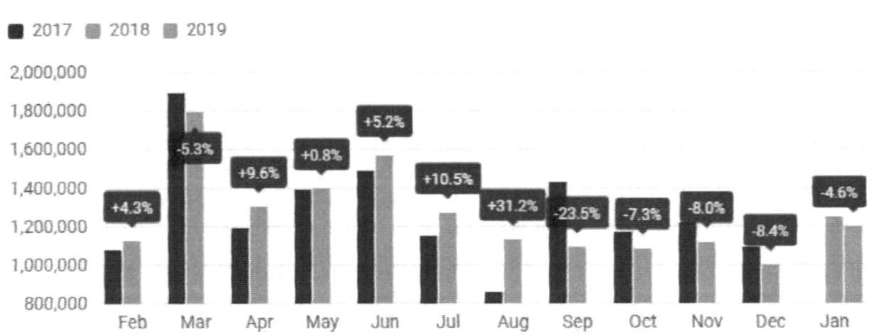

Figure 3. Car production in the UE, 2019 versus 2018. [Source: ACEA, 2019]

Paternosters lifts were popular during the first half of the twentieth century because they could carry more passengers than ordinary elevators. These were more common in continental Europe, especially in public buildings than in the UK. They are relatively slow elevators usually traveling at about 30 cm/sec (about 1 ft/sec) to facilitate starting and stopping. [11]

With the passage of time and the emergence of new technologies, this system has begun to grow and be developed in several areas, reporting on the first occurrence.

Lately, these systems have been developed in many areas and in various forms, based on the principle of operation of the first paternoster systems. [2]

IOP Conf. Series: Materials Science and Engineering **749** (2020) 012026 doi:10.1088/1757-899X/749/1/012026

2. Literature review

The automotive industry comprises a very wide range of companies and organizations in the design, development, manufacture, marketing, and sales of motor vehicles, being one of the largest economic sectors in the world

2.1. Series production in the automotive industry

Manufacturing or manufacturing process is meant the physical or chemical transformation of materials, substances or components into new products or components through a series of processes.

The British Encyclopedia [3] mentions that "manufacturing is an industry that makes raw material products by using manual labor or machines, usually routinely carried out with the division of labor."

Manufacturing as a system of design processes and the sequence of actions needed to produce the product as it was conceived and designed in order to solve a problem or meet a need.

As can be seen in figure 4, at the beginning of this year 2019, the European automotive market registered a slower start, with a decline of 4.6% over the previous year. [5]

This effect can be felt throughout the manufacturing sector in the automotive industry, and in the case of small producers, it is already becoming an alarming situation, for which they have to resort to various fireworks, such as reducing the number of employees, increasing the range of products, new smart technologies etc.

In Romania, according to the managers' opinions, the main causes limiting the processing activity are (figure 5):

- Insufficient internal demand (14% of total cases);
- Lack of workforce (14% of total cases);
- Financial issues (5% of total cases).

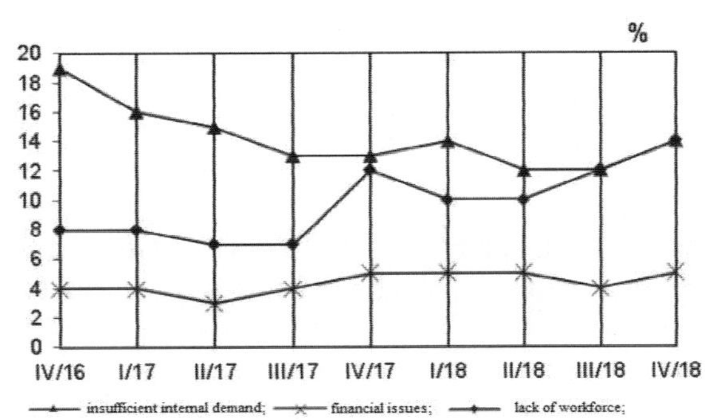

Figure 4. The main causes of production limitation
[Source: INS, January 2019]

2.2. General classification for parts of the automotive industry

Depending on the customer relationship criteria, the types of manufacturing can be structured into the following categories:

- manufacture on stock;
- custom manufacturing;
- mixed manufacturing.

There are many companies that combine the first two types and have mixed manufacture for strategic or competitive considerations. For example, the desire to be a unique manufacturer in its product line can lead a stock company to complete its catalog of offers by offering custom-made products. The desire to compete on a delivery rate criterion can motivate a custom-made manufacturing company to store key components and subassemblies by offering larger quantities of custom-made assembly. Receipt of an order initiates assembly and finishing product based on the customer's specifications.

Depending on the way of obtaining, the products from the automotive industry can be obtained:
- by injection of plastics;
- by machining (forging, casting, cutting processing, cold pressing etc.)
- by sewing;
- by quilting; "cașerare";
- by pasting etc.

2.3. Paternoster system

Paternoster is an excellent solution for optimal downtime. The particular reduction in collection time is seen in the sorting of batches, offering a quicker recovery

The paternoster type system is very used in the storage/logistics area, for the simple fact that space is saved, the handling time of the objects placed in the paternoster cabins/containers decreases and the efficiency of the management used is increased.

There are storage machines, vending machines, as storage systems and computer-controlled transfer that significantly improve storage efficiency and processing of products.

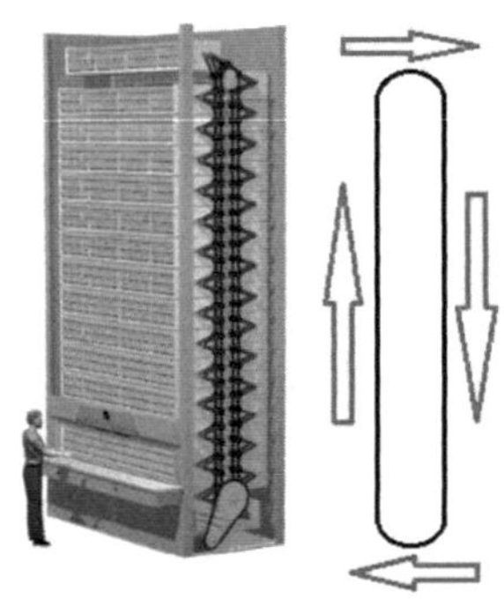

Figure 5. Operation of a paternoster
[Source: Paternosterkast.com]

As can be seen from figure 6, according to Paternosterkast, an automatic storage system saves up to 70% of the storage space, reduces choice errors and decreases the lifting time by more than 60%. The most efficient way to use the storage machine is to include warehouse management in the machine. This is suitable for use with a variety of storage systems such as storage machines, small parts racks, and pallet racks. [7]

These paternoster storage systems are currently used for both bulk and bulk products as well as bulky and heavy products.

Figure 6. Paternoster system for large products
[Source: Paternosterkast.com]

For example, in figure 6, there is presented a paternoster system used for storage/ logistics of products with large dimensions and masses, such as semi-finished products made of laminated bars or steel blocks. This system is equipped with 34 containers, and a container can be loaded up to a maximum weight of 4 tons, the entire system can be loaded with up to 136 tons of products. The same type of system can be seen in figure 5, but with smaller dimensions and can be used in case of small products.

Control of these systems is permanently made from a computer that can be connected to a managed internet source such as SAP.

2.4. Quality and risk management for rebut items

Manufacturing process management- MPM is a process of defining and managing the manufacturing processes that will be used to manufacture, assemble, and perform the final product inspection. [12]

MPM is a process by which the manufacturing industries will use different types of technology to help produce products with different machine locations and different ways in which assembly lines could be placed. Production engineers use CAD data to define the different processes through which the new product will be manufactured. The MPM system is used to develop the sequence of operations, machine location planning, production line balancing and 3D design of human and robotic operations, as well as the delivery of machine programs and electronic work instructions at workshops. Production engineers assess different manufacturing scenarios, material flows, calculate production costs and the effects of changes on production lines. [13]

Statistical process control is a quality control method that uses statistical methods to monitor and control a process that produces serial products, such as a manufacturing process. Using this control during the manufacturing process makes it possible to determine whether the process is being carried out normally by providing only compliant products. Obtaining the necessary information for a statistical control is done by sampling n-size samples at defined time intervals and verifying one or more quality characteristics of the objects or elementary sampling units in order to ascertain whether during the manufacturing process these characteristics do not change significantly.

When applying statistical control, it is necessary to be able to act on the process by making appropriate adjustments to ensure that a normal process is carried out.

When applying statistical process control, the main tools are the control sheets, the Pareto diagram, and the design of the experiments.

At statistical control, objects in the sample can be controlled by:

-quantitative, measurable characteristics such as dimensional dimensions; these variables are often distributed after the Gauss normal distribution, but a Poisson, binomial, etc. distribution is also possible;

- qualitative or attributive features such as the appearance of the product,
- the number of defects in the sample or the number of defects per sample elementary unit.

Another method used in quality analysis and control is the FMEA (Failure Mode and Effects Analysis), described as a procedure for product development and management of potential failure analysis operations under a fault probability classification system and which continues 3 main objectives:

- Identifying and assessing the risk of potential product failure, flexible manufacturing system, and failure effects;
- Identification of actions that could eliminate or reduce the likelihood of a potential breakdown and development of control plans;
- Documenting the flexible manufacturing system and risk management;

Another way of analysing product quality, from the design stage, is a more complex methodology, called DFMA (Design for Manufacture and Assembly), to provide guidance to the design team to simplify product structure, reduce manufacturing costs, assemble and quantification of improvements. In the practice of applying DFMA, the emphasis is on identifying, quantifying and eliminating unnecessary or redundant or ineffective elements in designing the product. [14]

DFMA focuses on three main directions:

- Reducing the number of components (parts, subassemblies) of an assembly;
- Redesigning simpler, modular components, easier to process and control, and even cheaper;

IOP Conf. Series: Materials Science and Engineering **749** (2020) 012026 doi:10.1088/1757-899X/749/1/012026

- Optimization of production and control technologies;

A product scrap, seen as a whole unit or unit is defined as a product resulting from the manufacturing process, which does not meet the quality set by standards, internal rules, contracts etc. and which can't be used directly for the purpose for which it was realized, representing an economic loss.

A product considered to be a rebuttal following the manufacturing process can be recovered using a reshaping method to be brought to the quality set by the qualitative and standard conditions or can be recovered without intervening with any change to it, finding the defect by selling at a low price to less demanding customers.

3. Discussion

The longer the production line, the more it requires a larger number of people, processes and, of course, a longer time. The more people are needed, the higher the mistakes, the scrapping rate.

Therefore, we can see once again how the company loses money by paying extra people's wages, storing stocks, raw materials by repairing defective products, etc. It is not at all an appropriate solution to simply replace old machines with new ones, but replacing old machines with new ones that do not require human involvement (or require minimal involvement or supervision, although a worker who stays and looks at a machine does not add value) is a much more effective solution and, most likely and important, is a long-term solution.

It can be said that lately, eliminating the human operator from as many places on the production line has become a prime concern and even a "trend" in the idea of stopping as much as possible from the errors of the human operator.

Lately, many companies are very focused on the "zero defects" concept, but as nothing in this world is perfect, rebate products are always on the line of production due to the desire to increase the volume of products and to be in the prime plan with what's happening in the market. Technologically, it can be said that in many cases it is possible to achieve the "zero defect" target. However, it is a difficult threshold to overcome, large firms with high investment potential, accept the best 98% of good products. A target of 99% involves extremely high costs, much higher than 2% non-conforming products.

Focusing heavily on increasing product volumes, many manufacturers neglect more or less non-compliant products. Very few wonders what happens to scrape that can mostly be reused after a reshuffle. If the costs of handling, storing, logistics of these scrapes were calculated, it would be economical and a substantial gain.

In general, the criteria for classification for defects are analogous to the criteria for classifying the quality characteristics:

- nature - functional and structural, appearance, packaging;
- measurement possibilities (measurable/ attributive);
- importance - critical, major, minor;
- after effect - allowing reshuffle;
- after the moment of occurrence, related to the product's lifetime: youth, maturity, old age;
- the character of occurrence: accidental; systematic;
- by cause, origin, the most important criterion for the removal of their occurrence;

Depending on the causes, defects fall into one of the six factors, namely: material, worker, machine, working method (technology), measurement, and environment.

3.1. Suggestions for capitalizing remanufactured scrap

Referring to scrapes that have appearance defects, but which retain their functional role, they can be capitalized without interfering with their modification, reshaping them.

For example, there are manufacturers that produce components for the interior of cars. These components result from machining operations such as injection moulding, hot pressing, sewing, gluing, taping etc. For various reasons, defects such as lack of material in certain areas of the product, a bulge in a certain area, seams defects (deformed stitch, ruptured sewing material, "rare mesh", material wrinkling, seams missing etc.), discoloration of the material after the soldering operation, burns and scratches resulting from the make-up operation etc.

By reporting on the automotive industry, these products that have certain appearance defects can be redeemed by selling the final product to less demanding customers or to companies that provide certain services (taxi, courier, etc.) but with a re-valued sales cost finding defects.

- Regarding scrapes that have functional defects, but which through a certain reshuffling of the product, I can restore the functional state for which they were created, can be capitalized if the repair costs and the allocated time are small.
- If we refer to scrapes that have both aesthetic and functional defects, it is not appropriate to remedy them due to high costs but can be redeemed by using under normal conditions that rebut for another purpose for which it was originally created.

Figure 7. Type of Paternoster by Maxi Store
[Source: Youtube.com]

For example, if we think of car safety belts, assuming that for some reasons they do not meet the qualitative standards set by the standards, these rebated belts can be reused for another purpose, such as the use of straps to secure goods in trucks or for palletizing goods.

In figure 7, there is shown a paternoster system, which is customized with different sizes and shapes of the racks in which the scrap can be stored and sorted, depending on the defect category settings that are introduced into the paternoster computer.

3.2. Possibilities of implementation of the redeemed scrap in the paternoster system

Taking into account that at present they are manufacturers in the automotive industry that use up to 90% of their waste, they are doing it in a classic way, and considering the possibility of implementing a paternoster management/ logistics system, would save from the storage space, reduce time and handling costs, manufacturers would have an overview of the scrap for a certain period of time.

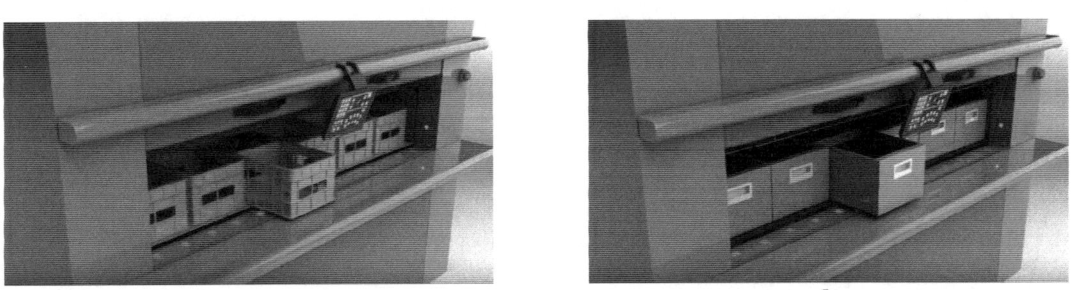

a **b**

Figure 8. Custom Paternoster by Maxi Store [Source: Youtube.com]

In figure 8, there are exemplified some compartment models that can be customized according to the customer's need and operated manually or automatically (horizontally) to the loading position, to facilitate operator ergonomics and reduce the dead times.

The rebuttable parts can be recorded with the bar code that is unique to each piece. Later, when it is desired to "archive" some rebuttable pieces, it is possible to scan the bar code and somewhere in an observation field to introduce the defects of the product after the evaluation of the Technical and Quality Control Department, then its paternoster system store it on the "district/container" that has a certain defect category.

On paternoster shelves/containers, defects can be determined from its software, defects categories and subcategories, and when a product with an aesthetic defect is introduced, for example, the system knows where it can store. They can be customized in many ways, for example, depending on the category and severity of the defect, to reassess the value of these scrapes.

Nowadays, the big manufacturers that are in trend with technology, use a system similar to the one mentioned in the consumable products in the idea of recycling materials and reduce the future costs of acquisition. For example, the cutting plates used for machining when worn out are returned by human operators in an automated system that sorts them according to the chemical composition of the tiles, later being sent to manufacturers and reused as raw material.

4. Conclusion

Due to the speed of today's technology, it has begun to grow and grow in any field, and the rise in the labour shortage increases the need for technology development.

By implementing such a paternoster system, one of the causes limiting the processing activity, namely the lack of labour, mentioned in subchapter 2.1, would be stopped.

In the following papers, some case studies will be approached regarding the implementation of a Paternoster type system on certain remnants of the automotive industry in Brasov.

The purpose of implementing these systems is to maximize the use of these types of rebuttal, to have a traceability of the production flow trace of the semi-finished parts, to simplify existing procedures and documentation for capitalizing on these products, to reduce costs for sorting, logistics, inventory, storage of said scrap, to automate this process.

References

[1] Benoit B 2015 *Is It Time for Germany's Doorless Elevators to Move On?* The Wall Street Journal accessed in December 2018
[2] "*Circulating Multi-Car Elevator System Exponential increase in carrying capacity*" accessed in January 2019
[3] *Encyclopedia Britannica* accessed in March 2019
[4] *Facts about the Automobile Industry* accessed in March 2019
[5] "*Hart's Cyclic Elevator Mansion House Chambers - J. and E. Hall*" *The Engineer:* 61. 26 January 1883
[6] "*In the footsteps of Peter Ellis*" accessed in January 2019;
[7] "*Lift Machine - TORNADO improves efficiency and saves space and time*" accessed in March 2019
[8] "*Paternosterkast, Bertello*" accessed in February 2019
[9] "*Paternoster. dictionarry.oed.com*" accessed in December 2018
[10] "*Passenger car registrations, 2019 versus 2018*" accessed in March 2019;
[11] Strakosch G R 1998 *The vertical transportation handbook* Wiley. pp.99–100 ISBN 978-0-471-16291-9 Accessed in November 2018
[12] Siciliano, Albert 2009 *Improving Manufacturing Process Management with Windchill* MPMLink
[13] Shalvi M 2003 *Manufacturing Process Management A New Era in Manufacturing*
[14] Valter N *Design pentru fabricatie si asamblare* Edit. Bren ISBN 987-606-610-224-7

IOP Conf. Series: Materials Science and Engineering **749** (2020) 012027 doi:10.1088/1757-899X/749/1/012027

Influence of drag finishing parameters on the cutting edge radius of solid carbide mills

M Vozár[a], B Pätoprstý, J Peterka, P Pokorný, T Vopát and V Šimna

Slovak University of Technology, Faculty of Materials Science and Technology, Trnava, Slovakia

E-mail: [a] marek.vozar@stuba.sk

Abstract. Paper deals with the cutting-edge preparation of milling tools by drag finishing. The aim of the experiment was to determine the optimal parameters of the process on a prototype device for edge preparation. The prototype device was based on a 3-axis milling centre and the drag finishing media used was Al_2O_3 granulating. Tools for the experiment were manufactured on a tool grinding machine Reinecker WZS 60 from the cemented carbide rod with chemical composition WC+Co (10%). After that, the tools were drag finished. Then, cutting edges radii were measured on a Zeiss SURFCOM 5000 surface and form measuring machine. The effect of process parameters such as spindle rotation speed, feed rate and time of the drag finishing on the size of the cutting-edge radius and shape was examined. The experiment design was described as well as measurement and evaluation of the output values. Purpose of the experiment and multi-criteria analysis described in this article was to determine the influence of process parameters on the cutting-edge radius sizes when drag finishing the cemented carbide mills.

1. Introduction

The issue of edge preparation of carbide cutting tools is of high importance from the standpoint of tool wear cycle and tool life. It has been stated by author Denkena that cutting-edge microgeometry has significant impact on the tool wear [4], surface quality of machined workpiece [3], as well as on the subsurface residual stresses [10]. Having optimized tool microgeometry also allows for more economic manufacturing, which is the reason why research in this field is important. Controlled application of cutting-edge preparation has several main purposes to add strength to the edge, to eliminate the previous defect on the edge, to increase tool life, to minimize the edge chipping and to prepare the tool surface for the coating [11]. The tool edge rounding affects significantly the developed mechanical and thermal loads during material removal [4].

Using various edge preparation methods different shapes of the cutting edge can be achieved. Sharp cutting edge can be turned into either rounded shape and chamfered shape. Moreover, there are several variants or combinations of these shapes, such as trumpet form and waterfall form of rounded edge or double chamfer, uneven chamfer or combination of chamfer and rounding [9]. The most used methods for cutting edge preparation of solid carbide tools or inserts are abrasive jet machining and brushing, in addition to those there are also other methods that can be used to achieve desired cutting edge microgeometry, such as ultrasonic machining, laser beam machining etc. [6, 9]. One of these methods is drag finishing. The method is widely used for finishing parts that need to have good

IOP Conf. Series: Materials Science and Engineering **749** (2020) 012027 doi:10.1088/1757-899X/749/1/012027

surface quality; its advantage is relatively short time needed to obtain the needed results [2]. However, with the modification of certain process parameters such as the type and size of granulate, feed rate or process time, it can also be used for stable cutting-edge preparation of solid carbide tools [4]. There are manufacturers of drag finishing machines specifically designed for edge preparation and surface modification of cutting tools, such as OTEC GmbH or Rösler GmbH.

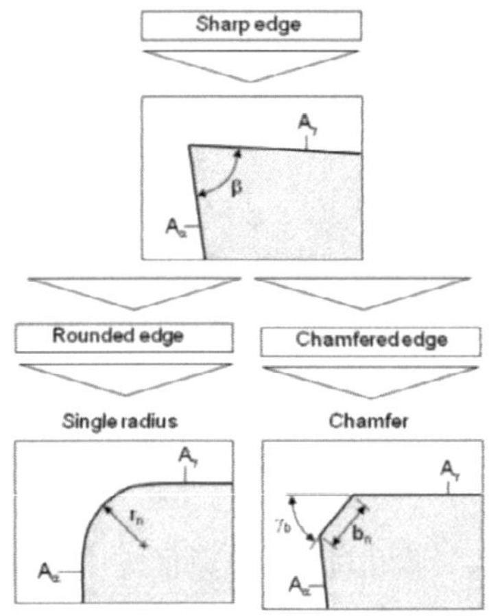

Figure 1 Cutting edge after preparation [9]

Using this method has positive impact on the coating adhesion on the finished surface, moreover, with modified parameters it can also be used to remove macro-particles and droplets from the coating, resulting in even lower roughness of the coated surface [1].

The purpose of the carried out experiments described in this article is to investigate the influence of process parameters of drag finishing on the cutting edge microgeometry of cemented carbide mills as well as to investigate the possibility to use machining centre for cutting edge preparation. Use of statistical tools such as multi-criteria optimization can help with determining optimal process parameters without extensive experimental testing, therefore saving time while allowing for better experiment results [5].

2. Materials and methods

Table 1 Tool parameters

Tool parameter	Value
Maximum depth of cut	18 mm
Cutting diameter	9,8 mm
Shank diameter	10 mm
Center diameter	5 mm
Cutting edge count	2
Flute helix angle	40°
Rake angle γ	8°
Flank angle α_1	8°
Flank angle α_2	22°

Before the mills could be drag finished, they first had to be manufactured, because there was a need to have tools with sharp cutting edge. For the purpose of the experiment the tools were manufactured from cemented carbide rod of 10 mm diameter. Specific grade of the cemented carbide was PCG F10 that is equivalent to K20-K30 by ISO. Volume of Cobalt in the material was 10% and achieved hardness was 1580 by Vickers. Material density was 14,35 $kg.m^{-3}$. [8].

The wire electrical discharge machining method was used to make workpieces with dimension Ø10h6 x 63 mm. Monolithic solid carbide mills were manufactured on the Reinecker WZS 60 tool grinding machine in the Centre of Excellence of 5-axis Machining at the Faculty of Materials Science and Technology of Slovak Technical University. CNC program was made according to the geometry of the tools in the NUMROTOplus simulation software. Specific tool parameters can be found in table 1. A number of different grinding wheels were used in order to achieve the correct shape of the mills.

2.1. Measuring the tools

After the mills were manufactured, their macrogeometrical as well as microgeometrical shapes were measured and compared. ZOLLER genius 3 universal optical tool measuring machine was used for the purpose of measuring tool macrogeometry. It was necessary to verify, if all the required parameters were manufactured correctly. Microgeometry of the cutting edge after grinding was measured on the CNC measuring station Zeiss Surfcom 5000. This shape measuring machine is capable of measuring both the contour and the roughness of the surface. It is equipped with laser interferometer which allows for a high precision measuring resolution up to 0,31 nm. Its measuring error is ± (0.2+L/1.000) µm with L being measured length in mm. Therefore, it is a viable measuring machine to obtain cutting edge microgeometry data before its preparation [12].

Cutting edge radius was measured on every manufactured mill, so that it could be compared to the edge shape after drag finishing. Edge radii values are in the table 2.

Table 2 Measured values of cutting-edge radius

Tool	Edge radius (rake face-outside) [µm]	Edge radius (rake face-centre) [µm]	Edge radius on the helix [µm]
T1	3,04	2,31	1,26
T2	2,12	1,92	3,59
T3	2,55	2,87	1,85
T4	2,02	2,19	2,23

2.2. Cutting edge preparation

Drag finishing of the manufactured mills was carried out on a prototype device based on the Dugard EAGLE 1000 3-axis machining centre. Abrasive media used for the drag finishing was Al_2O_3 granulate QZ 1-2 W with grain size ranging from 1.0 to 2.0 mm, which is used for preparation of edges with radii up to 30 µm.

Figure 2 Prototype machine for cutting edge preparation

IOP Conf. Series: Materials Science and Engineering **749** (2020) 012027 doi:10.1088/1757-899X/749/1/012027

Moreover, prepared edges and tool surface should have very low values of roughness parameter after finishing [7]. Multiple toolpaths alternatives were drawn in Autodesk PowerShape CAD software, based on the information from manufacturers of drag finishing machines. Toolpaths were exported into Autodesk PowerMill CAM software, so that their simulations could be run. After simulations, one of the strategies that were going to be used for the experiment was chosen based on the required feed rate and tool dynamics. NC code of the chosen toolpath was generated and exported, then loaded into the milling centre. After that, the mills were drag finished with multiple different parameters that are described in table 3. The reason of choosing the parameters was to achieve constant cutting edge radius after a certain time as well as ensure the repeatability of the drag finishing process.

Table 3 Drag finishing parameters

Tool no.	Spindle speed [min^{-1}]	Feedrate [$mm.min^{-1}$]	Time [min]
1	75	2100	15 30 45
2	1000	2100	
3	75	6000	
4	1000	6000	

3. Results and discussion

After all the mills were drag finished, they were measured on the SURFCOM 5000 surface and form measuring machine. Results of the measurement of the cutting edges of the face teeth (r_{n1} values) and helical teeth (r_{n2} values) of the mills are in the table 4.

Table 4 Measured values of cutting edges radii after drag finishing

Tool	Cutting edge preparation in time t1 = 15 min			Cutting edge preparation in time t2 = 30 min			Cutting edge preparation in time t3 = 45 min		
	Rake face (center)	Rake face (outside)	Helix	Rake face (center)	Rake face (outside)	Helix	Rake face (center)	Rake face (outside)	Helix
	[μm]	[μm]	[μm]	[μm]	[μm]	[μm]	[μm]	[μm]	[μm]
T1	6,79	5,66	7,13	7,02	6,69	8,2	7,14	6,7	8,4
T2	4,99	12,48	16,01	6,16	14,75	19,17	6,4	17,73	21,34
T3	6,59	8,14	7,91	7,97	8,16	8,5	8,03	8,22	9,23
T4	6,25	13,74	17,47	7,73	15,7	22,24	7,92	19,45	24,86

Measurement was performed as follows-the tool was clamped into holder, to measure rake face surface and helix surface too. On the machine was set a curve on which the touch of the tool would pass. The contour of rake face and helix surface was obtained. Secondly, the tangent lines were attached to rake and flank contour surfaces of milling tool. After that, the circle based on 3 points was drawn. Two points were located on tangent lines and third point was located on radius of tool (figure 3). The evaluation was performed with magnification of 10000 times.

Once the results were obtained, it was possible to move on to the evaluation in order to determine what input parameters of the drag finishing process have the most significant impact on the achieved cutting-edge radii.

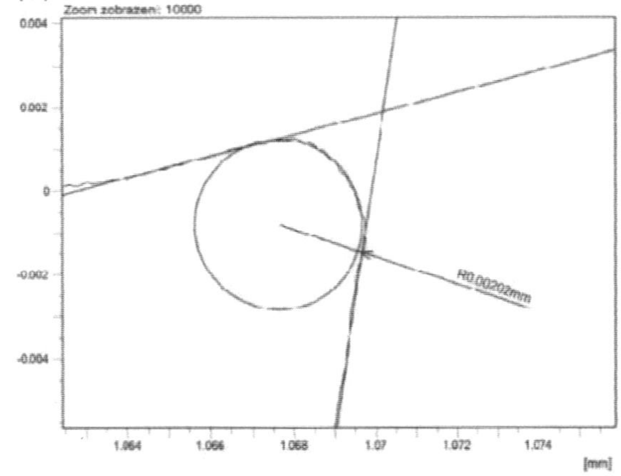

Figure 3 Measuring cutting edge radius

3.1. Size of the cutting-edge radius

The cutting-edge radius increases with the increasing time of cutting-edge preparation. The measuring was done on the different locations, on the rake face close to middle of the tool (centre), on the rake face close to edge of the tool (outside) and the helix surface on the tool. Every cutting-edge radius is different and the smallest is on the rake face close to middle of the tool. It was caused by zero rotation speed in the tool axis and this part of the tool performs movement based only on feed rate. The abrasive granulate was not so effective in the tool axis and effectiveness increased with more distance from the tool axis.

On the graphs can be seen, that with the higher spindle rotation speed, the cutting edge radius increases. In the first graph where the spindle rotation speed of the preparation was 75 min^{-1}, the cutting edge radius after 45 min of cutting edge preparation was only about 7 to 8 μm. But when the spindle rotation speed was 1000 min^{-1}, the cutting edge radius was almost three times higher than in first situation.

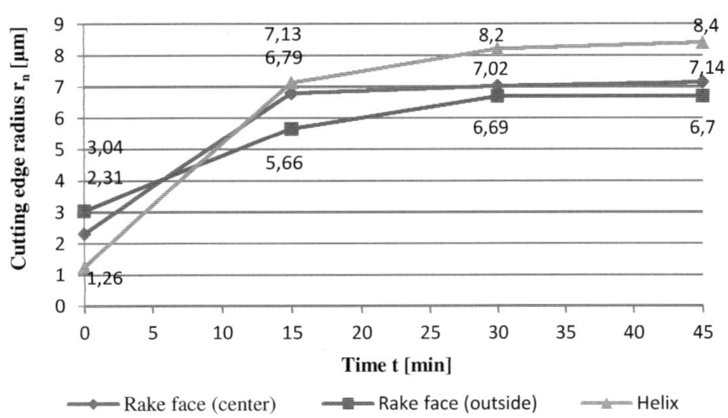

Graph 1 Dependence of cutting-edge radius r_n on time t on tool 1

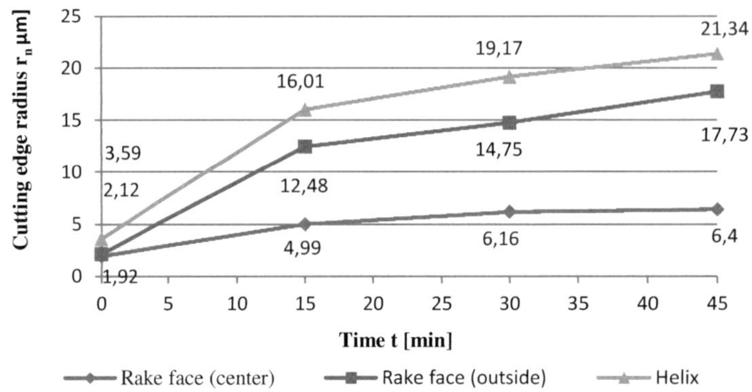

Graph 2 Dependence of cutting-edge radius r_n on time t on tool 2

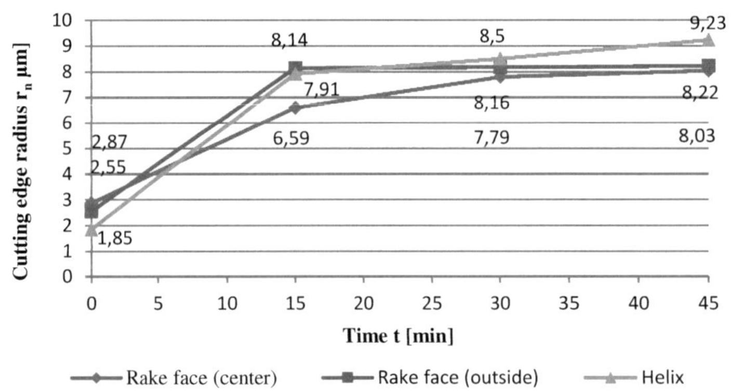

Graph 3 Dependence of cutting-edge radius r_n on time t on tool 3

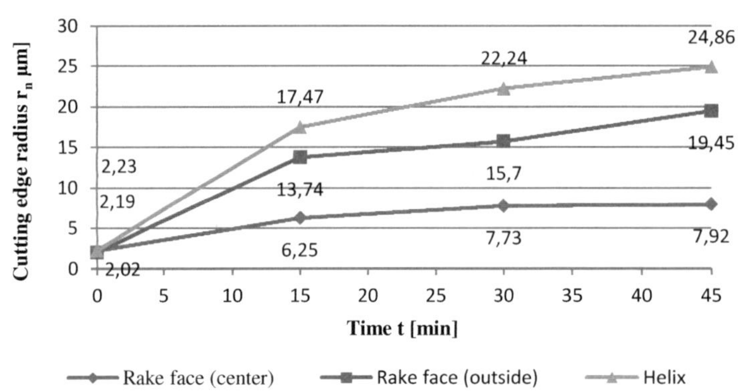

Graph 4 Dependence of cutting-edge radius r_n on time t on tool 4

After evaluating all graphs it can be seen that feed rate does not have a significant effect on the cutting edge radius. The values of cutting-edge radius with feed rate 6000 $mm.min^{-1}$, was only about 1µm higher than in first situation where was feed rate 2100 $mm.min^{-1}$. But, what can be seen is that, the higher spindle rotation speed has effect on uneven cutting edge radius on helix and rake surface, where for tool 2 the difference between cutting edge radius in centre and cutting edge radius on helix was 15µm and for tool 4 the difference was 16µm. Spindle rotation speed has the most significant effect on the cutting edge radius.

IOP Conf. Series: Materials Science and Engineering **749** (2020) 012027 doi:10.1088/1757-899X/749/1/012027

3.2. Multi-criteria analysis

When dealing with the multi-criteria analysis, the desired outcome is usually an optimization of the process based on the chosen weight of the criterion. The assigned weight is a numerical value that reflects the importance of the criterion from the author's standpoint. In the case of multi-criteria analysis, the partial weights should add up to one. Solving an issue through the multi-criteria analysis, it's recommended that all of the evaluated criteria are of one type - that is either minimum or maximum. In the case of the values obtained from the experiment, all are quantitative and measured in the same units-µm, there was no need to transform them in any way, and it was possible to proceed to their evaluation. Chosen weight values for each criterion were set as follows:

- Criterion of cutting-edge radius on the helix - $v_1=0.5$. This criterion is deemed as the most important, as it affects the dimensional parameters of the milling tool, namely the tool diameter and its tolerance. Controlling this parameter is important because it directly affects the precision of the machining process.
- Criterion of the deviation of the cutting-edge radius on the face teeth from the radius in the helical teeth - $v_2=0.4$. This criterion describes how uniform is the drag finishing process, higher deviation means worse repeatability.
- Criterion of cutting-edge radius on the face teeth - $v_3=0.1$. While it can still affect dimensional accuracy for axial machining to some extent, this criterion is not deemed as important as the previous ones, therefore it was assigned the lowest weight.

All the criteria above were evaluated as maximization criteria, used method was weighted sum model, where the values of the criteria are multiplied by their weight and then added together according to the following formula *(1)*:

$$w_i = r_{n2} \cdot v_1 + r_{nMAX} \cdot v_2 + r_{n1} \cdot v_3 \quad (1)$$

where: w_i is value of weighted sum of criteria of i variant
r_{n2} is cutting edge radius on the helical teeth [mm]
r_{n1} is cutting edge radius on the face teeth [mm]
v_{1-3} is weight of the criterion 1-3
r_{nMAX} represents the maximized criterion value of the edge radius difference between the teeth on the face and on the helix of the tool.

The calculated results of this method are listed in the table 5. The analysis was done in the statistical software Minitab.

Table 5 Results of the multi-criteria analysis

Order	$\sum w_i$	Variant	n [min^{-1}]	vf [mm.min^{-1}]	t [min]
1	12,66	12	1000	6000	45
2	12,17	10	1000	2100	45
3	10,14	6	1000	2100	30
4	10,07	8	1000	6000	30
5	9,74	4	1000	6000	15
6	9,04	2	1000	2100	15
7	7,41	7	75	6000	30
8	7,38	3	75	6000	15
9	7,39	11	75	6000	45
10	7,25	9	75	2100	45
11	70,2	5	75	2100	30
12	6,72	1	75	2100	15

The order of the variants was determined in regard to the weighted sum value, where the lowest number represents the best variant according to the chosen weight of the criterion.

Based on the results of the analysis, following can be stated:

- For variants with higher spindle speed $n=1000\ min^{-1}$, better results were achieved than for variants with $n=75\ min^{-1}$, this indicates higher productivity when drag finishing with higher spindle speeds.
- The longer the process time was, the better the results were in consideration with the assigned weight values.

Higher feed rates $v_f=6000\ mm.min^{-1}$ allowed for better results compared to the $n=75min^{-1}$.

3.3 Determination of the influence of the drag finishing parameters

The influence of the input parameters of the process on the radius of the cutting edge was determined by the DoE (Design of Experiments) method. The observed variable was radius of the cutting edge rn and influence of the following parameters: spindle rotation speed (n), feed rate (vf) and time (t). In the figure 2, there is a Pareto chart of standardized effects of the factors.

The influence of the factors and their interactions can be determined based on this chart, the most significant influence have the factors spindle rotation speed (n), time (t) and combination of these factors as well.

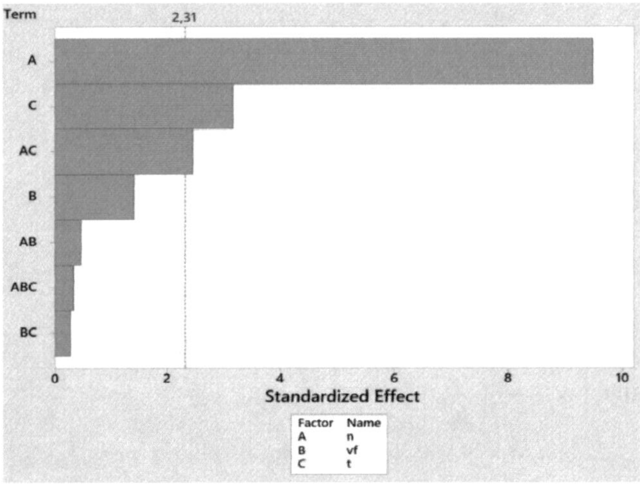

Figure 4 Pareto chart of standardized effects

4. Conclusion

After drag finishing solid carbide tools in an Al_2O_3 granulate on an experimental device based on the Dugard Eagle 1000 3-axis machining centre, the radius of the cutting edge was measured on the Zeiss Surfcom 5000 surface and form measuring machine. The drag finishing process was carried out with set values of three different parameters-spindle rotation speed, feed rate and time. Various combinations of values for these parameters were tested in the experiment, in order to determine the influence of each parameter, as well as their combinations. Twelve different variants of the drag finishing process were tested and after every test the radius of the cutting edge was measured.

There was investigated that the greatest influence on cutting edge radius has spindle rotation speed, the size of cutting-edge radius increases with the rise of the time. However, with the higher spindle rotation speed increases the difference of cutting-edge radius in different location on the cutting tool. The uneven cutting-edge radii were reached on the tools. The feed rate influences the size of cutting-edge radius only marginally, because the comparison between tool 1 and tool 3 where was different feed rates was influenced only by 1μm. The size of cutting-edge radius was most influenced by spindle rotation speed and when the spindle rotation speed was 1000 min-1 size of the cutting edge was 3 times higher than when the spindle rotation speed was 75 min-1. However, the difference between size of the cutting-edge radius on rake face and helix with usage of previously mentioned condition was 15 to 16μm. This causes uneven size of cutting-edge radius.

The obtained data were then used in a multi-criteria evaluation of the drag finishing process. Depending on the assigned weight that reflected the importance of the measured values, the data were used for a weighted sum model that allowed for sorting out the process parameters and their combinations according to their effectiveness. Furthermore, design of experiments (DoE) model was used so that the results of the previous statistical model could be verified. Based on the results of the multi-criteria analysis, the parameters with most significant influence on the drag finishing process were identified, as well as with what values of these parameters were achieved the best results of the edge preparation as far as reliability and repeatability of the process goes. The most important parameters of the drag finishing process according to the multi-criteria analysis were found out to be spindle rotation speed and process time.

Optimal values of the process parameters were determined to be following: spindle rotation speed of *1000 min*$^{-1}$, feed rate of *6000 mm.min*$^{-1}$ and time of *45 min*. All of these values were the highest ones used in the experiment; therefore, it could be assumed that the higher the process parameters values are, the better the results of drag finishing will be. Experiment results will be used as a basis for future research of high-speed drag finishing edge preparation method.

Acknowledgements
The article was written with the support of the Project of VEGA grant agency of the Ministry of Education, Science, Research and Sport of the Slovak Republic and Slovak Academy of Sciences, no. 1/0097/17: "The research of novel method for cutting edge preparation to increase the tool performance in machining of difficult-to-machine materials", and APVV Project of Slovak Research and development Agency of the Ministry of Education, Science, Research and Sport of the Slovak Republic, no. APVV-16-0057: "Research into the Unique Method for Treatment of Cutting Edge Microgeometry by Plasma Discharges in Electrolyte to Increase the Tool Life of Cutting Tools in Machining of Difficult-to-Machine Materials"

Authors are also grateful to Tomáš Štefíček and Martin Počuch, who participated in this research.

References

[1] Akmal M, Layegh K S E, Lazgolu I, Akgün A, Yavaş Ç 2017 *Friction Coefficients on Surface finish of AlTiN Coated Tools in the Milling of Ti6Al4V* Procedia CIRP 58 pp.596-600
[2] Barletta M, Pietrobono F, Rubino G and Tagliaferri V 2014 *Drag finishing of sensitive workpieces with fluidized abrasives* J. Manuf. Proc.16 pp.494-502.
[3] Denkena B, Koehler J and Rehe M 2012 *Influence of the Honed Cutting Edge on Tool Wear and Surface Integrity in Slot Milling of 42CrMo4 Steel* CIRP Procedia CIRP 1 pp.190-195
[4] Denkena B, Lucas A and Bassett E 2011 *Effects of the cutting edge microgeometry on tool wear and its thermo-mechanical load.* CIRP Ann. Manuf. Technol. 60 73-76
[5] Gopalsamy B, Mondal B and Ghosh S 2009 *Taguchi method and ANOVA: An approach for process parameters optimization of hard machining while machining hardened steel* J. Sci. Ind. Res. 68 pp.686-695
[6] Krebs E et al. 2017 *High-quality cutting edge preparation of micromilling tools using wet abrasive jet machining process* Production Engineering
[7] OTEC 2018 Process media. Available on the internet: https://www.otec.de/en/products/media
[8] PCG Precision Carbide Germany 2018 Catalogue for Carbide Rods Available on the internet: http://www.pcg-gmbh.de/assets /skizze_28122018181733 2018-11_PCG-Katalog_en.pdf
[9] Rodríguez C J C 2009 *Cutting edge preparation of cutting tools by applying micro-abrasive jet machining and brushing* Kassel: Kassel University Press GmbH
[10] Shen Q, Liu Z Q, Hua Y, Zhao J F, Lv W Y and Mohsan A U H 2018 *Effects of Cutting Edge*

Microgeometry on Residual Stress in Orthogonal Cutting of Inconel 718 by FEM Materials 11 Issue 6 pp.1-16

[11] Yussefian N Z 2012 *Cutting edge microgeometry modeling & electro-erosion honing* McMaster University

[12] ZEISS 2018 *Contour and Surface Measuring Machines* Available on the internet: https://www.zeiss.com/metrology/products/systems/form-and-surface/contour-measurement/surfcom-family.html

IOP Conf. Series: Materials Science and Engineering **749** (2020) 012028 doi:10.1088/1757-899X/749/1/012028

Aspects of the use solar energy valorification for industrial and public lighten area

V M Brezoczki[a] and G M Filip

Technical University of Cluj-Napoca, Faculty of Engineering, Department of Mineral Resource, Material and Environment Engineering, Baia Mare, Romania

E-mail: [a] brezoczkivaleria@gmail.com

Abstract. The paper presents a viable success alternative for solar energy valorification to generate electricity for industrial lighten system and city streets lighting system in Baia Sprie town. Electricity provided by solar power plants since 2016 and used for lighting the industrial area, urban parking, street lighting, partially replacing conventional energy. Within the Romanian Regional Sustainable Development Strategy there are several measures meant to obtain electricity from renewable energy sources which do not produce negative effects for the environment, which means that the expansion of the photovoltaic plant is desirable. In accordance with the Development Strategy of the town of Baia Sprie, the increase of the plant's productions capacity through the extension of the plant surface for can be used by private consumers in different industries (woodworking, insulating windows, tinsmithing, conservation of forest fruits etc.).

1. Some considerations on solar energy

Conventional energy sources based on oil, coal, and natural gas have proven to be highly effective drivers of economic progress, but at the same time damaging to the environment and to human health, especially by increasing the atmospheric levels of CO_2 even bellow $50\mu g/m^3$. So all over the world they search for alternative and renewable energy resources for electricity production.

The potential of alternative resources such as biomass, wind, solar, hydropower, and geothermal is enormous, and can provide sustainable energy services. It is becoming clear that future growth in the energy sector depends on the renewable energy [1], [2].

One of the renewable energy sources this paper refers to is solar energy used for the production of electricity. Solar energy represents the radiant energy produced by the Sun as a result of nuclear fusion, sent to Earth through space as energy quantum (photons) which interact with the atmosphere and the Earth's surface, i.e. energy produced directly through the transfer of radiant energy from the Sun. This energy can be captured and used as heat in thermo-solar applications or can be transformed into electricity by means of the photovoltaic cells.

Solar energy could account for 8-15% of global electricity in 2050, depending on factors such as market demand, energy policy, manufacturing costs and technological advances. It may also play a role in providing power to the roughly 1.6 billion people worldwide without access to the electricity grid [3].

IOP Conf. Series: Materials Science and Engineering **749** (2020) 012028 doi:10.1088/1757-899X/749/1/012028

Solar cells, also called photovoltaic (PV) cells, convert sunlight directly into electricity, which is called the *PV effect*, discovered in 1954, when scientists at Bell Telephone discovered that silicon (an element found in sand) created an electric charge when exposed to sunlight [4].

Traditional solar cells are made from silicon, are usually flat-plate, and generally are the most efficient. Second-generation solar cells are called thin-film solar cells because they are made from amorphous silicon or non-silicon materials such as cadmium telluride. Thin film solar cells use layers of semiconductor materials only a few micrometers thick. Because of their flexibility, thin film solar cells can double as rooftop shingles and tiles, building facades, or the glazing for skylights.

Third-generation solar cells are made from a variety of new materials besides silicon, including solar inks using conventional printing press technologies, solar dyes, and conductive plastics. Some new solar cells use plastic lenses or mirrors to concentrate sunlight onto a very small piece of high efficiency PV material.

Because the lenses must be pointed at the sun, the use of concentrating collectors is limited to the sunniest parts of the country.

The cost of electricity produced from PV systems is quite higher than from most other competing technologies, but these costs are expected to continue to decline steadily improving in efficiency and increasing rapidly in sales. These potential reductions in cost, combined with the simplicity, versatility, reliability, and low environmental impact of PV systems, should help them become increasingly important sources of economical premium-quality power over the next 20 to 30 years.

Despite the fact that solar energy is renewable and easy to produce, the main problem is that the Sun does not offer constant energy anywhere on the Earth. Moreover, due to the Earth's revolution around its own axis, and thus of the day-night alternance, solar energy can be used for generating electricity only for a limited amount of time during the day. Another limitation of this type of energy is in relation to weather, viz. during cloudy days the potential of capturing the solar energy decreases noticeably due to the screening of the Sun, thus limiting the applications of this form of renewable energy.

Europe holds a leading position in the production of power produced from solar energy, yet not sufficient enough to reach the objective listed under the European Climate Change Programme, which provisions that by 2020, 20% of the power required for consumption in Europe should be from renewable sources.

From the perspective of solar energy, Romania falls under the European zone B (the first class, class A, contains countries such as Spain, Portugal, Greece, or Italy). In Romania, there are 4 sunny areas that can be defined, from a maximum of 1600 kWh/m^2 in the south of the country, i.e. in Dobrogea, to 1250 kWh/m^2 in the northern part of the country, figure 1.

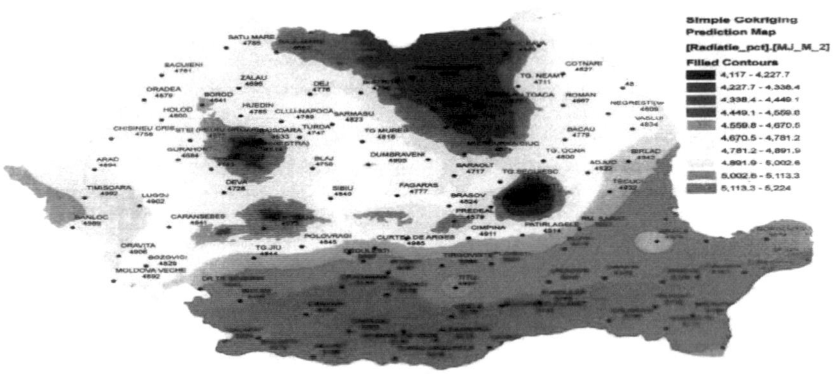

Figure 1. Romanian solar map [4]

Romania's geographical position is favourable for the use of solar energy from March to November.

IOP Conf. Series: Materials Science and Engineering **749** (2020) 012028 doi:10.1088/1757-899X/749/1/012028

2. The use of solar energy in the NW part of Romania, in the town of Baia Sprie

The areas in the north and north-west of Romania are not the most favourable ones as regards the use of solar energy as renewable energy source, yet mounting certain solar panels or smaller solar energy stations in sunny areas may somewhat contribute to the reduction of the power consumption from primary energy resources, and more specifically to the reduction of pollution generated by thermal power stations.

This option of using solar energy as renewable resource was chosen by a small town in the NW of Romania, i.e. the town of Baia Sprie.

2.1. General aspects regarding the geographical location, the climate and the town administration

Located in the North-Western Development Region of Romania, Maramureş County, the town of Baia Sprie is situated about 9 km from the county capital city Baia Mare. The town lies at the southern base of the mountainous plateau Oaş, Gutâi, Ţibles, at an altitude of 380m. [6]

Through its positioning, at the south-eastern foot of the Gutâi Mountain, Baia Sprie is considered a developed town in the contact area of two important relief units between the wide hill depression of Transylvania and the Oriental Carpathians. The town was one of the most important mining centres for non-ferrous metals and gold the in north-western part of Romania.

The climate is specific for the sub-mountainous areas, with an annual average temperature of 9-11°C, abundant rainfall which creates a dense hydrographical network.

The town has an administrative surface of 9602 ha and a population of 17,076 inhabitants [7], including the neighbouring towns of Chiuzbaia, Satu Nou de Sus and Tăuţii de Sus.

2.2. The public lighting system and the main power consumers

The public lighting system in Baia Sprie is made up of: switch-on points, distribution boards, junction boxes, subterranean or aerial low tension electric lines, foundations, lampposts, grounding installations, consoles, street lamps, accessories, conductors, insulators, clamps, coatings, command equipment, automation and measurement equipment used for public lighting.

The public lighting network of the town of Baia Sprie is 55 km long. Due to the fact that the town stretches over an extended area, it was required to have a dense network of roads, which make up 178 km all together. Along this system of public roads, public lighting requires great consumption of electricity, leading to major expenses from the local public budget [8].

A detailed analysis of the public electricity consumers in Baia Sprie over the period 2008-2009 highlighted the main consumers, as well as the recorded consumption values (table 1).

Table 1. Public consumers of electricity [6]

Year	Public lighting	Schools, kindergartens	Town hall	Culture House	Hostel	Sports hall	Hospital	Total
2009KW/h	775.938	104.922	26.469	9.498	12.407	4096	49.119	982.449
2008KW/h	837.745	111.792	25490	9.088	10.868	5120	42.258	1.042.361

Of this detailed analysis one can conclude that the main consumer of electricity in Baia Sprie is the public lighting system, whose annual recorded consumption is approximately 80% of the total consumption for the recorded year of the entire town. The entire quantity of conventional electricity is used form the National Electricity System (NES), Romania's national system of producing and distributing electricity, without any extra support of energy produced from renewable resources: geothermal, solar or wind energy.

2.3. Solutions regarding the reduction of electricity consumption

The high expenses incurred with the public lighting electricity consumption required certain solutions regarding the considerable reduction of this consumption and thus of the expenses from the local budget. At a first stage, along 2010, a series of measures were taken with governmental support. These measures were aimed at replacing the 2347 streetlights with 483 LED lights and 27 energy-efficient light bulbs, modernising all command and switch-on stations, extending the network with about 400 new appliances, and 4000 metres of network cable. The result of this approach was the reduction of the annual consumption by 20%, yet the solution was not a true success due to the limitations of the support from the local budget [6].

Another viable option, in agreement with the European tendencies regarding the harnessing of renewable energy, has become a priority. Thus, a series of studies regarding the solar potential of the area around the town of Baia Sprie has been carried out. Moreover, a series of scenarios regarding the factors which influence the production of electricity from solar energy, the optimal tilt angle of the photovoltaic panels, their type and number, the surface they cover, the right positioning, the size of the electricity production capacity etc. have also been analysed.

All these aspects were aimed at achieving an optimal production of electricity by harnessing solar energy in order to ensure sustainable energy for the public consumers in the town of Baia Sprie, whose energy consumption is soaring.

2.4. Baia Sprie Solar power plant

When calculating the energy the photovoltaic system would be required to produce, it was considered that the development of the town would lead to the increase of electricity consumption reaching a value of 1,250,000 KW/h (1250 MWh), and when estimating the electricity production capacity, this projected consumption was considered.

The studies lead to the design and implementation of a solar power plat (figure 2) in Baia Sprie in the autumn of 2015, with the following technical specifications:

- 896 photovoltaic panels with monoaxial solar orientation (figure 3);
- the length of use of these panels can be 25-30 years;
- 4 power inverters with a power of 60 kW each (figure 4);
- 250 kVA power converter station to which the inverters are connected (figure 5);
- the annual electrical output is 265 MW;

The solar power plant activity is monitored in real-time on the Internet. The solar power plant functions 24/7, 365 days/year, except during the night.

The solar panels transform the captured solar energy into continuous current, then, with the help of the power inverters, it is transformed into alternative tree-phase electric power and transported to the power converter station. All the electricity produced is uploaded into the national electricity system (NES), from where it goes to the public consumers that are connected to it.

Figure 2. The solar power plant [9]

Figure 3. Photovoltaic panels [9]

Figure 4. Power inverters [9]

Figure 5. Power converter station [9]

The monthly variation of energy produced in 2016, as compared to the projected energy, is represented in figure 6.

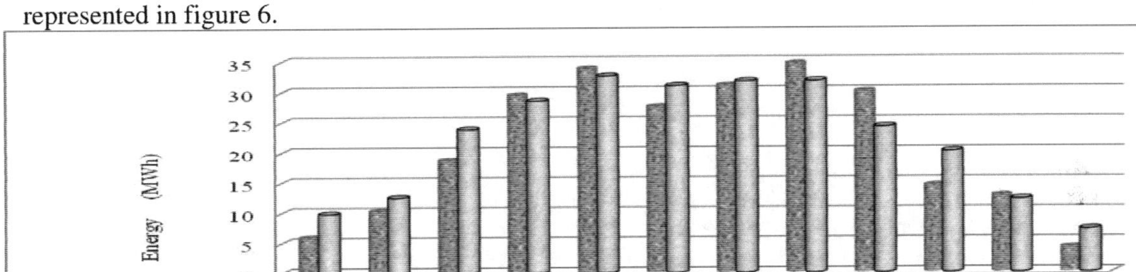

	January	February	March	April	May	June	July	August	September	October	November	December
Generated (MWh)	6	10,3	18,7	29,5	33,9	27,6	31,1	34,7	30,1	14,4	12,7	4,1
Prognosed (MWh)	9,8	12,5	23,8	28,5	32,6	31	31,7	31,8	24,2	20,1	12,1	7,1

Figure 6. Monthly variation of energy produced compared to projected energy for 2016 year

The greatest difference between the projected electricity and the generated electricity were recorded in January, February, March, October, November and December, when the meteorological conditions were unfavourable for capturing solar energy (cloudy sky, abundant snowfall, invisible sky due to fog, glaze, and heavier rainfall than the multiannual average). These differences vary from a minimum of 2.2 MWh, in February, to a maximum of 6.7 MWh.

In December, April, May, August and September, due to favourable meteorological conditions, the generated electricity recorded significantly higher values as compared the projection for those months.

For 2016, the projected quantity of energy was 265.1 MWh, and the energy produced by the solar power plat was 253.1 MWh, i.e. 12 MWh less. This difference was determined by a series of technical problems (5 in total in 2016) which occurred in the operation of the grid of the National Electricity System, which made it impossible to insert the electricity generated by the solar plant.

At the same electricity generating capacity, over the year 2017, without December having finished, estimating the generated electricity identical to the one from the previous year, i.e. of 4.1 MWh, the solar plant produced 265,92 MWh, exceeding the annual projected value by 265 MWh. The data are centralized in figure 7.

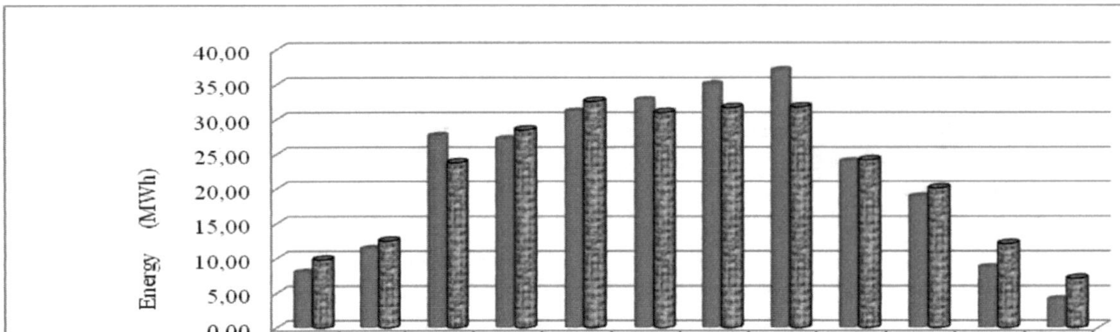

	January	February	March	April	May	June	July	August	September	October	November	December
Generated (MWh)	7,99	11,35	27,65	27,18	31,15	32,79	35,07	37,07	24,00	18,87	8,70	4,1
Prognosed (MWh)	9,8	12,5	23,8	28,5	32,6	31	31,7	31,8	24,2	20,1	12,1	7,1

Figure 7. Monthly variation of energy produced, compared to the projected energy for 2017

3. Conclusions

The Baia Sprie solar power plant is the first of its kind in Maramures. The presented solar power plant generates enough electric energy for covering the total electric consumption of the industrial and public lighting system in Baia Sprie, between May and October, but for the period between November and April it can cover only 20.24% of the electric consumption necessity. During the aforementioned period, the day is shorter, and the weather is not quite favourable for solar power use, so the electrical production decreases.

Within the Romanian Regional Sustainable Development Strategy there are several measures meant to obtain electricity from renewable energy sources which do not produce negative effects for the environment, which means that the expansion of the photovoltaic plant is desirable.

In accordance with the Development Strategy of the town of Baia Sprie, the increase of the plant's productions capacity through the extension of the plant surface for can be used by private consumers in different industries (woodworking, insulating windows, tinsmithing, conservation of forest fruits, etc.).

References

[1] Duke, R D and Kammen, D. M. 1999 *The Economics of Energy Market Transformation Initiatives*, The Energy Journal 20 (4) pp.15-64
[2] http://www.ren21.net/wp-content/uploads/2016/06/GSR_2016_Full_Report.pdf
[3] http://www.rsc.org/campaigning-outreach/global-challenges/energy/
[4] https://www.nrel.gov/workingwithus/re-photovoltaics.html
[5] http://energystreet.ro/en/photovoltaic/romania-solar-map/
Baia Sprie Territorial Administrative Unit 2010, *Green Local Energy Quality Policies in the Public Administration of Baia Sprie* pp.2-3
[6] National Institute of Statistics 2016, *Population of Romania by localities* http://www.insse.ro/cms/ro/content/popula%C5%A3ia-rom%C3%A2niei-pe-localitati-la-1-ianuarie-2016
[7] Baia Sprie Territorial Administrative Unit 2013, *Local Strategy for Sustainable Development of Baia Sprie 2014-2020*
[8] Personal album, 2016

IOP Conf. Series: Materials Science and Engineering **749** (2020) 012029 doi:10.1088/1757-899X/749/1/012029

Reactive Power Consumption Analysis of a Medium Sized Factory

O Chiver[a], L Neamt, C Barz, E Pop and Z Erdei

Technical University of Cluj-Napoca, Faculty of Engineering, Department of Electrics, Electronics and Computers Engineering, Baia Mare, Romania

E-mail: [a] olivian.chiver@cunbm.utcluj.ro

Abstract. The paper achieves a study on the reactive power consumption in case of a medium sized factory. The factories as others industrial consumers usually have the power factor lower than the neutral one (in Romania was 0.92 and now is 0.9); that is why these types of consumers have to pay for reactive power. The reason for the power factor is small (sometimes much lower than the neutral) is not always just because of inductive receptors (as transformers, induction motors, induction furnaces, etc.) but also due to some technical-organizational issues. Many of these issues could have been avoided if more attention had been paid to the proposals made by the power installations designer in the design stage and it would not have been considered only the aspect of investment costs, which, though not decreased very much, provoke much more operating costs. This paper will emphasize these features that cause additional costs with reactive power for a medium sized factory who has more substations with two transformers in each of them. It also proposes, after an analysis of the current situation, technical-organizational solutions to minimize the reactive power consumption.

1. Introduction

Usually, industrial consumers such as factories have many receivers such as induction motors, transformers or induction furnaces, which make these consumers resistive-inductive, and therefore the power factor at the point of electrical energy measure will result in a sub-unit. In Romania, if the power factor was less than 0.92 (now 0.9), the electricity supplier billed the corresponding reactive energy. In addition to this additional charge, also energy losses in consumer installations are higher as a result of the flow of reactive current through them. In order to avoid this situation, industrial consumers are interested in applying measures to increase the power factor to a higher value than the neutral value, as close as possible to the unit.

Two main types of measures are possible to apply: technical-organizational measures and/or inductive reactive energy compensation by producing capacitive reactive energy with the help of dedicated installations. The first category includes those measures that can be applied and lead to an increase in the power factor without additional investments in capacitive reactive energy generation. Thus, reducing the no-load time operation of the inductive power receivers (e.g.: induction motors, transformers, etc.), increasing their load, reducing the supply voltage of the induction motors during periods of low load, and achieving good quality repairs (especially for the electric motors) are measures that once applied will result in an improvement in the power factor. Also in this category can be mentioned the possibility of optimal use of synchronous motors compensation capacity. It should be noted that all these measures, except the latter, can only increase the power factor, and there is no impediment to their application. Synchronous motors (if found among the consumer's receivers) allow both inductive and capacitive power generation, the adjustment being made continuously, but the power is limited by the load factor.

In terms of the reactive energy compensation with dedicated types of equipment, there are three possibilities: the use of capacitor batteries [1], SVC (static VAR compensator) [2] or rotating synchronous compensators.

Nowadays, rotary synchronous compensators are no longer used due to the disadvantages related to maintenance costs, active power losses, and the initial cost of the investment. The main advantage of these compensators is that they allow continuous regulation of reactive energy, both inductive and capacitive.

The condenser batteries, made with different power levels, are the most common means of compensation for both low and medium power consumers. They are also used for high voltage side compensation. The main advantages of these installations are given by low power consumption, relatively low cost, very cheap maintenance. As a disadvantage, it can be remembered that they allow reactive power adjustment only in steps, cannot compensate for the capacitive regime and it can provoke disturbances at the commutation [3]. In addition, the reactive power supplied depends on the voltage level, which is another disadvantage.

SVCs have the main advantages: they can compensate for any regime (inductive or capacitive), provide fine power regulation, good behavior with dynamic load [4-5] and can be used on the low, medium or high voltage side. The main disadvantage is that they are still quite expensive. However, one or other option can be advantageous in different situations, a thorough analysis is usually required.

The paper is structured in the following sections: section 2 - analysis of the current situation; section 3 - technical and organizational solutions; section 4 - conclusion.

2. Analysis of the current situation

2.1. Data registered by the electricity supplier

The analyzed industrial consumer has its own transformer station being powered by the 110 kV grid. From the 110/20 kV, 10 MW transformer, four 20 kV cells are fed, figure 1.

One is the measuring cell, one is the reserve. From the other two cells, leaves two underground cables (L1 and L2) which feed four transforming substations in parallel, figure 2. The substations are each with two identical transformers of 20/0.4 kV, except for station 1, where we have two 20/1 kV transformers. It can be noticed that only in substation 1 the transformers can be put in parallel (substations 3 and 4 are the same as substation 2, only the powers of the transformers differ), respectively all the consumers can be fed at one moment only from one of the transformers. In the other three substations, at this time it cannot operate with the transformers in parallel, or so that all consumers are powered at a given moment only by one of the transformers.

The analysis started with the data recorded at the measuring point. The values recorded in the past three and a half months have been considered. Thus, for the first month, the consumption was according to figure 3. For these consumes of energy, the values of the power factor are in figure 4.

It can be seen that all values of the power factor are less than 0.92, except for a day when the factory was stopped for maintenance work. This stop is usually done once a year. In the next figures, the same consumption is shown for the other two and a half months.

The data of the electricity supplier only provides information on the general electricity consumption. They are sufficient only if the power factor compensation at the measuring point is achieved. In order to carry out more in-depth analysis, and possibly, propose other solutions, further measurements are required. So, the next step consisted in performing the measurements and records in the substations of the consumer.

IOP Conf. Series: Materials Science and Engineering **749** (2020) 012029 doi:10.1088/1757-899X/749/1/012029

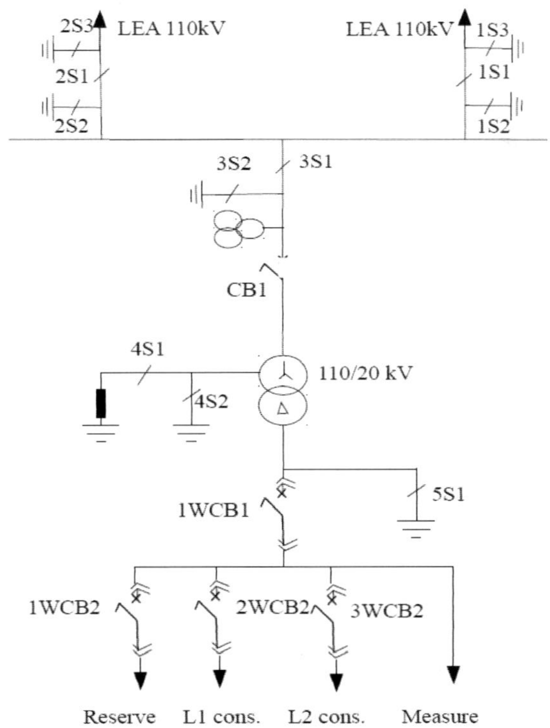

Figure 1. Electrical diagram of the 110/20 kV power station

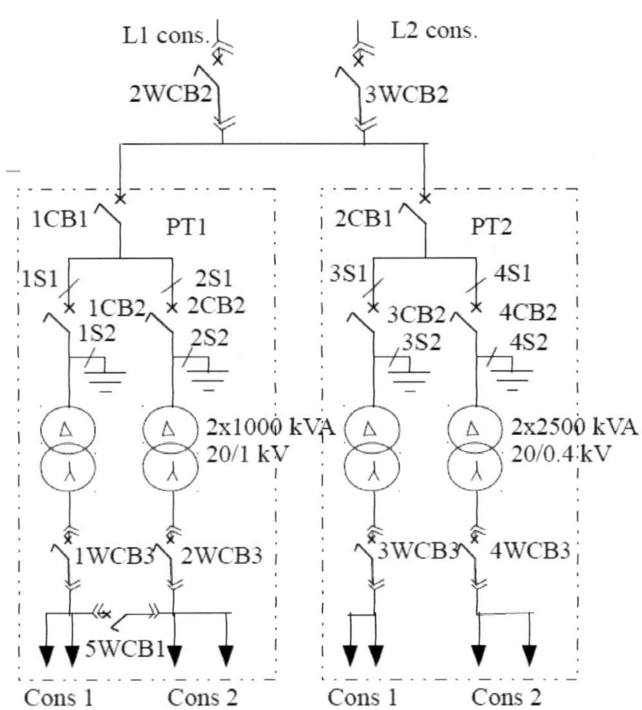

Figure 2. Electrical diagram of substations 1 (left) and 2 (right)

IOP Conf. Series: Materials Science and Engineering **749** (2020) 012029 doi:10.1088/1757-899X/749/1/012029

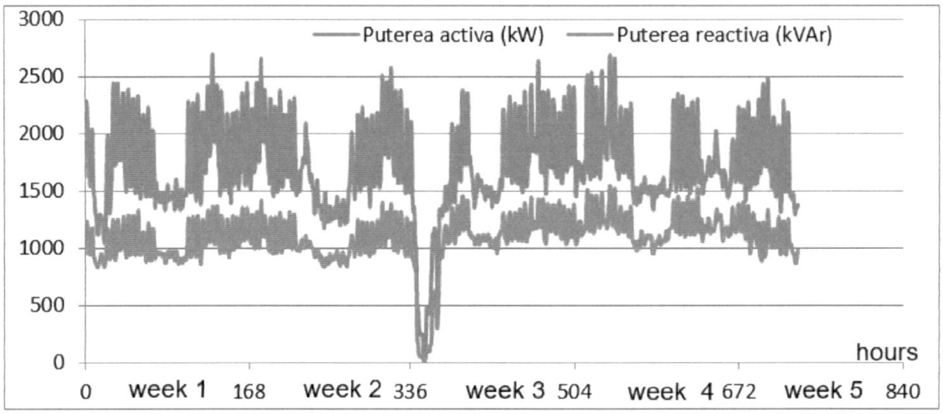

Figure 3. Active (kW) and reactive (kVAr) power (month 1)

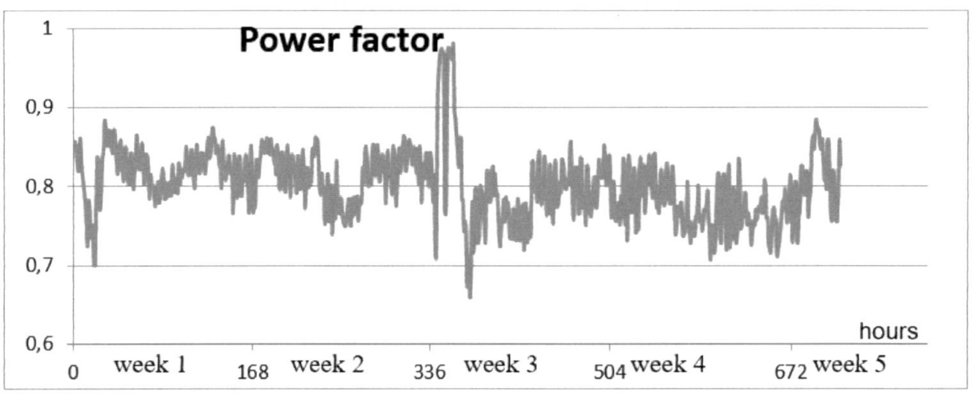

Figure 4. Power factor (month 1)

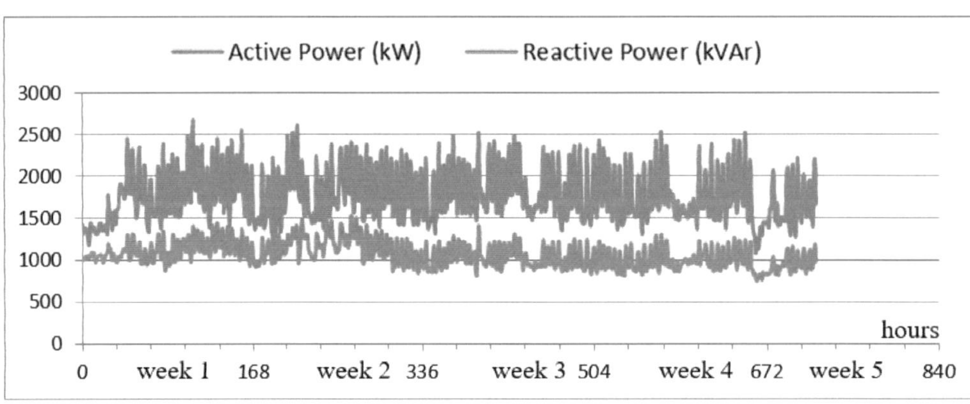

Figure 5. Active (kW) and reactive (kVAr) power (month 2)

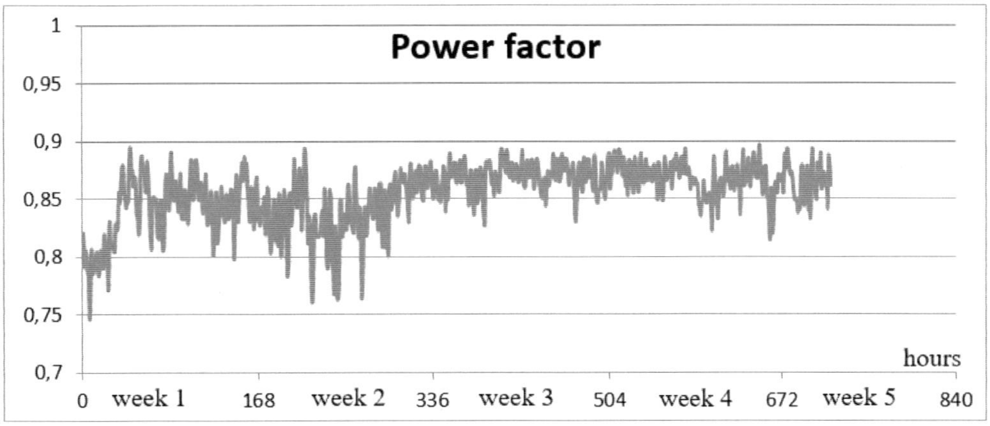

Figure 6. Power factor (month 2)

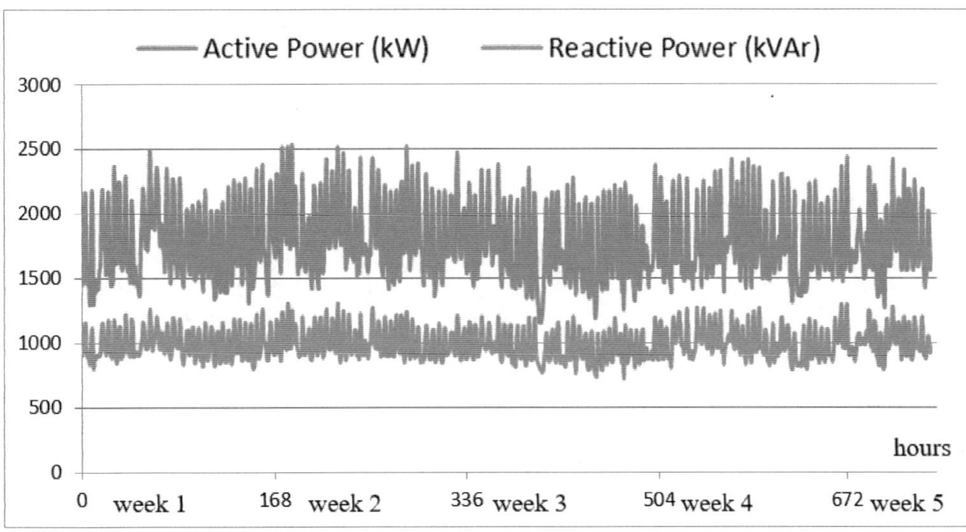

Figure 7. Active (kW) and reactive (kVAr) power (month 3)

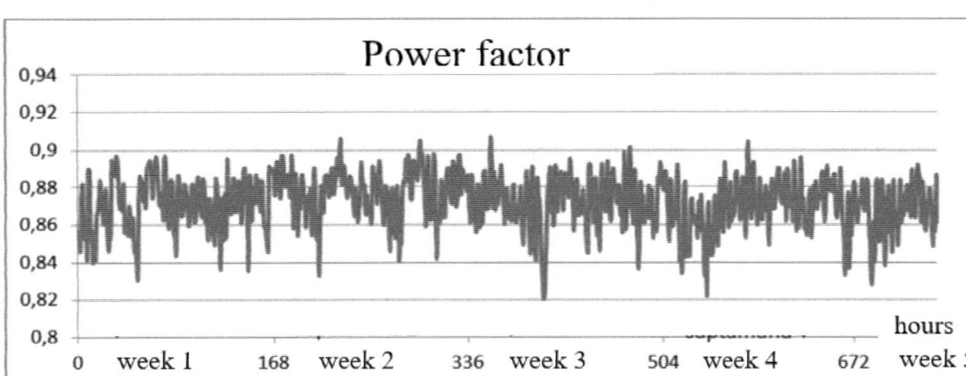

Figure 8. Power factor (month 3)

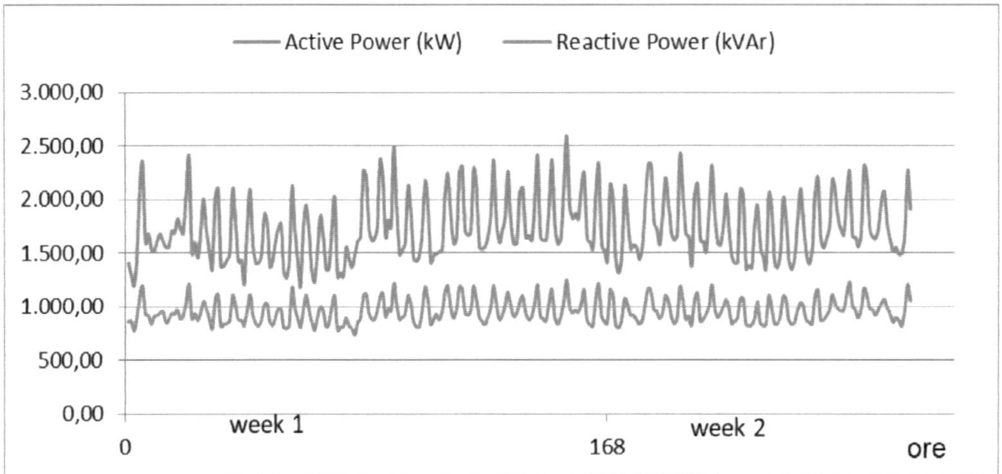

Figure 9. Active (kW) and reactive (kVAr) power (half a month)

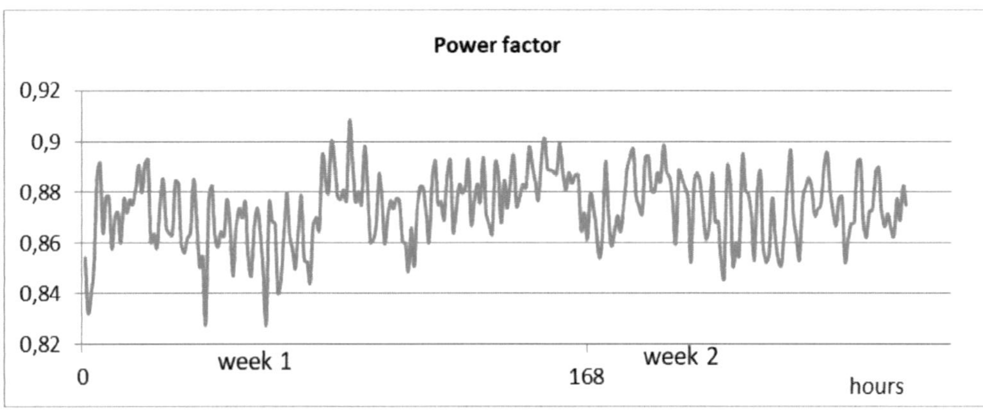

Figure 10. Power factor (half a month)

2.2. Measured and recorded data in the substations

The measurements and records were done with Fluke 435 power quality and energy analyzer, on the secondary bars of each transformer. Only in one of the substations (substation 4), the measurements were realized on different receiver groups because we did not have access on the secondary bars of the transformers.

Substation 1, transformer 1 (1000 kVA, 20/1 kV), registration time - 24 hours

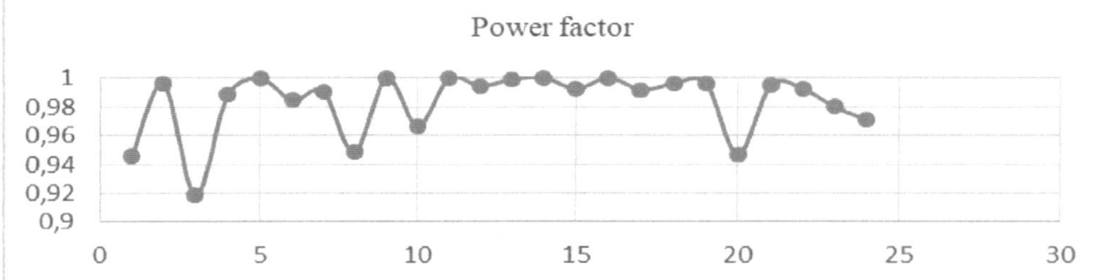

Figure 11. Power factor (Substation1, transformer 1)

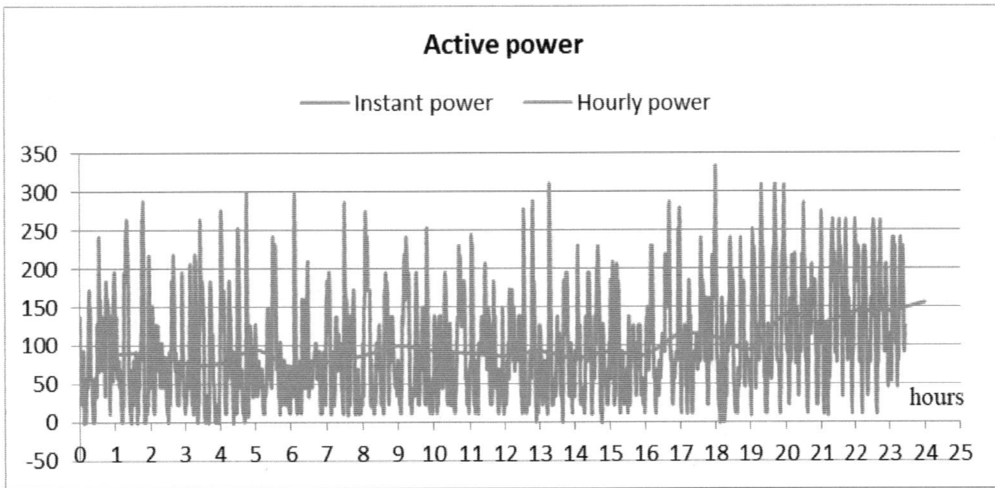

Figure 12. Active power (Substation 1, transformer 1)

Substation 1, transformer 2 (1000 kVA, 20/1 kV), registration time - 27 hours

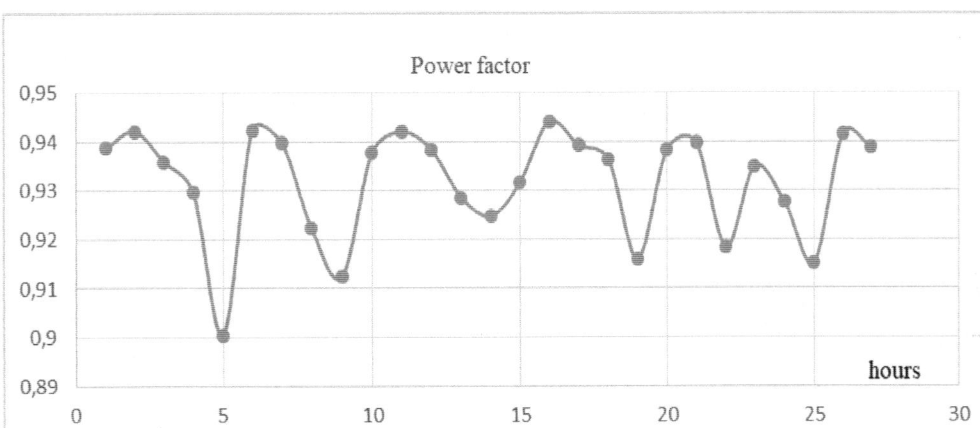

Figure 13. Power factor (Substation 1, transformer 2)

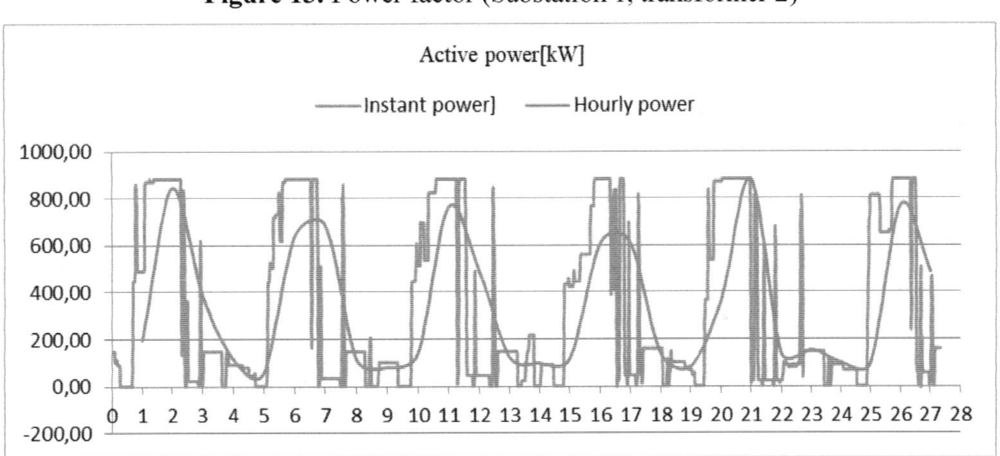

Figure 14. Active power (Substation 1, transformer 2)

IOP Conf. Series: Materials Science and Engineering **749** (2020) 012029 doi:10.1088/1757-899X/749/1/012029

Substation 2, transformer 1 (2500 kVA, 20/0.4 kV), registration time - 27 hours

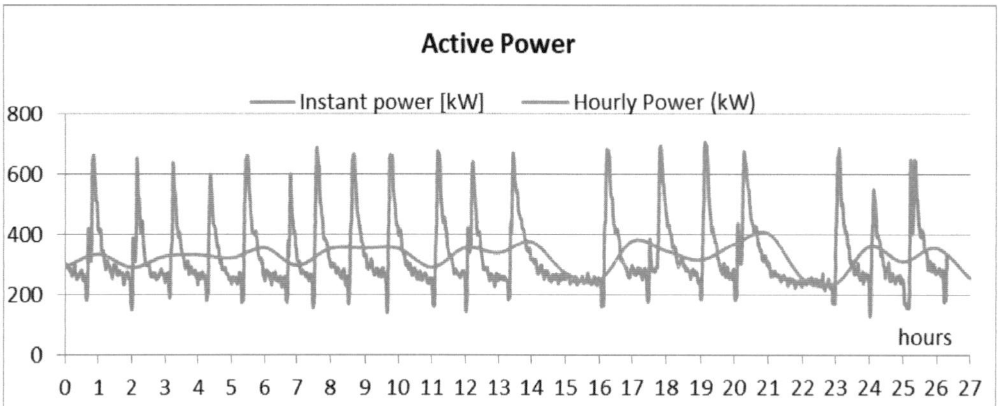

Figure 15. Active power (Substation 2, transformer 1)

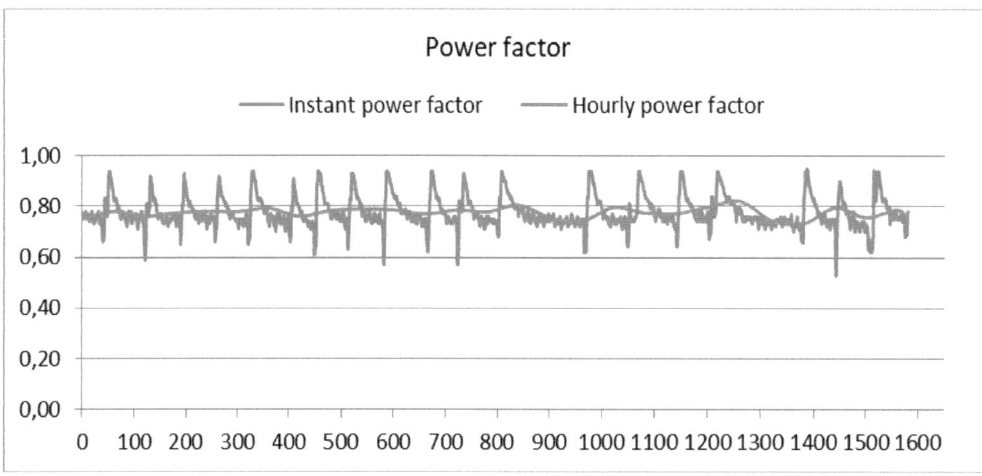

Figure 16. Power factor (Substation 2, transformer 1)

Substation 2, transformer 2 (identical with transformer 1), registration time – 32 hours

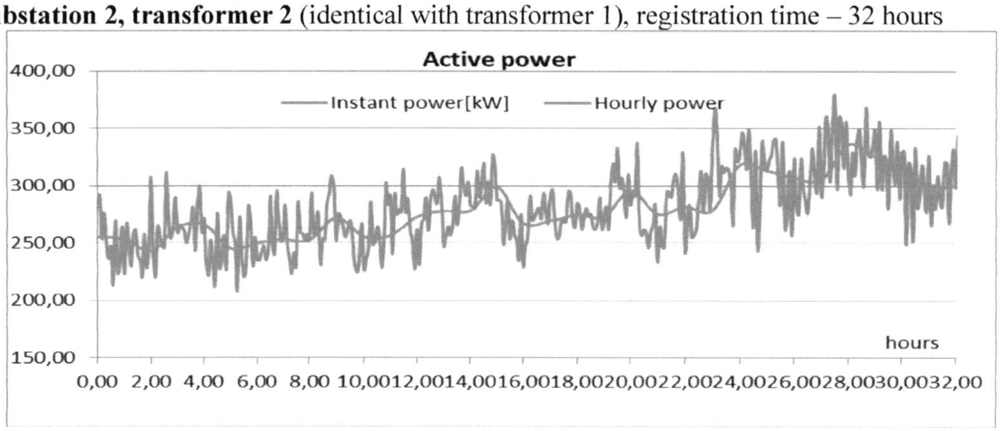

Figure 17. Active power (Substation 2, transformer 2)

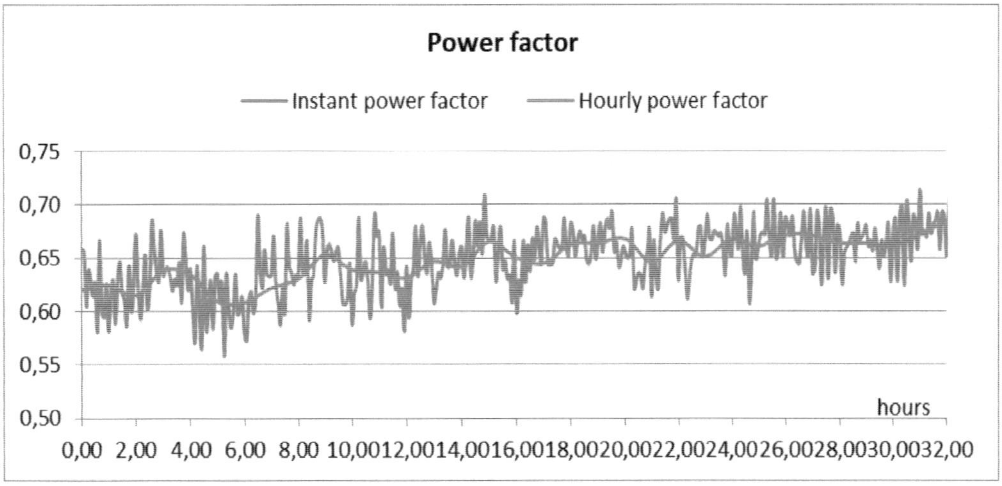

Figure 18. Power factor (Substation 2, transformer 2)

Substation 3, transformer 1 (both transformers are like in the substation 2), registration time – 23 hours

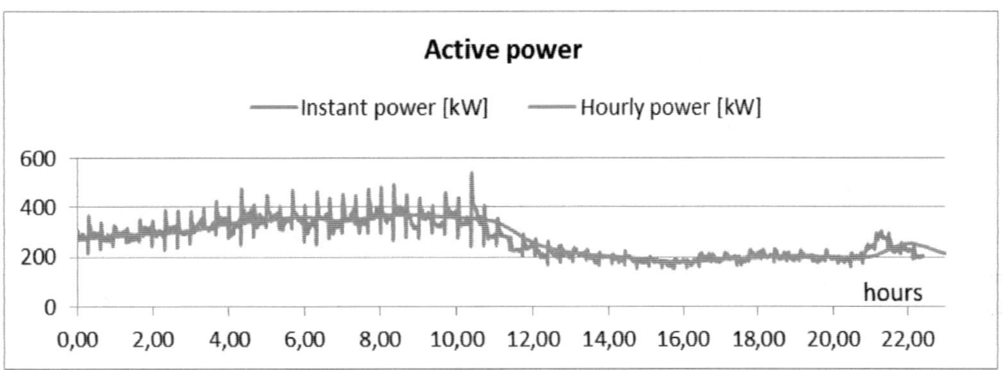

Figure 19. Active power (Substation 3, transformer 1)

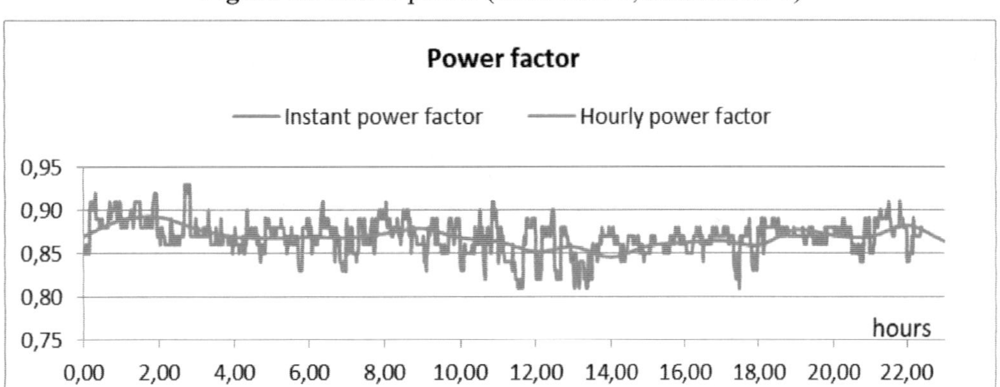

Figure 20. Power factor (Substation 3, transformer 1)

Substation 3, transformer 2, registration time – 28 hours

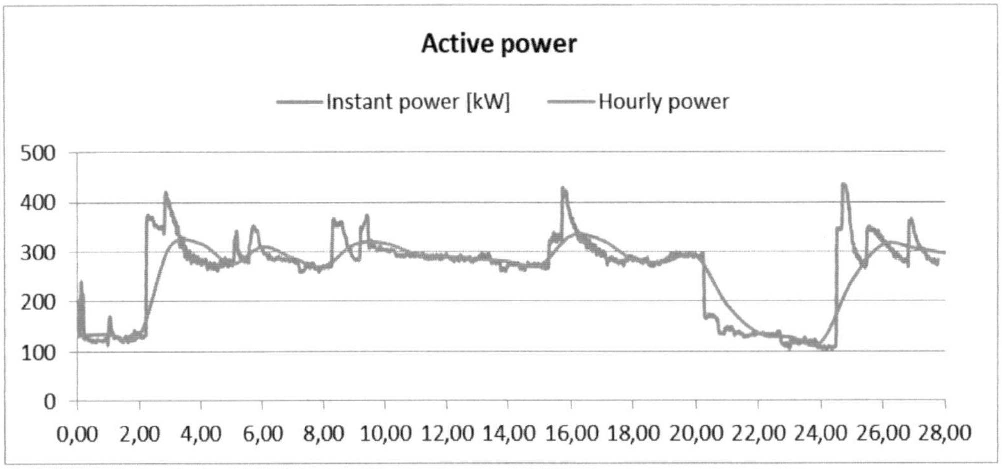

Figure 21. Active power (Substation 3, transformer 2)

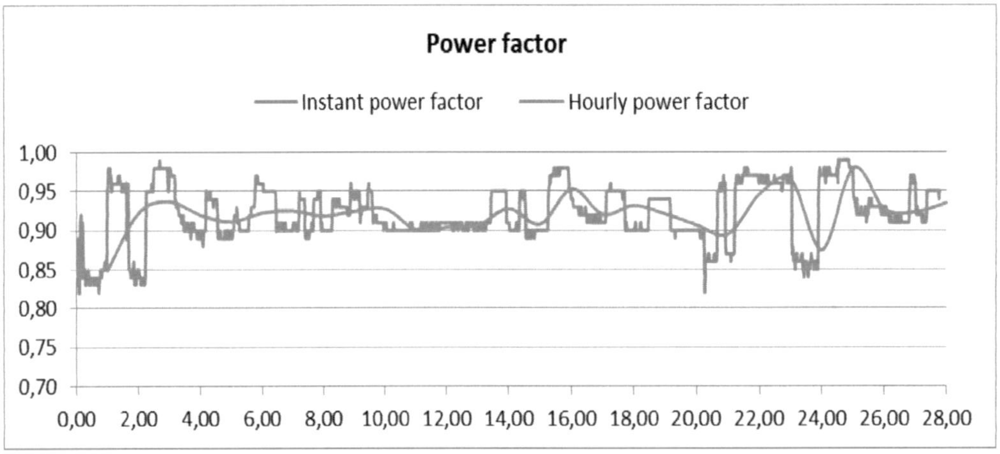

Figure 22. Power factor (Substation 3, transformer 2)

Substation 4 (two transformers 1600 kVA)

Making measurements in this substation was difficult because there was no access to the secondary bars of the transformers. On the two transformers, which cannot be parallel operated, several consumers are grouped on each transformer. Measurements were made on the cables that feed six main consumer groups. Only on two of them recorded significant values: one a maximum of 140 kW but at a power factor greater than 0.92; on the other, a maximum of 135 kW, with a short duration, the power factor being around 0.9. The other four records showed insignificant power up to 35 kW.

3. Technical and organizational solutions

3.1. Substation 1

In this substation the power factor was relatively high (higher than 0.9) and there is no reason for investment in reactive power compensation. On the other hand, there is a rapid fluctuation of reactive energy from inductive to capacitive, which would complicate the type of the compensation system. It has to be taken into account that the capacitive energy can compensate the reactive energy of the other substations.

It is also important to state that in this substation, all consumers can be supplied from a single transformer, but the consumed power does not allow this (being periods when the total power exceeds the rated power of a single transformer). Consequently, based on the data recorded in this substation, no technical or organizational measures are required for the optimization of energy consumption.

3.2. Substation 2

The first transformer is periodically loaded with power from 200 kW to 700 kW, the average power is about 400 kW. The maximum apparent power was 750 KVA, about 30% of rated power. The power factor is around 0.8 at the secondary bars, but in primary, it is much lower because of the reduced load (smaller than 15%).

The other transformer is loaded with a maximum apparent power of 550 KVA (22% of the rated power). Of course, in this situation the power factor is low.

The best solution for this substation is to make a connection between the secondary bars of the two transformers, this will allow the operation with a single transformer at a given time. By connecting in parallel the two transformers (using two couplers) will be possible to carry out maintenance work at each transformer without interrupting the activity. On the other hand, operation with a single transformer will automatically increase the power factor in this substation.

3.3. Substation 3

Also, in this substation, the situation is just like the previous one. The two 2500 KVA transformers, each was loaded with a maximum apparent power of 600 KVA (24%) and 450 KVA (18%) respectively. It is necessary to make a coupling between the secondary bars of the two transformers, and not powering one of them. This is mandatory at this time considering also the other benefits.

3.4. Substation 4

Although in this substation the transformers are of lower power (1600 KVA) their load is very low, and the lack of the possibility to operate with a single transformer is an obvious disadvantage.

4. Conclusion

This paper examines how the electricity is consumed by a medium sized factory. The exposed situation highlights the special importance that should be accorded to the design of the electricity utilization installation. In the analyzed case, the lack of the possibility of putting all the consumers from a substation in parallel is the main misconception that involves a lot of disadvantages that can hardly be solved later on. Besides the fact that the safety of operation is low due to this design error, the compensation of the reactive energy is also hard to solve and is only achievable with additional costs.

In this paper, we presented only the solutions referring to the improvement of the existing equipment exploitation, without proposing static compensation solutions.

References

[1] Deaconu S I, Popa G N and Babau R M 2014 *Study, design and industrial implementation of capacitive power factor controller for large load fluctuations in steel industry* International Conference and Exposition on Electrical and Power Engineering (EPE) 2014 Iasi Romania October 16-18 pp.962-967

[2] Lei Wang L, Lao K W, Lam C S, Wong M C 2017 *Delta-connected static var compensator (SVC) based hybrid active power filter (SVC-HAPF) and its control method* 43rd Annual Conference of the IEEE Industrial Electronics Society IECON 2017 Beijing China Oct. 29-Nov. 1 pp.6384-6388

[3] Tusaliu P, Ciontu M and Tusaliu V 1996 *Disturbances at the commutation of the condensers batteries as part of the national Electrical Energetic System* Proceedings of 8th Mediterranean Electrotechnical Conference on Industrial Applications in Power Systems

Computer Science and Telecommunications (MELECON 96) Bari Italy May 16 pp.1594-1597

[4] Mateo V, Gole A M and Ho C N M 2017 *Design and implementation of laboratory scale static var compensator to demonstrate dynamic load balancing and power factor correction* 2017 IEEE Electrical Power and Energy Conference (EPEC) Saskatoon SK Canada Oct. 22-25 pp.1-6

[5] Bui V, Truong D and Ho D 2017 *Improve power oscillation stability in a grid connected wind power system by using a Static Var Compensator* International Conference on System Science and Engineering (ICSSE) Ho Chi Minh City Vietnam,July 21-23 2017 pp.200-203

[6] Wysocki W and Szlosek M 2011 *Compensation of reactive power as a method for reducing energy losses: On the example of calculations and measurements of load flow through the distribution transformer in one of the polish distribution network* 11th International Conference on Electrical Power Quality and Utilisation Lisbon, Portugal Oct. 17-19 2011 pp.1-5

[7] Vlasenko R, Bialobrzheskyi O and Gladyr A 2018 *Analysis of three-phase parallel active power filter operation mode with arc furnace active power's fluctuations* 2018 IEEE 3rd International Conference on Intelligent Energy and Power Systems (IEPS) Kharkiv Ukraine Sept. 10-14 pp. 80-84

IOP Conf. Series: Materials Science and Engineering **749** (2020) 012030 doi:10.1088/1757-899X/749/1/012030

Concept for a precise academic Gripper

S Epple[1a], R Jung[1], C Barz[2] and V Nasui[3]

[1] University of Applied Sciences Kempten, Research Centre, Kaufbeuren, Germany
[2] Technical University of Cluj-Napoca, Faculty of Engineering, Department of Electrics, Electronics and Computers Engineering, Baia Mare, Romania
[3] Technical University of Cluj-Napoca, Faculty of Engineering, Department of Engineering and Technology Management, Baia Mare, Romania

E-mail: [a] steffen.epple@hs-kempten.de

Abstract. A new, precise, scalable multi-purpose gripper system for academic research purposes is proposed. The multi-purpose gripper is intended to be easily adapted to different tasks and object sizes. The gripper system is driven by two separate motor-gear-lead spindle units, operated by a small CNC control. Software running on the CNC unit is completely open source. A novelty for a gripper in this constellation is an interlock-line that may for example be used for a safety door, allowing to stop movement of the gripper fingers and thus being capable to avoid the bruise of an operator's finger or arm part. This paper describes the steps mechanical parts were designed and simulated to allow a lightweight concept leaving some payload even for relatively small industrial robots used in academic research. Stability of the gripper fingers was tested by pressing the gripper finger towards each other by hand while "blocking" the movement with a finger showing no visible bending effect of the gripper parts. To test the concept of the gripper system, a first prototype is under construction. Parts of the gripper system were simulated, 3D-printed with poly-lactic-acid (PLA) and mounted to a small laboratory robot. The CNC-gripping function will be furthermore optimized and tested at the prototype.

1. Introduction

Precise gripping of very different object sizes in different fields of automation becomes more and more essential. Existing parallel gripper designs allow either small objects to be gripped or large objects to be gripped. This situation leads to many time consuming tool changes, only to adapt to a different object size. When one piece manufacturing is intended, this could lead to time losses and as a result to a decrease of productivity [1].

Many designs still use air pressure activated mechanisms [13], which need a mechanical stop to be adjusted to the object size in order not to crush down objects which cannot withstand the full gripping force. Alternatively the air pressure is regulated. At low air pressures this can lead to the effect that the gripper barely starts moving due to the static friction of the air seals of the piston.

Common electric designs use one motor and allow current control to limit gripping forces. Position control is not available, which means that the gripper fingers move towards each other until the current limit of the motor is reached.

All these concepts have in common that they do not allow precise gripping over a wide range of different object sizes and weights.

2. Goals of the concept

Several goals were attempted to reach with this concept:

Main goal is to gain flexibility in usage and convenience in programming: The proposed concept allows moving each gripper finger independently and position controlled by using one CNC axis per gripper finger. This allows asymmetric gripping relative to the tool centre point as well as very small movements of the gripper fingers.

The second goal is to alleviate the drawback of existing electrically driven grippers to do always a full cycle. This is slow compared to pneumatic driven grippers.

In combination with precise positioning of the gripper, this concept allows short cycle times since it is not mandatory to do a full open-close-cycle with every grip. A partial opening, just to fit the object to grip between the gripper fingers, is the intended method of operation.

The third goal is to allow scalability: The working principle can be adapted to much smaller robots e.g. for performing small parts assembly tasks or to much larger robots, e.g. for palletizing cardboard boxes of very different sizes. The size of the presented prototype was chosen to allow usage by a 1.2kg payload robot.

The last goal is to allow easy reproducibility. This is achieved by using 3D printed parts, which can be fused on any 3D printer large enough to fit the parts onto its build plate, as well as using standard parts.

3. Working principle of the gripper mechanics

High resolution linear drives have been realized in many different varieties: Direct belt drives are used widely in 3D printers, rack and pinion drives are used in scanners and small laboratory robots.

In this project we decided to use small geared stepper motors in combination with a lead spindle and nut mechanism. This kind of mechanism theoretically allows to reach high positioning resolution (1,8 μm per half-step) at reasonable force and low weight.

The gripper fingers are mounted on a single miniature rail (Type: MGN9).

The connecting parts between the spindle nut and the slide bar (MGN9H) of each gripper finger are specially designed and simulated parts which were also 3D printed.

The gripper fingers and connecting parts are screwed on top of each other onto the slide bars.

4. Conceptual approach

Figure 1 shows the basic mechanical concept of the gripper. At the time the pictures were taken, the linking parts were not mounted. In the outlook, a picture of the complete gripper system is shown.

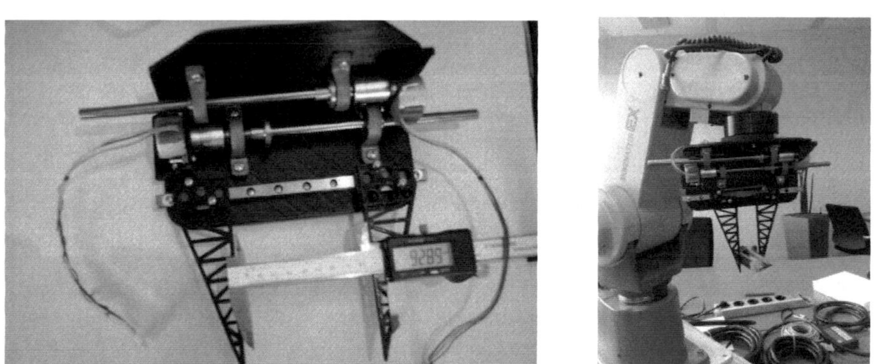

Figure 1. Basic mechanical concept of the prototype

5. Safety analysis

Regarding the construction of the gripper, the hazard of crushing the experimenter's fingers is very unlikely since expected forces are worst case limited by the rigidity of the rail. In the unlikely case of

IOP Conf. Series: Materials Science and Engineering **749** (2020) 012030 doi:10.1088/1757-899X/749/1/012030

the motors not stalling and the force at the gripper finger getting too high, the rail will bend and render itself non-hazardous. In this case, no actions need be taken.

Although the gripper baseplate and motor brackets are 3D printed, sharp edges would lead easily to injuries (scratches) of the experimenter's hands and forearms. As an avoiding measure of safety, edges of 3D printed parts were rounded wherever applicable.

6. The control

One of the major disadvantages of electronics is obsolescence and components are hard to replace.

The big advantage of the small stepper motors used is that they do not need large stepper amplifier stages. Combining the advantage with the disadvantage leads to an available and optimized solution: The Arduino CNC shield [3] with an Arduino Uno and a set of A4988 driver boards combined with the GRBL 0.9 software [2] is a readily available and a cost efficient solution for tiny CNC controls.

GRBL is widely in use by many non-commercial CNC mills and in a modified form also used as motion-control backbone of open source firmware for popular 3D printers.

The CNC G-Code control commands used for the gripper are M0 (move, fast approach position), M1 (move with working speed to position) and G92 (set axis to zero).

For controlled stop of the active cycle the GRBL specific feed hold command "!" and for continuing the cycle start/resume "~"-commands are used.

For setup, the $0 and $1 commands are necessary to define the x and y steps/mm.

7. Simulations

Before fusing the parts on a 3D printer, a simulation was performed to see if these parts can withstand the maximum foreseeable forces. For each part several optimizations steps including simulation, 3D printing and system integration tests were performed. All drawings and simulations were carried out using FreeCAD [5] and its integrated FEM workbench [6].

Figure 2. Van-Mises-stress simulation for the motor bracket

Figure 2 shows the van-Mises-stress result for a simulated force of 50N at each of the two small mounting holes of the flange (upper part of figure 2). The aim is to detect weak points under the assumption of a torque higher than the motors can exert. The forces around these holes are less critical than the forces at the mounting holes (coloured red in lower part of Figure 2) of the bracket to the baseplate. The inner surfaces of these two holes define fixed boundary values for the stress simulations.

Figure 3 shows the van-Mises stress simulation results for an assumed force of 100N at the outermost tip of a gripper finger. "Fixed points" boundary conditions were the inner surfaces of the four small mounting holes in oblong order in the upper area of the part. Simulated Material is PLA, using the properties provided by FreeCAD FEM workbench [6].

Figure 3. Van-Mises-stress simulation for a gripper finger

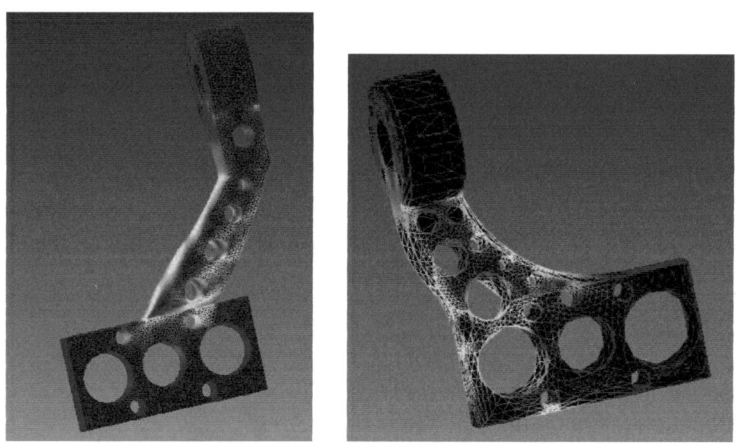

Figure 4. Van-Mises stress simulation results for both linking parts

Figure 4 shows the van-Mises stress simulation results for the linking parts between the brass nut and the gripper fingers. Gripper finger and linking part are screwed on top of each other onto the sliding part of the linear rail. The inside surfaces are assumed “fixed” for the simulation. A force of 100N is assumed as boundary condition to act in spindle direction to the side surface, where the brass nut is screwed to the linking parts.

8. Experimental results

Different experiments have been conducted to see if the mechanics work as intended:

Moving one gripper finger in small steps and measuring the distance of the move was compared to the expected distance. It turned out that the number of steps/mm calculated did not fit exactly. This was expected and corrected.

The mechanical resolution of the movement of the sliding brackets expected is 1.8μm/half step. The experimentally obtained resolution is 1.93μm in average over a 45mm move. This equals 518.067 half steps/mm instead of calculated 512 half steps/mm. [12].

It was observed, that near the end of the spindle the same number of steps resulted in a shorter movement than in the middle of the same spindle. It is not yet clear whether the spindle is the source for this inaccuracy or whether mechanical friction and errors in parallelism between guide rail and spindle is responsible for the observed effect. This point needs further investigation, although it is a common scale error effect at the extremes of travel [11].

The measured backlash when changing the direction is approximately 0.5mm.

Depending on the gripping force, the springiness observed is in the range of ±0.3mm.

Moving one gripper finger to the other until the motor stalls requires a backward movement of 2.5mm to take out the pressure between the gripper fingers. For a maximum gripping force the gripper fingers can be closed approximately 2.5mm smaller than the size of the object without losing steps.

Moving both gripper fingers with a gripped object in one direction (off-centre) works as expected. The stepper motors move perfectly synchronous, the observed difference of the "gap" between the fingers is approximately 0.02mm, which is the range of the measurement resolution of the digital calliper gauge. The off-centering capability depends on the initial gripper opening width. Each gripper finger can be moved for approximately 25mm over the tool centre point and 45mm in opening direction.

This prototype is able to grip object sizes in the range from 0mm to 90mm.

9. Conclusions

- The expected positioning resolution is within expected range and tolerance.
- The absolute positioning exactness depends on the adjustment of the zero point. The nonlinearity of the linear motion in this use case is within the acceptable tolerance and does not significantly influence the functionality of the gripper.
- Grippers can be built with 3d printed components up to certain rigidity. Therefore a simple but rigid fused filament printer is sufficient, but it must be adjusted very precisely.
- The springiness deriving from the 3D printed parts in contrast to the use of classical metal parts is beneficial for the function in two ways: firstly it allows larger tolerances of the structure and secondly it enhances the function of gripping. With this springiness the "force steps" introduced by stepping motion are smoothed. This is very beneficial when handling sensitive parts, e.g. brittle parts or partially assembled conglomerates of parts.

10. Outlook

This gripper prototype represents a work in progress. The next steps in research are:

- Integrating the cameras into the gripper control.
- The introduction of functional safety [4] to avoid gripping human body parts, especially fingers by using the on-gripper-camera(s) and observing the picture(s) for skin-color as described in [7].
- Camera controlled slip detection and appropriate counter action, not using extra sensors as e.g. in [8].
- The automation of the gripping process by using cameras for multidimensional re-positioning of the gripper relative to an object, using a different approach than [10].
- Supporting assembly tasks by using the on-gripper-cameras as shown in figure 5. In this figure cameras and gripper are slightly twisted to the viewer for better visibility. In "working position" the cameras' view is perpendicular to the moving direction of the gripper fingers and the gripper with attached camera beam will be mounted to a robot.
- The bending effect resulting from the springiness of the gripper fingers could be used for force measurement when calibrated and evaluated from the camera picture. Since springiness depends on the construction, an optimized gripper finger could be specifically designed for a task where this kind of measurement might be needed. E.g. to measure counter-reactions of parts when they get gripped and squeezed might be a possible quality control application.
- Source of the observed nonlinearity of the spindle drive must be investigated.

IOP Conf. Series: Materials Science and Engineering **749** (2020) 012030 doi:10.1088/1757-899X/749/1/012030

Figure 5. Principle setup for further research. Source for red and green cubes [9]

References

[1] Jalba C K et. al 2016 *Optical Gripper* DOI 10.1088/1757-899X/200/1/012054
[2] Sonny J 2017 grbl [online][accessed: 2018-05-30] https://github.com/grbl/grbl/wiki
[3] Kruger B 2015 *Arduino CNC Shield*-100% GRBL Compatable [online] [Accessed: 2018-05-30] https://blog.protoneer.co.nz/arduino-cnc-shield/
[4] Standard ISO 13849-1:2015 *Safety of machinery-safety related parts of control systems*
[5] Unknown author *FreeCAD-Welcome!* [online] [Accessed: 2018-05-30] https://www.freecadweb.org/
[6] Unknown author *FEM Module* 2018-02-01 [online] [Accessed: 2018-05-30] https://www.freecadweb.org/wiki/index.php?title=FEM_Module
[7] Khamar S et. al. 2015 *Comparative Study of Skin Color Detection and Segmentation in HSV and YCbCr Color Space*, DOI 10.1016/j.procs.2015.07.362
[8] Ion I et. al. 2015 *Intelligent Sensing and Motion Control in Robotic Manufacturing Systems*, Applied Mechanics and Materials DOI 10.4028/www.scientific.net/AMM.762.299
[9] iDig3Dprinting: XYZ 20mm Calibration Cube, Published on January 19 2016 www.thingiverse.com/thing:1278865
[10] Comsa et. al. 2012 *An Overview of Library Automation* DOI 10.4028 www.scientific.net/AMM.162.583
[11] Miller J E et. al. 2016 *Improved machine tool linear axis calibration through continuous motion data capture* DOI 10.1016/j.precisioneng.2016.08.010
[12] Bell C 2014 *Maintaining and troubleshooting your 3D printer* ISBN 9781430268093 DOI 10.1007/978-1-4302-6808-6
[13] https://schunk.com/de_en/gripping-systems/category/gripping-systems/schunk-grippers/parallel-gripper/

IOP Conf. Series: Materials Science and Engineering **749** (2020) 012031 doi:10.1088/1757-899X/749/1/012031

The Use of S Parameters in Two-Port Analogue Networks Stability Analysis

Z Erdei[1,a], M Iordache[2], D Niculae[2], M Stanculescu[2] and A Grib[2]

[1] Technical University of Cluj-Napoca, Faculty of Engineering, Department of Electrics, Electronics and Computers Engineering, Baia Mare, Romania
[2] Politehnica University of Bucharest, Department of Electrical Engineering, Bucharest, Romania

E-mail: [a] zoltan.erdei@cunbm.utcluj.ro

Abstract. If electrical systems are considered to consist of two-port linear analogue circuits (networks) in harmonic state, then $\underline{S}$ parameters can be used to establish the stable functioning states corresponding to such systems. This paper presents, on one hand, the practical use of $\underline{S}$ parameters to increase the efficiency of information transmission and propagation and of electromagnetic energy wireless transfer from the emitter-receiver point of view. On the other hand, the active power efficiency transfer from the wireless system input to the loads connected at its output, it's taken into account. Starting from the correct formulation of $\underline{S}$ parameters, their generation is done automatically, based on modified nodal equations corresponding to the power transfer wireless systems. The magnetically coupled coils, used for wireless power transfer can be considered as a two-port circuit. To estimate the parameters of the magnetically coupled coils, advanced programs for electromagnetic field numerical computation is used: Cadence, ADS, Ansoft Extractor Q3D, COMSOL, Feko etc. In this paper there are also defined several parameters, function of $\underline{S}$ parameters, a series of constants which are further used to compute the stability factor. To generate: the reflection coefficient, the stationary waves ratio (SWR), the input and output impedances, the input and output active power transmission efficiencies, the power gains, various stability coefficients, the centres and the radius of the stability circles, the $\underline{\boldsymbol{S}}$, $\underline{\boldsymbol{T}}$, $\underline{\boldsymbol{Z}}$ and $\boldsymbol{Y}$ matrices and the Smith diagrams, dedicated functions have been designed and implemented and the Microwave toolbox from MATLAB have been used. By comparing the simulated results with the experimental ones and with those presented in literature, the obtained error is acceptable.

1. Introduction

The characterization of linear or non-linear circuits which are functioning at low signals, in harmonic state, at small and medium frequency can be done by using the following parameters: Z, Y, H, T etc. When these circuits are operating at high and very high frequencies, these parameters can no longer be used, because they imply circuits with certain branches to be open or short-circuited such that the measurements to take place and such that the currents and the voltages from the circuit to be determined. At lower frequencies, the transfer and impedance matrices are commonly used, but at microwave frequencies they become difficult to measure and therefore, the scattering matrix description is preferred. The S parameters can be measured by embedding the two-port network (the Device Under Test (DUT)) in a transmission line whose ends are connected to a network analyser. A typical network analyser can measure S-parameters over a large frequency range, for example, the HP 8720D vector network analyser (VNA) covers the range from 50 MHz to 40 GHz. Frequency resolution is typically 1 Hz and the results can be displayed either on a Smith chart or as a conventional gain versus frequency graph, [1-5].

The scattering (repartition) parameters-denoted by $\underline{S}$-are complex quantities, frequency dependant, associated to a two-port linear system, in harmonic state. At the beginning, $\underline{S}$ parameters have been used in electrical long line theory. To define them, the incident (direct) voltage waves, the reflected (inverse) or transmitted ones have been used. In general, $\underline{S}$ parameters can be defined in transmission information systems, such as the microwave systems (waveguides), where these parameters can be studied using the electric circuit theory [4-10]. Because there are a lot of possibilities to introduce these parameters [1, 10, 11], their physical interpretation and understanding can become difficult.

$\underline{S}$ parameters do not have a direct correspondent in the electric circuit theory, because they do not contain circuit elements in which the propagation exists as in the waveguides. Anyway, in the circuit theory reasoning similar to that for microwave circuits has been extended, by introducing the concept of *power waves*, terminology coming from their significance that is connected to the dependence between the active power absorbed by a load connected to a port and its working frequency [3, 6]. Knowledge and understanding of $\underline{S}$ parameters are especially important for those applications that operate at high frequencies and include active and passive components from the integrated circuits [2], including microelectromechanical systems (MEMS), as well as the wireless power transfer systems (WPTS) [12].

For automatic generation of $\underline{S}$ parameters, for different configurations and structures of the two-port circuits and of the wireless power transfer systems, one can use either the modified nodal equations generated by the Circuit Symbolic Analysis Program (CSAP), [1, 13, 14], or state equations generated by Symbolic State Equation Generation Program (SYSEG), [13, 14]. Identifying the parameters of the analogue two-port circuits and the magnetically coupled coils used for wireless power transfer, for the frequencies between 30 kHz and 40 MHz, the most advanced and efficient electromagnetic field numerical computation programs are used: Cadence [15], ADS [16], ANSYS-Ansoft Extractor Q3D [17], Feko [2, 18] etc. In general, the two-port linear analogue circuits in harmonic state and the magnetically coupled coils can be described by a number of equivalent circuit parameters, such as: $\underline{\boldsymbol{T}}$-the transfer coefficients matrix, $\underline{\boldsymbol{Z}}$-the impedance matrix, $\underline{\boldsymbol{Y}}$-the admittances matrix and the matrix of $\underline{\boldsymbol{S}}$ parameters. Literature [4, 13, 14] presents procedures that allows the transition from one matrix to another.

This paper, based on electrical circuit theory, presents: starting from the correct formulation of S parameters [1], the automatic generation of matrix S and of the scattering parameters and then, based on this matrix, the computation of the following matrices: Z, Y and T. Furthermore, for a passive, liner, two-port circuit, in harmonic state, there are being determined: the input and output complex impedances (Z_{in} and Z_{out}), the reflection coefficients (from the generator $\underline{\Gamma}_G$, from the load $\underline{\Gamma}_L$, from the input $\underline{\Gamma}_{in}$ and from the output $\underline{\Gamma}_{out}$), the voltage transfer function. The relationships between these quantities and S parameters are deduced. It is presented the practical use of these parameters to increase the efficiency of information transmission and propagation and of electromagnetic energy wireless transfer from the emitter-receiver point of view. Also, the active power efficiency transfer from the wireless system input to the loads connected at its output, it's taken into account. Based on the maximum active power transfer theorem, it is computed the maximum active power transmitted to by a two-port system to the load and, function of S parameters, the most used in practice performance parameters are determined: the transducer power gain (G_t) the available power gain (G_a), the power gain (G_p, also called the operating gain), the maximum available power gain and the maximum stable power gain [5-9].

To generate: the reflection coefficient, the stationary waves ratio (SWR), the input and output impedances, the input and output active power transmission efficiencies, the power gains, various stability coefficients, the centres and the radius of the stability circles, the $\underline{\boldsymbol{S}}$, $\underline{\boldsymbol{T}}$, $\underline{\boldsymbol{Z}}$ and $\boldsymbol{Y}$ matrices and the Smith diagrams, dedicated functions have been designed and implemented and the Microwave toolbox from MATLAB have been uses. The results obtained by simulation were compared to those reported in the literature and with the experimental ones.

2. Wireless power transfer systems stability

For any passive linear two-pot network (figure 1), the scattering parameters can be used to study the stability conditions of electrical systems whose functioning in harmonic state can be described with a passive linear two-port system, such as, for example WPTS [5-9, 19-23].

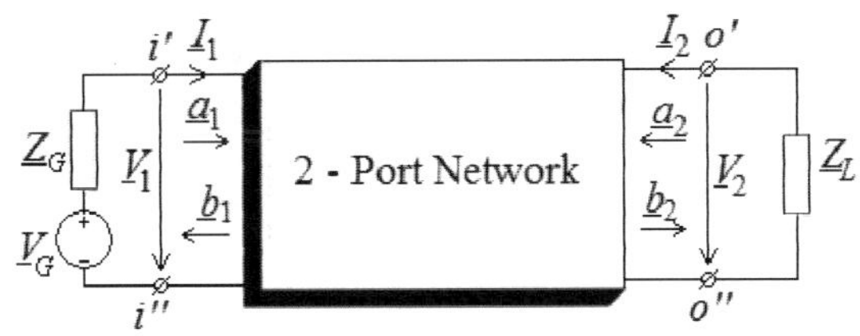

Figure 1. Passive linear two-port network connected to generator and to the load.

In the discussions concerning the stability conditions of a two-port network, function of S parameters, the following definitions and constants are used [10]:

$$\underline{\Delta} = det(\underline{S}) = \underline{S}_{11}\underline{S}_{22} - \underline{S}_{12}\underline{S}_{21}\ ;\ K = \frac{1-|\underline{S}_{11}|^2 - |\underline{S}_{22}|^2 + |\underline{\Delta}|^2}{2|\underline{S}_{12}\underline{S}_{21}|} \text{ – Rollett stability factor;}$$

$$\begin{cases} \mu_1 = \dfrac{1-|\underline{S}_{11}|^2}{|\underline{S}_{22} - \underline{\Delta}_{11}\underline{S}_{11}^*| + |\underline{S}_{12}\underline{S}_{21}|} \\ \mu_2 = \dfrac{1-|\underline{S}_{22}|^2}{|\underline{S}_{11} - \underline{\Delta}_{11}\underline{S}_{22}^*| + |\underline{S}_{12}\underline{S}_{21}|} \end{cases} \Rightarrow \text{Edwards - Sinsky stability parameters ;} \quad (1)$$

$$B_1 = 1 + |\underline{S}_{11}|^2 - |\underline{S}_{22}|^2 - |\underline{\Delta}|^2;\ B_2 = 1 + |\underline{S}_{22}|^2 - |\underline{S}_{11}|^2 - |\underline{\Delta}|^2;$$

$$\underline{C}_1 = \underline{S}_{11} - \underline{\Delta}\underline{S}_{22}^*; D_1 = |\underline{S}_{11}|^2 - |\underline{\Delta}|^2;\ \underline{C}_2 = \underline{S}_{22} - \underline{\Delta}\underline{S}_{11}^*; D_2 = |\underline{S}_{22}|^2 - |\underline{\Delta}|^2.$$

Between the constants from (1), the following identities can be simply deduced:

$$\begin{aligned} & B_1^2 - 4|\underline{C}_1|^2 - B_2^2 - 4|\underline{C}_2|^2 = 4|\underline{S}_{12}\underline{S}_{21}|^2(K^2-1) \\ & |\underline{C}_1|^2 = |\underline{S}_{12}\underline{S}_{21}|^2 + \left(1 - |\underline{S}_{22}|^2\right)D_2 \\ & |\underline{C}_2|^2 = |\underline{S}_{12}\underline{S}_{21}|^2 + \left(1 - |\underline{S}_{11}|^2\right)D_1 \end{aligned} \quad (2)$$

The reflection coefficient from the generator $\underline{\Gamma}_G$ has the expression:

$$\underline{\Gamma}_G \stackrel{d}{=} \frac{\underline{a}_1}{\underline{b}_1}\Bigg|_{\substack{\underline{V}_G = 0, \underline{V}_G = -\underline{Z}_G \underline{I}_1 \\ \underline{V}_L = -\underline{Z}_L \underline{I}_2}} = \frac{\underline{Z}_G - Z_0}{\underline{Z}_G + Z_0} \Rightarrow \underline{Z}_G = Z_0 \frac{1 + \underline{\Gamma}_G}{1 - \underline{\Gamma}_G}, \quad (3)$$

where Z_0 is the real-valued positive reference impedance. In practice, the reference impedance is chosen to be $Z_0 = 50$ ohm.

IOP Conf. Series: Materials Science and Engineering **749** (2020) 012031 doi:10.1088/1757-899X/749/1/012031

The reflection coefficient from the load can be defined as:

$$\underline{\Gamma}_L \overset{d}{=} \left.\frac{\underline{a}_2}{\underline{b}_2}\right|_{\substack{\underline{V}_G=0, \underline{V}_G=-\underline{Z}_G \underline{I}_1 \\ \underline{V}_L=-\underline{Z}_L \underline{I}_2}} = \frac{\underline{Z}_L - Z_0}{\underline{Z}_L + Z_0} \Rightarrow \underline{Z}_L = Z_0 \frac{1+\underline{\Gamma}_L}{1-\underline{\Gamma}_L}, \tag{4}$$

The centres and the radius of the stability circles of the source (generator) and of the load, have the following expressions:

$$\underline{c}_G = \frac{\underline{C}_1^*}{D_1};\ r_G = \frac{|\underline{S}_{12}\underline{S}_{21}|}{|D_1|} \tag{5}$$

for the source stability circle,

$$\underline{c}_L = \frac{\underline{C}_2^*}{D_2};\ r_L = \frac{|\underline{S}_{12}\underline{S}_{21}|}{|D_2|} \tag{6}$$

for the load stability circle.

As a consequence of the last two equations (2) and of the definition relationships (5) and (6), the quantities c_G, r_G, c_L and r_L satisfy the following relations:

$$1-|\underline{S}_{11}|^2 = \left(|\underline{c}_L|^2 - r_L^2\right)D_2;\ 1-|\underline{S}_{22}|^2 = \left(|\underline{c}_G|^2 - r_G^2\right)D_1 \tag{7}$$

By using the equations given in (7), the stability parameters μ1 and μ2 can be computed with:

$$\mu_1 = \left(|\underline{c}_L| - r_L\right) sign(D_2);\ \mu_2 = \left(|\underline{c}_G| - r_G\right) sign(D_1) \tag{8}$$

For example, the stability coefficient μ1 has the following expression:

$$\mu_1 = \frac{1-|\underline{S}_{11}|^2}{|\underline{C}_2| + |\underline{S}_{12}\underline{S}_{21}|} = \frac{D_2\left(|\underline{c}_L|^2 - r_L^2\right)}{|D_2|\left(|\underline{c}_L| + r_L\right)} = \frac{D_2}{|D_2|}\left(|\underline{c}_L| - r_L\right) \tag{9}$$

One can easily prove that, to compute the input and output reflection coefficients, one can use the also the following formulas:

$$\begin{aligned} \underline{\Gamma}_{in} &= \underline{S}_{11} + \frac{\underline{S}_{12}\underline{S}_{21}\underline{\Gamma}_L}{1-\underline{S}_{22}\underline{\Gamma}_L} = \frac{\underline{S}_{11} - \Delta\underline{\Gamma}_L}{1-\underline{S}_{22}\underline{\Gamma}_L}; \\ \underline{\Gamma}_{out} &= \underline{S}_{22} + \frac{\underline{S}_{12}\underline{S}_{21}\underline{\Gamma}_G}{1-\underline{S}_{11}\underline{\Gamma}_G} = \frac{\underline{S}_{22} - \Delta\underline{\Gamma}_G}{1-\underline{S}_{11}\underline{\Gamma}_G}. \end{aligned} \tag{10}$$

A two-port network is unconditionally stable, if any positive impedance of the generator and of the load with resistive (real) parts R_G and R_L always determine positive input and output impedances with resistive (real) parts R_{in} and Rout.

On the other hand, the unconditional stability requires that for any generator and for any load with $|\underline{\Gamma}_L| < 1$ and $|\underline{\Gamma}_G| < 1$, it results that $|\underline{\Gamma}_{in}| < 1$ and $|\underline{\Gamma}_{out}| < 1$. A two-port network is potentially or conditionally unstable if for $|\underline{\Gamma}_L| < 1$ and $|\underline{\Gamma}_G| < 1$, it results that $|\underline{\Gamma}_{in}| \geq 1$ and/or $|\underline{\Gamma}_{out}| \geq 1$ [15-28].

The load stability region represents the set of all reflection coefficients $\underline{\Gamma}_L$ for which it results that $|\underline{\Gamma}_{in}| < 1$. The generator stability region is the set of all reflection coefficients $\underline{\Gamma}_G$ for which $|\underline{\Gamma}_{out}| < 1$.

IOP Conf. Series: Materials Science and Engineering **749** (2020) 012031 doi:10.1088/1757-899X/749/1/012031

For the unconditionally stable case, the stability regions of the generator (source) and of the load contain the interiors of the unit circles with $|\underline{\Gamma}_L| < 1$ or $|\underline{\Gamma}_G| < 1$. Anyway, for potential instability, only the portions of the unit circles can maintain the regions stable, so $\underline{\Gamma}_G$, $\underline{\Gamma}_L$ will give stable input and output impedances.

From equations (1)-(10) one can easily prove, the following connexions between the stability regions and the stability circles [5-8]:

$$1-|\underline{\Gamma}_{in}|^2 = \frac{|\underline{\Gamma}_L - \underline{c}_L|^2 - r_L^2}{|1-\underline{S}_{22}\underline{\Gamma}_L|^2} D_2,$$
$$1-|\underline{\Gamma}_{out}|^2 = \frac{|\underline{\Gamma}_G - \underline{c}_G|^2 - r_G^2}{|1-\underline{S}_{11}\underline{\Gamma}_G|^2} D_1. \qquad (11)$$

From (11) it results that the stability region is defined according to the conditions:

$$1-|\underline{\Gamma}_{in}|^2 > 0 \Leftrightarrow \left(|\underline{\Gamma}_L - \underline{c}_L|^2 - r_L^2\right) D_2 > 0 \qquad (12)$$

Function of the sign of D2, the conditions from (12) is equivalent to the interior or to the exterior of the load stability circle, having the centre cL and radius rL:

$$|\underline{\Gamma}_L - \underline{c}_L| > r_L, \text{if } D_2 > 0; \; |\underline{\Gamma}_L - \underline{c}_L| < r_L, \text{if } D_2 < 0 \qquad (13)$$

called the load stability region.

The boundary $|\underline{\Gamma}_L - \underline{c}_L| = r_L$ corresponds to $|\underline{\Gamma}_{in}| = 1$. The set complementary to this region corresponds to the instability region with $|\underline{\Gamma}_{in}| > 1$. In a similar manner, the source stability region can be defined as follows:

$$|\underline{\Gamma}_G - \underline{c}_G| > r_G, \text{daca } D_1 > 0; \; |\underline{\Gamma}_G - \underline{c}_G| < r_G, \text{daca } D_1 < 0 \qquad (14)$$

called the source (generator) stability region.

In order to have unconditional stability, the stability regions must contain the unit circle in its entirety. If D2 > 0, then the unit circle and the stability circle should not overlap at all, as presented in figure 2. Geometrically, the distance between the points A and O (figure 2) is (OA) $= |\underline{c}_L| - r_L$. The non-overlapping of these circles implies that (OA) > 1, or equivalently, $|\underline{c}_L| - r_L > 1$ [5-8].

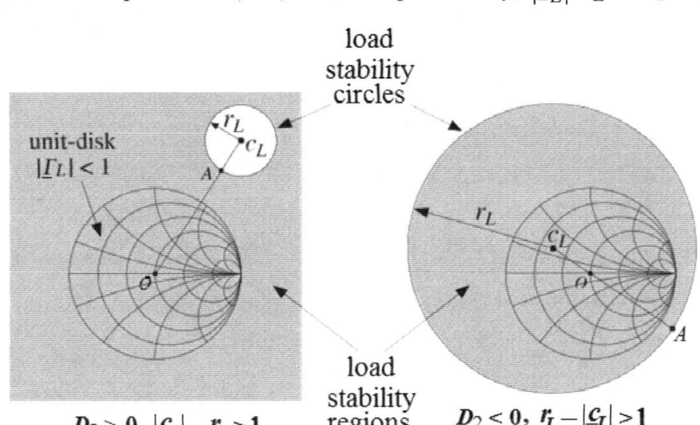

Figure 2. Load stability regions in the unconditionally stable case [10]

IOP Conf. Series: Materials Science and Engineering **749** (2020) 012031 doi:10.1088/1757-899X/749/1/012031

The stability conditions (14) can be combined in $sign(D_2)|\underline{c}_L| - r_L > 1$, or equivalently μ1 > 1, in accordance with equation (8). Geometrically, the parameter μ1 represents the distance (OA). So, the condition for the input unconditional stability is:

$$\mu_1 > 1 \text{ - unconditional stability condition} \tag{15}$$

Edwards and Sinsky showed that the only condition necessary and sufficient for the unconditional stability of a two-port network is that for which input and output impedance is μ1 > 1 (or, alternatively, μ2 > 1) [7, 14]. Obviously, the source stability regions are similar to those presented in figure 2.

If the stability condition is not fulfilled, i.e. μ1 < 1, then the portion of the unit circle that belongs to the stability circle will be stable and will lead to stable input and output impedances. Figure 3 depicts such a potential instability case. If D2 > 1, then μ1 < 1 which is equivalent to $|\underline{c}_L| - r_L < 1$ and if D2 < 0, then $r_L - |\underline{c}_L| < 1$. Anyway, the unit circle is partially overlaps the stability circle, as showed in figure 3. The portion from the unit circle that is not inside the stability region corresponds to an unstable Zin. There are several unconditional stability criteria which are equivalent to the only criterion μ1 > 1. All these criteria require that Rollett stability factor K to be greater the one (i.e. K > 1), as well as one other condition. Any of the following criteria are necessary and sufficient for the unconditional stability, [7, 30]:

$$\begin{aligned} &K > 1 \text{ and } |\underline{\Delta}| < 1; K > 1 \text{ and } B_1 > 0; K > 1 \text{ and } B_2 > 0; \\ &K > 1 \text{ and } |\underline{S}_{12}\underline{S}_{21}| < 1 - |\underline{S}_{11}|^2; K > 1 \text{ and } |\underline{S}_{12}\underline{S}_{21}| < 1 - |\underline{S}_{22}|^2. \end{aligned} \tag{16}$$

All criteria from (16) are equivalent to μ1 > 1. In particular, from (16) results that the unconditional stability requires that $|\underline{S}_{11}| < 1$ and $|\underline{S}_{22}| < 1$. These conditions are necessary but not sufficient for stability. A common situation in practice is to have a potentially instable two-port, but with $|\underline{S}_{11}| < 1$ and $|\underline{S}_{22}| < 1$. In these cases, equation (6) implies $D_2\left(|\underline{c}_L|^2 - r_L^2\right) > 0$, but the lack of stability requires $\mu_1 = sign(D_2)\left(|\underline{c}_L|^2 - r_L^2\right) < 1$. As a consequence, if D2 > 0, then we must have $|\underline{c}_L|^2 - r_L^2 > 0$ and $|\underline{c}_L| - r_L < 1$, that determines the inequality $|\underline{c}_L| < r_L < |\underline{c}_L| + 1$. The geometric distance (OA) $= |\underline{c}_L| - r_L$ satisfies the inequality 0 < (OA) < 1, such that the stability circle overlaps the unit circle but it does not contain its centre. On the other hand, if D2 < 0, the two conditions imply that $|\underline{c}_L|^2 - r_L^2 < 0$ and $r_L - |\underline{c}_L| < 1$, which results in $|\underline{c}_L| < r_L < |\underline{c}_L| + 1$. This situation is presented in the right part of figure 3 [7, 28-31].

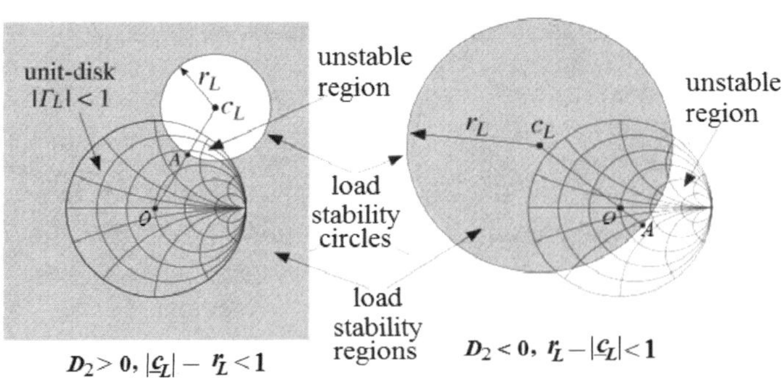

Figure 3. Load stability regions in potentially unstable case [7, 10]

The geometric distance (AO) = $r_l - |c_L|$ satisfies again the inequality 0 < (OA) < 1, but now the unit circle canter is inside the stability circle, which is also the stability region.

3. Examples

We consider the circuit represented in figure 5. In this case, the coupling circuit is a two-port passive network. Voltage controlled oscillators are important components in almost every digital communication system. In recent years, the microwave oscillators are used to control the phase in microwave antenna arrays as an alternative to electronic beam steering methods. A voltage controlled oscillator (VCO) is a circuit that produces an oscillatory output [1-8, 11-16]. The frequency of the output signal depends on the level of an input voltage signal supplied to the VCO.

The active parts of the two Van der Pol oscillators are modelled by two voltage-controlled nonlinear resistors (see figure 4) [3, 6-8, 14].

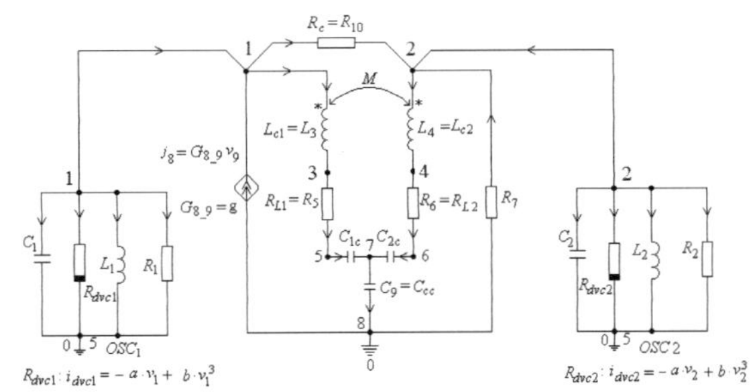

Figure 4. Two parallel resonant circuits coupled through a two-port passive network

The nonlinear characteristics of the two voltage-controlled nonlinear resistors are expressed as it follows:

$$\begin{aligned} i_{nel1} &= -av_1 + bv_1^3 \\ i_{nel2} &= -av_2 + bv_2^3, \end{aligned} \tag{17}$$

where (-a) is the negative conductance necessary to start the oscillation and b a parameter used to model the saturation phenomenon.

Therefore, in the sinusoidal behaviour (neglecting the third harmonic) the two nonlinear resistors can be modelled by two linear resistors with the conductances:

$$G_{01} = -a + \frac{3b}{4}A_1^2;\ G_{02} = -a + \frac{3b}{4}A_2^2 \tag{18}$$

where A1 and A2 are magnitudes of voltages v1 and v2.

To compute the parameters S11 and S21 (S12 and S22) it is used the circuit from the figure 6.a and 6.b where the two voltage controlled nonlinear resistors have been replaced with linear resistors having the values Rdvc1=-150 Ω, Rdvc2=-280 Ω, and the input and load resistors have the values Ri =RL=R0=Z0=50 Ω.

Performing Spice simulation (or MATLAB simulation), for the numeric value parameters: C1 = 10 pF, C2 = 20 pF, L1 = 2.8095 nH, L2 = 1.2678 nH, R1 = 3500 Ω, R2 = 3500 Ω, Rc = 200 Ω, C1c = C2c = Cc = 6.5 pF, Ccc = 6.5 pF, Lc1 = Lc2 = Lc = 0.6 nH, M = 0.21 nH, R7 = 250 Ω, RL1 = RL2 = 1.352 Ω, g = 0.0002 S, values Ri =RL=R0=Z0=50 Ω a = 0.0085 S, and b = 0.00071 S/V2 , we get the magnitudes values A1 = 1.8987 V, A2 = 2.9585 V, and the synchronization frequency fs = 1.1512 GHz, for the initial conditions v1(0) = 2.0 V and v2(0) = 1.0 V.

Thus, the circuit in figure 4 can be analysed by the complex representation method [14].

IOP Conf. Series: Materials Science and Engineering **749** (2020) 012031 doi:10.1088/1757-899X/749/1/012031

Running the routines implemented in MATLAB, for the above-mentioned example, there have been obtained results as follows.

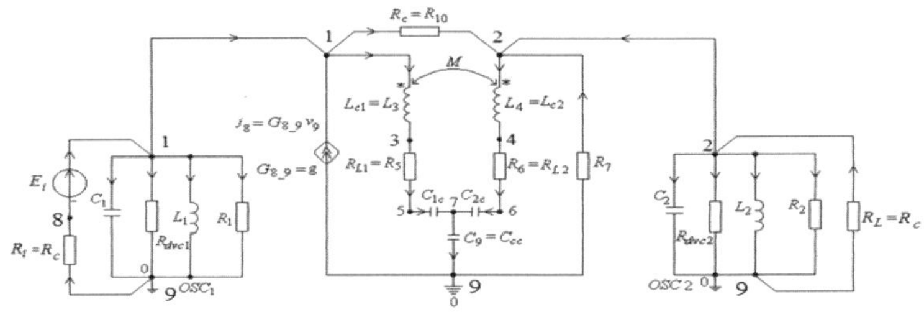

Figure 5.a) The circuit used for $\underline{S}_{11}$ and $\underline{S}_{21}$ computation

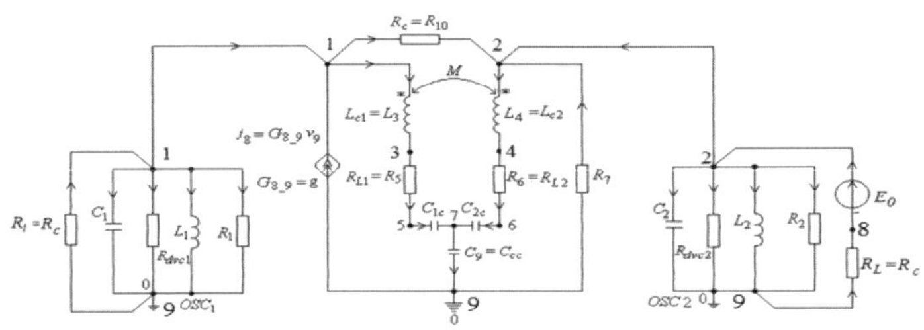

Figure 5.b) The circuit used for the $\underline{S}_{12}$ and $\underline{S}_{22}$ computation

For frequency f = 2.2 GHz, the parameters and constants mention in section 2 have been computed.
The matrix of parameters S(n_1) =
-0.96678-0.21332j-0.96678-0.21332j
-0.96678-0.21332j-0.96678-0.21332j
Figure 6 depicts the variations function of frequency of η_{21}-b, $\eta_{21\_S_{21}}$ -*r, η_{12} -k and $\eta_{12\_S_{12}}$ -m, and figure 7 presents the variations function of frequency of the transducer power gain Gt, the available power gain Ga, the power gain Gp, the maximum available gain GMAG and the maximum stable gain GMSG.

Frequency variations of the S-parameters modules are shown in figure 8. Because the two coupled oscillators are identical, the variations with the frequency of the S12 and S21 parameters modules are identical. Due to the symmetry of the two oscillators, the variation with the frequency of the yield $\eta_{21\_S_{21}}$ is identical to the one of the yield $\eta_{12\_S_{12}}$.

Frequency variation curves of the source standing wave ratio SWR_G and of the load standing wave ratio SWR_L have been shown in figure 9.

In figure 10 and 11 are presented the variations function of frequency for the stability coefficient $k_{stab} = |\underline{\Gamma}_G - \underline{c}_G| - r_G$ and the constant D1 ($k_{stab} = |\underline{\Gamma}_L - \underline{c}_L| - r_L$ and the constant D2). From figure 10 (figure 11) we notice that the analysed system, consisting of two magnetically coupled coils, is stable for the entire frequency interval, because the following inequalities are satisfied: $|\underline{\Gamma}_G - \underline{c}_G| > r_G, if\ D_1 > 0;\ |\underline{\Gamma}_G - \underline{c}_G| < r_G, if\ D_1 < 0$ ($|\underline{\Gamma}_L - \underline{c}_L| > r_L, if\ D_2 > 0;\ |\underline{\Gamma}_L - \underline{c}_L| < r_L, if\ D_2 < 0$).

Frequency variation curves of the source and load reflection coefficients modules ($\underline{\Gamma}_G$ and $\underline{\Gamma}_L$) have been shown in figure 12 and the ones of the input and output reflection coefficients modules ($\underline{\Gamma}_{in}$ and $\underline{\Gamma}_{out}$) have been presented in figure 13.

The stability factor K1 and constant B1 versus frequency are shown in figure 14 and in figure 15 is presented the frequency variations of the stability factor K1 and constant B2.

IOP Conf. Series: Materials Science and Engineering **749** (2020) 012031 doi:10.1088/1757-899X/749/1/012031

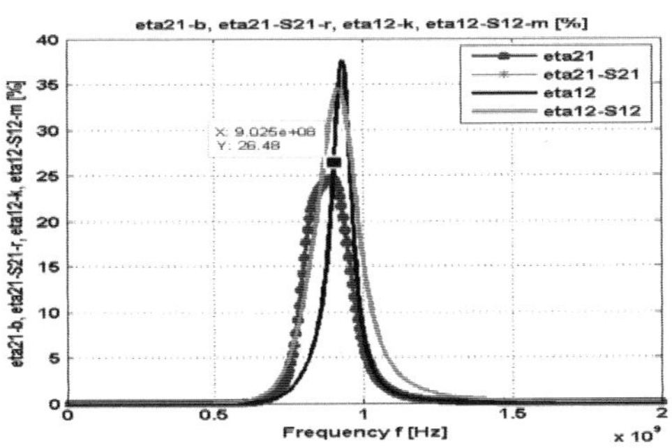

Figure 6. η_{21}-b, $\eta_{21\_S_{21}}$-*r, η_{12}-k and $\eta_{12\_S_{12}}$-m efficiencies vs frequency.

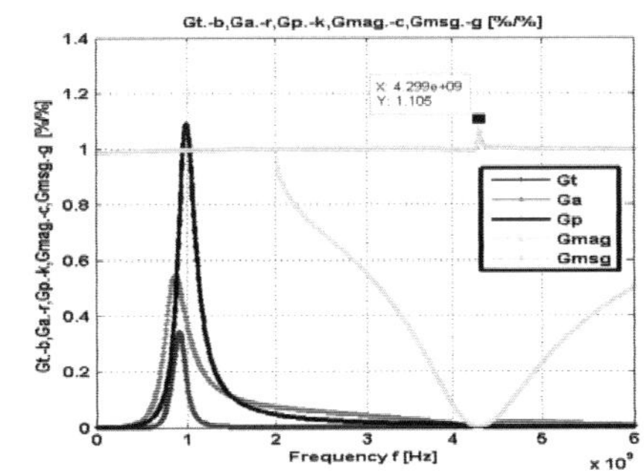

Figure 7. G_t-b, G_a-.r, G_p-.k, G_{mag}-c and G_{msg}-g parameters vs frequency.

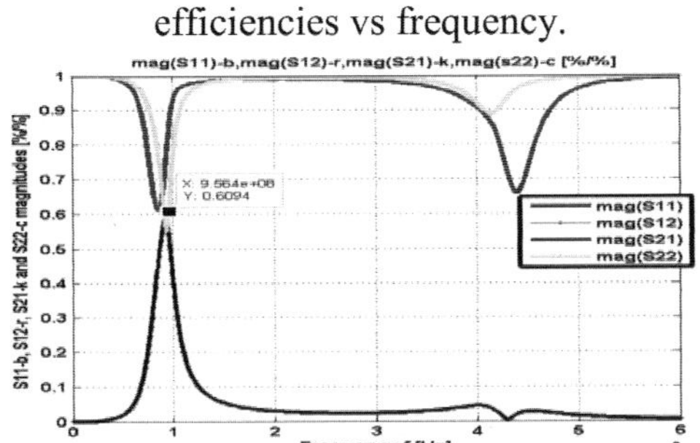

Figure 8. Parameter S magnitudes vs frequency

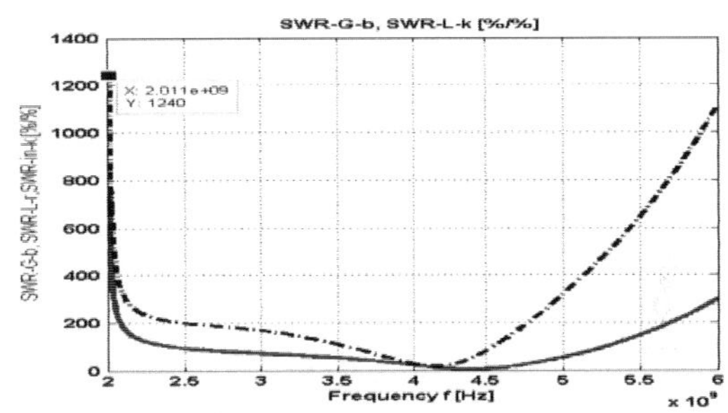

Figure 9. SWR-G-b and SWR-L-k parameters vs frequency

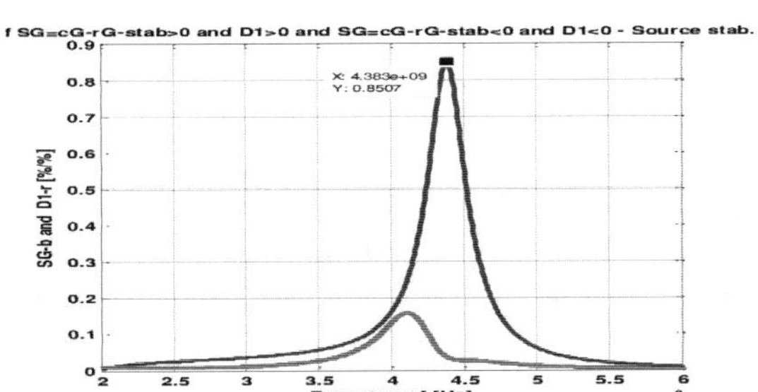

Figure 10. cG-rG-stab-b and D1-r factors vs frequency

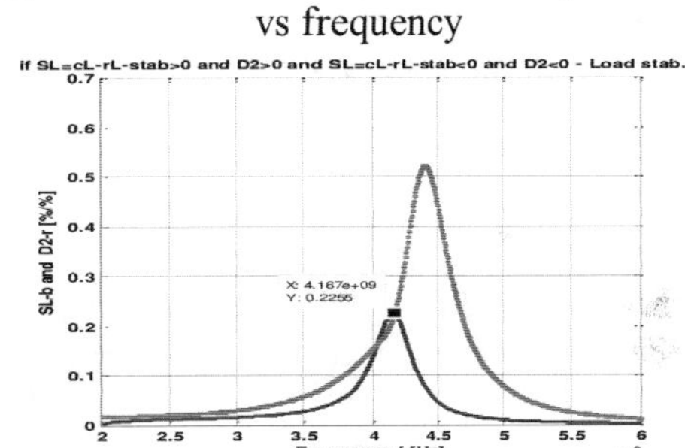

Figure 11. cL-rL-stab-b and D2-r factors vs frequency

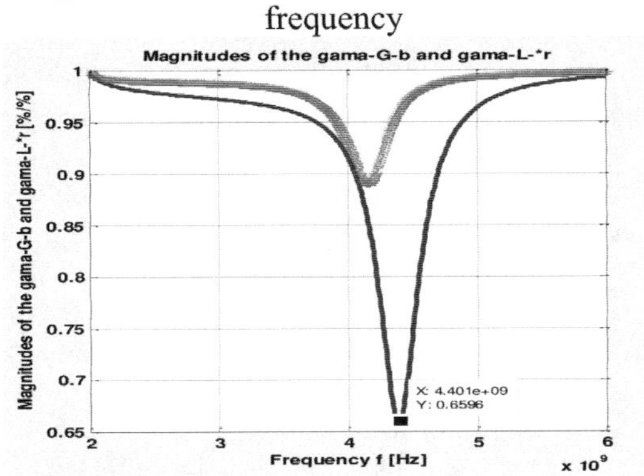

Figure 12. $\underline{\Gamma}_G$-b and $\underline{\Gamma}_L$-b-r magnitudes vs frequency

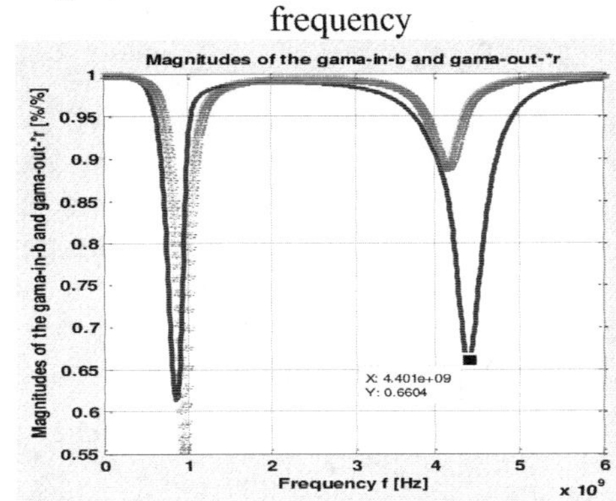

Figure 13. $\underline{\Gamma}_{in}$-b and $\underline{\Gamma}_{out}$-b-r magnitudes vs frequency

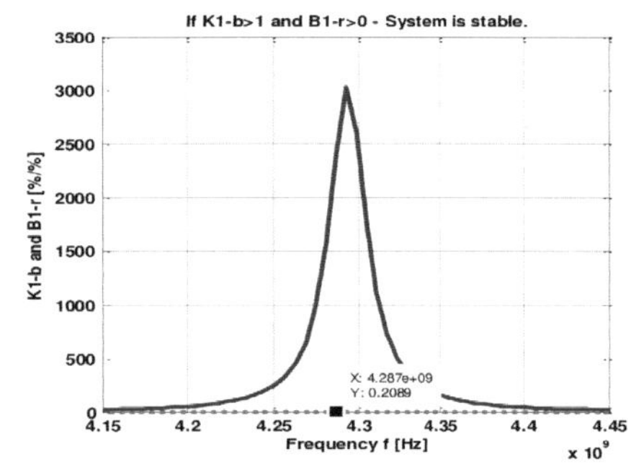

Figure 14. K_1-b and B_1-r constants vs frequency

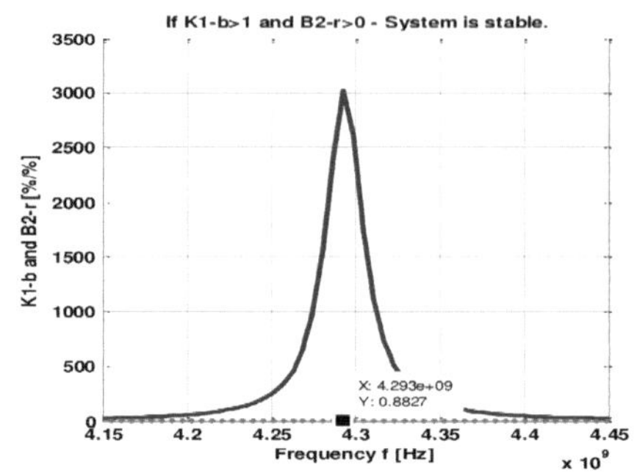

Figure 15. K_1-b and B_2-r constants vs frequency

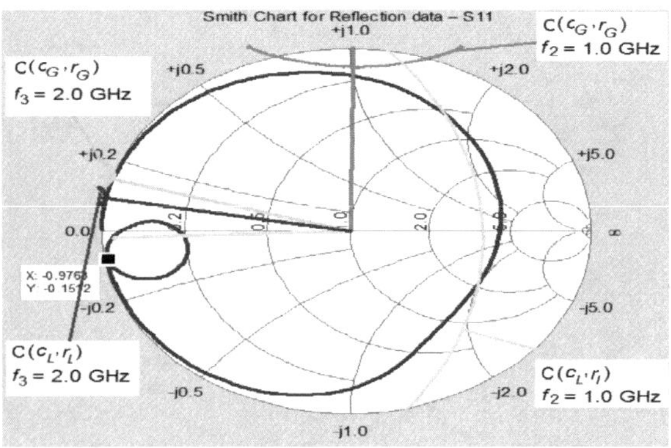

Figure 16. Smith Chart for $\underline{S}_{11}$, stability source circles and stability load circles for two frequencies

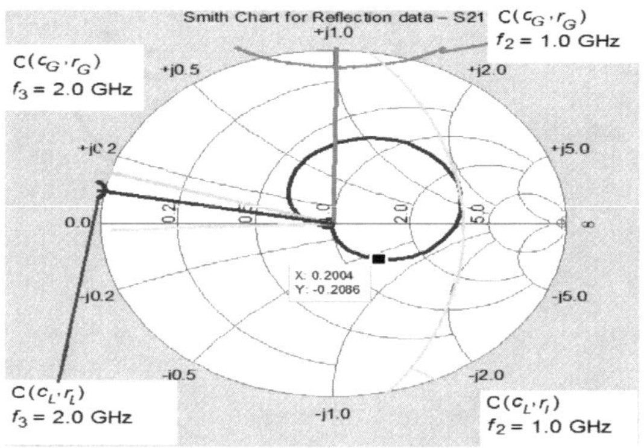

Figure 17. Smith Chart for $\underline{S}_{21}$, stability source circles and stability load circles for two frequencies

Figure 14 and 15, show the variations with the frequency of parameter S11, stability source circles and stability load circles for two frequencies (of parameter S21, stability source circles and stability load circles for two frequencies) on the Smith chart. The studied system consisting of two coupled oscillators is being symmetric, so the Smith charts of S12 and S21 are identical. The variations with the frequency of parameter and stability source circles for four frequencies (of parameter and stability source circles for four frequencies) on the Smith chart are shown in figure 16.

4. Conclusions

This paper presents the use of S parameters for analysing the stability of two-port analogue networks, with important applications for the WPTS. The analysis is done by starting from the correct formulation of these parameters, based on electrical circuit theory, and then generating automatically the matrix corresponding to the scattering parameters. Based on this matrix S, the following matrices are further computed: Z, Y and T. By considering the two magnetically coupled coils (the resonators)-used in WPTS-as a passive, linear two-port network, in harmonic state, we determined the following: the input and output impedances; the reflections coefficients; the voltage transfer function; the stability coefficients (K-Rollet stability factor, μ1 and μ2-Edwards-Sinsky stability factors), the centres and the radius of the load, source stability circles, various constants necessary to define the stability factors. It is also presented the practical use of S parameters to increase the efficiency of information transmission and propagation and of electromagnetic energy wireless transfer from the emitter-receiver point of view. Based on the maximum power transfer theorem, the maximum active power transmitted by a SWTP to the load is computed and, function of S parameters, the most used in practice performance coefficients are calculated: Gt, Ga, Gp, etc. There were established the necessary and sufficient conditions that two magnetically coupled coils (equivalent to a passive linear two-port network) to be unconditionally stable. It has been designed and implemented a procedure to generate, on the Smith diagram, the stability circles corresponding to the source and to the load.

Finally, were elaborated and implemented the computation procedures for: the reflection coefficients, for stationary wave ratios, for input and output impedances, for efficiency, for various stability coefficients, for centres and radius corresponding to the load and source stability circles, for the S, T, Z and Y matrices and for the Smith diagrams. For these procedures, were elaborated dedicated functions implemented in MATLAB. By comparing the simulated results with the experimental ones and with those presented in literature. The accuracy of the plots, obtained based on simulation is confirmed by the results obtained using the dedicated software ADS, which has specific routines to generate S parameters and the other quantities necessary to determine the stability regions for the WPTS.

References

[1] Stănculescu M, Iordache M, Niculae D, Iordache L and Bucată V 2016 *S Parameter Computation and Their Use for Electromagnetic Energy Wireless Transmission* IJCT-International Journal of Computers & Technologies Vol. 12 No June 2016 pp.37097-7108

[2] Frickey D A 1994 *Conversions between S, Z, Y, H, ABCD, and T parameters which are valid for complex source and load impedances* IEEE Trans. on Microwave Theory and Techniques No. 42 pp.205-211

[3] Marks R B and Williams D F 1995 *Comments on conversions between S, Z, Y, H, ABCD, and T parameters which are valid for complex source and load impedances* IEEE Trans. on Microwave Theory and Techniques No. 43 pp.914-915

[4] Iordache M, Dumitriu L, Niculae D and Zainea G 2015 *Power Transfer by Magnetic Induction Studied by Coupled Mode Theory* Book chapter in the book *Wireless Power Transfer* Editor J I Agbinya ISBN 978-87-92329-23-3 River Publishers Series in Communications Denmark Second Edition ISBN 9788793237629 EAN 9788792329233 pp.1-69

[5] Carson R 1982 *High-Frequency Amplifiers* WileyNew York NY

[6] Wai-Kai Chen 1991 *Active Network Analysis* World Scientific Publishing Co

[7] Edwards M L and Sinsky J H 1992 *A New Criterion for Linear 2-Port Stability Using a Single Geometrically Derived Parameter IEEE Transactions Techniques* December 1992

[8] Lombardi G and Neri B 1999 *Criteria for the Evaluation of Unconditional Stability of Microwave Linear Two-Ports: A Critical Review and New Proof, IEEE Transactions on Microwave Theory and Techniques* June 1999

[9] Elsherbeni A, Nayeri P and Reddy C J 2015 *Antenna Analysis and Design Using FEKO Electromagnetic Simulation Software* Pub. by SciTech Publishing an imprint of the IET

[10] Orfanidis S J 2008 *Electromagnetic Waves and Antennas* http://www.ece.rutgers.edu/orfanidi/ewa/

[11] Marks R B and Williams D F 1992 *A general waveguide circuit theory*, Journal of Research of the National Institute of Standards and Technology No. 97 pp.533-562

[12] Jiang H C and Wang Y E 2008 *Capacity Performance of an Inductively Coupled Near Field Communication System*, in Proceedings of the IEEE International Symposium of Antenna and Propagation Society Jul. 5-11 pp.1-4

[13] Imura T, Okabe H and Hori Y 2009 *Basic Experimental Study on Helical Antennas of Wireless Power Transfer for Electric Vehicles by using Magnetic Resonant Couplings* Proceedings of Vehicle Power and Propulsion Conference September 2009 IEEE Xplore 978-1-4244-2601-4/010/2010 pp.936-940

[14] Iordache M 2015 *Symbolic, Numeric-Symbolic and Numeric Simulation of Analogue Circuits-User Guides* MATRIX ROM Bucharest ISBN 978-606-25-0126-6 621.3.049.77

[15] Iordache M, Dumitriu L 2014 *Computer-Aided Simulation of Analoque Circuits-Algorithms and Computational Techniques* Editura POLITEHNICA Press Vol. I and II Bucharest 2014 ISBN 978-606-515-567-1 ISBN 978-606-515-585-5

IOP Conf. Series: Materials Science and Engineering **749** (2020) 012031 doi:10.1088/1757-899X/749/1/012031

[16] http://www.keysight.com/en/pc-1297113/advanced-design-system-ads?cc=US&lc=eng
[17] Agilent Technologies AGILENT TECHNOLOGIES *Large-Signal S-Parameter Simulation* Sept 2004
[18] ANSOFT Q3D EXTRACTOR, User Guide, www.ANSOFT.com
[19] https://www.feko.info/support
[20] Chen W, Chinga R A, Yoshida S, Lin J, Chen C and Lo W 2012 *A 25.6 W 13.56 MHz wireless power transfer system with a 94% efficiency GaN class-E power amplifier* Microwave Symposium Digest (MTT) IEEE MTT-S International August 2012 pp.1-3
[21] Brown W C 1996 *The history of wireless power transmission* Solar Energy Vol56 No1 pp.3-21
[22] Kurs A, Karalis A, Moffatt R, Joannopoulos J D, Fisher P and Soijacic M 2007 *Wireless Power Transfer via Strongly Coupled Magnetic Resonances*, Massachusetts Institute of Technology 2007 Science Vol. 317 No. 5834 July 2007 pp.83-86
[23] Harrington R F 1993 *Field Computation by Moment Methods* Wiley-IEEE Press ISBN: 978-0-7803-1014-8
[24] Pennock S R, Shephered P R 1998 *Microwave Engineering with Wireless Applications* McGraw-Hill Profesional ISBN: 0070497222
[25] Sah A K and Pant D P 2014 *Analysis and Optimization of Wireless Power Transfer Link* in Proceedings of International Conference on Technology and Innovation Management and IOE Graduate Conference Oct. 10-11 pp.82-93
[26] http://www.nathaniyer.com/ (quick Smith)
[27] http://www.tonnesoftware.com/supersmith.html (Tone superSmith)
[28] Woods D 1976 *Reappraisal of the Unconditional Stability Criteria for Active 2-Port Networks in Terms of S parameters IEEE Transactions on Circuits and Systems* February 1976
[29] Carson R 1982 *High-Frequency Amplifiers* Wiley New York, NY
[30] Barz C, Latinovic T, Balan A, Pop-Vadean A, Pop P P, *Using HMI Weintek in command of an industrial robot arm* International Conference on Applied Sciences Oct 02-04 2014 (ICAS2014) *IOP Conf. Ser.: Mater. Sci. Eng.* **85** (2015) p.1-6, DOI:10.1088/1757-899X/85/1/012003
[31] Berdie A, Osaci M, Muscalagiu I, Barz C, *A combined approach of AHP and TOPSIS methods applied in the field of integrated software systems* International Conference on Innovative Ideas in Science (IIS2016) 10-11 Nov 2016 Baia Mare Romania *IOP Conf. Ser.: Mater. Sci. Eng.* **200** (2017) 012041 pp.1-6 DOI:10.1088/1757-899X/200/1/012041

IOP Conf. Series: Materials Science and Engineering **749** (2020) 012032 doi:10.1088/1757-899X/749/1/012032

Angle position influences over external magnetic field of a solenoid in wireless power transfer

B Iuga[a] and R A Tirnovan

Technical University of Cluj-Napoca, Faculty of Electrical Engineering, Department of Electroenergetics and Management, Cluj-Napoca, Romania

E-mail: [a] bogdan_i2008@yahoo.com

Abstract: The general form of the Biot-Savart law is referring to a closed loop of wire, but it can be integrated over a finite length solenoid to calculate the external magnetic field in a specific point, at a rated distance from the centre. The law states that the point of measure must be on the longitudinal axis of the solenoid. In real cases the point of measure has angle orientation in relation to the axis which results in modifications of the field value.

1. Introduction

A finite length of conducting wire coiled around an axis results in a closed loop and a multiple number of closed loops give a solenoid, or a coil. This is the basic principle of calculating the exterior magnetic field generated by a solenoid using the Biot-Savart law.

The magnetic field has the strongest value inside any solenoid, especially if it has a core made of conductive material. If the coil does not have a conductive core, the maximum magnetic field generated is located on the half distance between the length edges [1]. External magnetic field is calculated integrating the general form of Biot-Savart law over the total length of the solenoid, but is important that the point of measure to be on the transversal axis of solenoid.

Jean-Baptiste Biot and Félix Savart formulated the law around 1820, which is the fundamental formula for calculating the magnetic field. It corresponds to Coulomb's law for calculating the electric field. The Biot-Savart law is an equation describing the magnetic field generated by an electric current as a vector that has magnitude direction and length [2].

The purpose of this paperwork is to show the calculation of the exterior magnetic field produced by a solenoid using Biot-Savart law and the location impact of the point of measure. In real cases, the point of measure is not on the centre axis, but it has an angle orientation, which brings to modifications in the value of the field.

2. Mathematical Analysis

The basic principle used in finding the value of magnetic field is to split the solenoid into small parts known as elementary components.

The general form of Biot-Savart law is referring to a finite length of conductive wire [3]:

$$d\vec{B} = \frac{\mu_0}{4\pi} \cdot \frac{I \cdot d\vec{S} \times \hat{r}}{r^2} \tag{1}$$

IOP Conf. Series: Materials Science and Engineering **749** (2020) 012032 doi:10.1088/1757-899X/749/1/012032

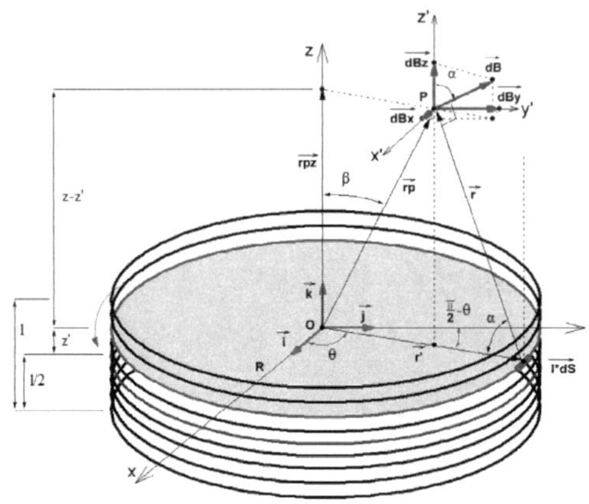

Figure 1. Vectorial components of exterior magnetic field of a solenoid

A solenoid with finite length l and the centre at l/2 will have the point of measure P located at z distance from the centre on the Oz axis.

First, the general formula is derived for a single spire (closed loop). This is possible by taking into consideration any loop dz`, located at a random distance z' from the centre of solenoid and (z-z`) distance from the point of measure and calculating the locations of corresponding vectors [4] [5].

$$d\vec{B} = \frac{\mu_0 \cdot I \cdot R}{4\pi \cdot (R^2 + (z - z`)^2)^{3/2}} \cdot \left((z - z`) \cdot \cos\theta \cdot \hat{i} + (z - z`) \cdot \sin\theta \cdot \hat{j} + R \cdot \hat{k}\right) \cdot d\theta \quad (2)$$

Because the point of measure has a deviation from the central axis, the angular dependence, cosβ, has been introduced [4] [5]:

$$d\vec{B} = \frac{\mu_0 \cdot I \cdot R}{4\pi \cdot (R^2 + (z - z`)^2)^{3/2}} \cdot \left( \cos\theta \cdot \frac{(z - z`)}{\cos\beta} \cdot \hat{i} + \sin\theta \cdot \frac{(z - z`)}{\cos\beta} \cdot \hat{j} + R \cdot \hat{k} \right) \cdot d\theta \quad (3)$$

where: R is the radius of loop
I is the current passing through solenoid

The current I that pass through solenoid will distribute uniformly on all the spires, so that leads to a new derived form:

$$dI = I \cdot (n \cdot dz`) \quad (4)$$

and

$$n = \frac{N}{l} \quad (5)$$

where N is the total spire number of solenoids.

After splitting the result into Cartesian axis xOyOz, represented by the corresponding versors $\hat{i}, \hat{j}, \hat{k}$ the only component that will have value is B_z as the others are null because of sinθ and cosθ:

$$B_z = \frac{\mu_0 \cdot I \cdot R^2}{4\pi \cdot \left(R^2 + \frac{(z - z`)^2}{\cos^2\beta} \right)^{3/2}} \quad (6)$$

The contribution of this element will be derived in function to I to find the field value for all the loops [2] [3]:

$$dB_z = \frac{\mu_0 \cdot R^2}{4\pi \cdot \left(R^2 + \frac{(z-z`)^2}{\cos^2 \beta} \right)^{3/2}} \cdot dI = \frac{\mu_0 \cdot R^2}{4\pi \cdot \left( R^2 + \frac{(z-z`)^2}{\cos^2 \beta} \right)^{3/2}} \cdot (n \cdot I \cdot dz`) \quad (7)$$

The dz` element is characterized by R and the (z-z`) which is the relative distance of the spire to the point of measure.

The result is integrated over the entire length of the solenoid, which leads to a possible final form of the magnetic field value.

$$B_z = \frac{\mu_0 \cdot R^2 \cdot n \cdot I}{2} \cdot \int_{-\frac{l}{2}}^{\frac{l}{2}} \frac{1}{\left[R^2 + \left(\frac{z-z'}{\cos \beta} \right)^2 \right]^{3/2}} \cdot dz' \quad (8)$$

3. Results

The basic principle used in finding the value of magnetic field is to split the solenoid into small parts known as elementary components.

$$u = \frac{z-z'}{\cos \beta} \quad (9)$$

and the derived form becomes:

$$dz' = -\cos \beta \cdot du \quad (10)$$

After replacement the integral has a simpler form, consisting of values known on left side and an integral with complex number on right side:

$$B_z = -\frac{\mu_0 \cdot R^2 \cdot n \cdot I}{2} \cdot \int_{-\frac{l}{2}}^{\frac{l}{2}} \frac{\cos \beta}{\left(R^2 + u^2 \right)^{3/2}} \cdot du \quad (11)$$

A trigonometric form is desired to simplify more the equation, so the complex function number derived becomes:

$$u = R \cdot \tan \rho \quad (12)$$

and his derived form:

$$du = R \cdot \sec^2 \rho \cdot d\rho \quad (13)$$

With this solution, the magnetic field form becomes:

$$B_z = -\frac{\mu_0 \cdot R^2 \cdot n \cdot I}{2} \cdot \int_{-\frac{l}{2}}^{\frac{l}{2}} \frac{\cos \beta \cdot \sec^2 \rho \cdot R}{\left(R^2 + \tan^2 \rho \right)^{3/2}} \cdot (\sec^2 \rho \cdot R) \cdot d\rho \quad (14)$$

Using the basic trigonometric formulas:

$$\tan^2 \rho + 1 = \sec^2 \rho \quad (15)$$

IOP Conf. Series: Materials Science and Engineering **749** (2020) 012032 doi:10.1088/1757-899X/749/1/012032

and

$$\frac{1}{\sec\rho} = \cos\rho \tag{16}$$

a new simple form is given:

$$B_z = -\frac{\mu_0 \cdot R^2 \cdot n \cdot I}{2} \cdot \int_{-\frac{l}{2}}^{\frac{l}{2}} \cos\rho \cdot d\rho \tag{17}$$

After computing the notes, the final form of the magnetic field, with angle position influences is given:

$$B_z = \frac{\mu_0 \cdot R^2 \cdot n \cdot I}{2} \cdot \frac{\cos\beta}{R^2} \cdot \left. \frac{\frac{z'-z}{\cos\beta}}{\sqrt{R^2 + \left(\frac{z-z'}{\cos\beta}\right)^2}} \right|_{-l/2}^{l/2} \tag{18}$$

4. Conclusions

The magnetic field is null on the Ox and Oy axis because of the dependence to the angle θ and the Oz component, also it decreases exponential in relation to Euclidian distance from the source to the point of measure.

If the point of measure is not on the central transversal axis of the solenoid, the magnitude of the field is measured with the help of the corresponding axial component. In this case, the cosβ that appears increases the Euclidian distance, which causes a substantial drop of the field.

Also, if the angle between and reaches π/2 radians (the point of measure is perpendicular to the solenoid) the field value drops to 0 [5]:

$$d\vec{B} = \cos\alpha \cdot d\overrightarrow{B_z} \tag{19}$$

All expressions presented will help develop a general law, which can calculate the exterior magnetic field at a given point and to understand why the axial components are canceled. The differences between ideal and real cases help heaving a better knowledge of the field value and what is it impact of the orientation angle.

References

[1] Dumitru C, Dumitru L 2004 *Bazele electrotehnicii* Bucuresti
[2] Jackson J D 1999 *Classical Electrodynamics* (3rd ed.) New York: Wiley Chapter 5
[3] Massachusetts Institute of Technology *Magnetic Fields: Creating magnetic field-Ampere's Law* Chapter 9
[4] Bălan V 1999 *Algebră liniară, geometrie analitică* Editura Fair Partners
[5] Olteanu M 2005 *Analiza Matematica - notiuni teoretice si probleme rezolvate* Ed. Printech
[6] Maxwell J C 1861 *On Physical Line of Force* Philosophical Magazine and Journal of Science

IOP Conf. Series: Materials Science and Engineering **749** (2020) 012033 doi:10.1088/1757-899X/749/1/012033

Research quantitative indicators of the potential of solar energy in the Carpathian region of Ukraine

O M Mandryk[1,a], N R Moskalchuk[1], L M Arkhypova[2], M M Pryhodko[3] and O V Pobigun[2]

[1] Ivano-Frankivsk National Technical University of Oil and Gas, Institute of Environmental Engineering, Department of Ecology, Ivano-Frankivsk, Ukraine
[2] Ivano-Frankivsk National Technical University of Oil and Gas, Institute of Architecture, Construction and Tourism, Department of Tourism, Ivano-Frankivsk, Ukraine
[3] Ivano-Frankivsk National Technical University of Oil and Gas, Institute of Architecture, Construction and Tourism, Department of Land Management and Cadastre, Ivano-Frankivsk, Ukraine

E-mail: [a] konsevich@ukr.net

Abstract. The paper presents the results of theoretical research on the resource potential of solar energy calculated by different methods for administrative areas of the Carpathian region; The advanced methodology of the estimation of the influence of solar power plants on the natural environment is proposed, which concerns the determination of the significance of the residual effects of renewable energy on the environment and is carried out after screening and mitigation measures and is based on the three identified parameters of influence: spatial, temporal and intensity of influence. The results of statistical analysis of observational data of meteorological stations of the studied region are presented. The results of experiments that study are dependent on power generation by photovoltaic panels from meteorological elements of weather conditions; it is proved that the amount of energy received significantly depends on the cloudiness and air humidity. At the same time, it has no direct dependence on atmospheric pressure and wind speed indices. The paper presents the results of the cartographic processing of quantitative indicators of the potential of solar energy for the arrangement of the objects providing renewable energy in the Carpathian region of Ukraine.

1. Introduction

It is known that in Ukraine, as well as in many countries in the world, there is an energy due to the lack of conventional energy sources [9]. For present moment it is believed that the use of renewable energy sources is one of the ways to solve this problem. The topicality of the scientific project is apparent and related to the fact that the world community regards the use of alternative and renewable energy as one of the most promising solutions to the growing problems of energy supply [1]. One of the most promising alternative sources of energy is solar energy. Considering the dynamics of reducing the cost of electricity, generated from the sun it is strategically important for Ukraine to develop these directions of „future", especially in such a difficult economic situation [6].

Efficiency and expediency of the use of sunny energy many works are sanctified to [1, 6, 7, 8]. Ponderable results are got in Institute of physics of semiconductors of NUN of Ukraine, in the Kyiv,

IOP Conf. Series: Materials Science and Engineering **749** (2020) 012033 doi:10.1088/1757-899X/749/1/012033

Odesa, Uzhhorod and Chernivtsi universities, on some industrial enterprises ("Pillar", "Quasar"), in other Ukrainian laboratories.

All assert simply, that application is in Ukraine of alternative energy sources, foremost, sunny energy, no doubt will give a benefit [8].

On the whole the modern state of questions of analysis and estimation of sun energy resources is in-process considered. The basic types of works are distinguished and analysed in relation to the natural resources of refurbishable energy, scientific achievements in relation to natural resources for development of alternative energy, that contain general and special description of refurbishable energy sources, methods of their use but estimations.

Analysis undertaken studies showed that today all more authors dedicate the labours to systematization and perfection of unconventional energy sources and complex calculation of options that give an opportunity on their basis to get energy (e.r. Solove, 2007).

Very often the necessary are given about sunny energy are absent and for this purpose use analytical dependences that with sufficient exactness help to decide the put tasks. Being and description of such dependences labour is sanctified (e.r. Kondratev, 1981).

Other source about the level of insolation of concrete territory are results of long-term research of NASA (eosweb.larc.nasa.gov), what is included by the satellite measuring, design of atmosphere and surface researches. In this model during the design of atmosphere such factors are used, as climatic zones, reflectance, cloudiness, precipitations and aerosols in an atmosphere.

2. Problem statement

The aim of the work - to justify the advisability of implementing facilities for using the solar energy of the Carpathian region and also to prove that for the development of solar energy in the Carpathian region there are not objective resource obstacles. At a recent time one can observe an increasing interest in the use of photovoltaic panels to produce electricity by converting solar energy. Along with other devices to generate electricity from renewable energy sources, photovoltaic panels have several advantages, including simplicity of design and construction, small weight and size, long-term life (e.r. Duffie, 1980). While the major disadvantages of its use are their low efficiency, instability of electrical energy due to weather conditions and the dependence of output power from the angle of incidence of sunlight on light-absorbing panel. In order to use solar installations, one must have data on the distribution of the intensity of solar radiation in the area. The administrative Carpathian region includes Ivano-Frankivsk, Lviv, Zakarpattya and Chernivtsi areas.

Thus, there is a need to solve the following tasks:

- Accumulation of data on solar radiation intensity in each region;
- Data processing and determination of probable characteristics of the intensity of solar radiation distribution.

3. Presentation of the fundamental material

The regional average annual values of the total solar radiation within the whole Carpathian region on the horizontal (underlying) surface are presented on the Map Info map built in GIS by us in Figures 1, 2, containing 10 colour gradations. The results indicate a gradual decrease in the quantitative indicators in the direction of the south-east-northwest.

The efficiency of light-absorbing panels operation in the Carpathian has been studied on the basis of Ivano-Frankivsk National Technical University of Oil and Gas since April 2015. There are daily conducted measurements of the amount of produced electricity and meteo-indexes in Table 1.

IOP Conf. Series: Materials Science and Engineering **749** (2020) 012033 doi:10.1088/1757-899X/749/1/012033

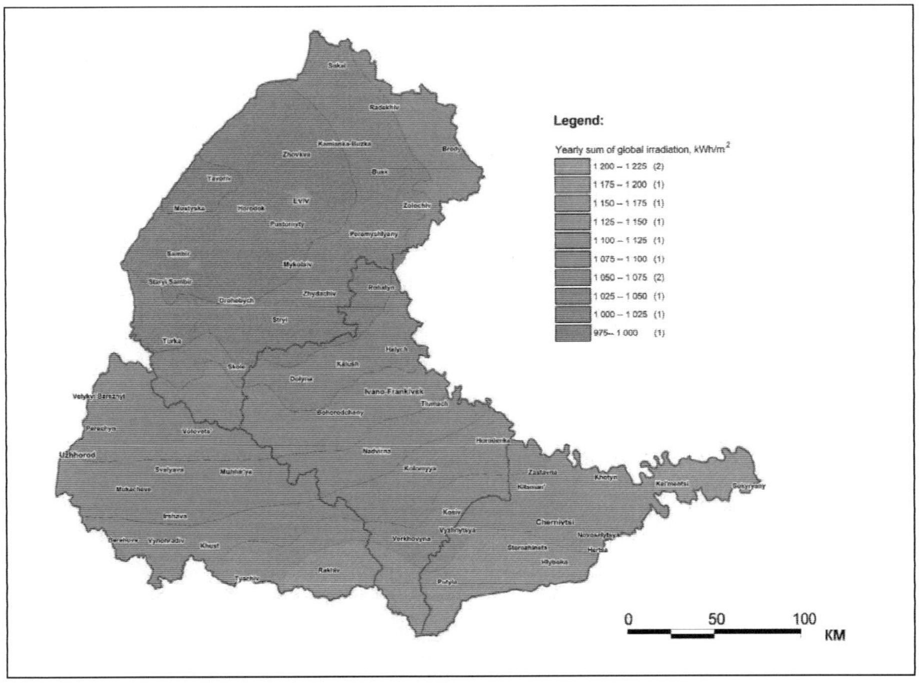

Figure 1. Total solar radiation on a horizontal surface in the Carpathian region of Ukraine

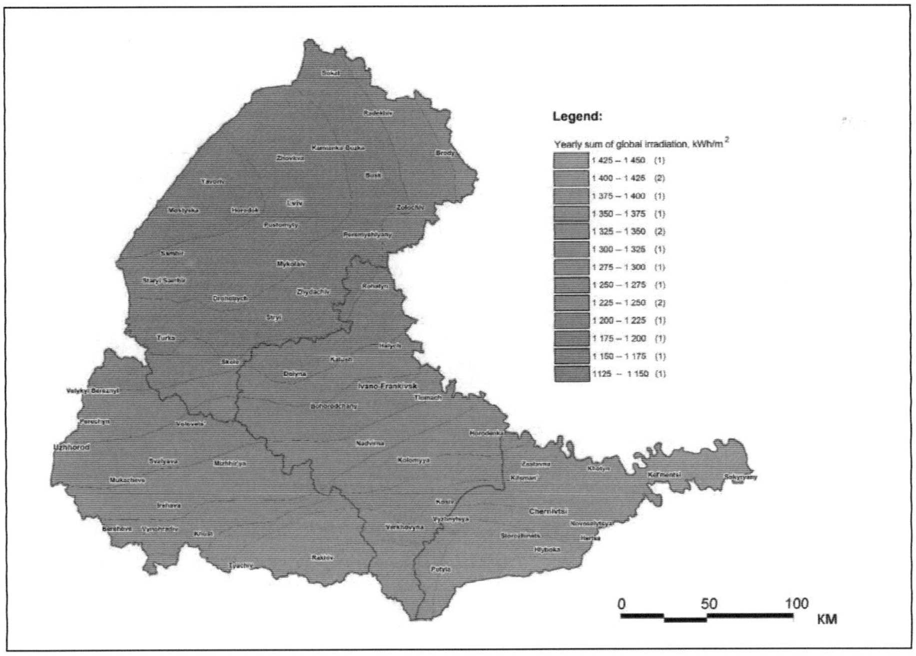

Figure 2. Total solar radiation on a vertical surface in the Carpathian region of Ukraine

IOP Conf. Series: Materials Science and Engineering **749** (2020) 012033 doi:10.1088/1757-899X/749/1/012033

Table 1. Experimental measurements at the photoelectric station of IFNTUOG (April 2015)

Date	E-tot (kWh)	E-day (kWh)	Cloudiness (%)	Probability of precipitation %	Air temperature (°C)	Atmospheric pressure, mmHg	Humidity (%)	Wind speed (m/s)
01.04.	6305	18,2	70	37	11	726	30	10
02.04.	6314	9,4	63	17	7	755	31	6,4
03.04.	6320	6,1	88	48	3	756	54	7,7
04.04.	6328	7,6	67	29	4	759	66	4,2
05.04.	6335	7,3	67	21	6	761	45	2,4
06.04.	6342	6,9	99	70	4	764	75	3,6
07.04.	6348	6,1	87	46	7	767	50	6,3
08.04.	6356	7,7	83	53	7	767	67	5,2
09.04.	6371	15,2	0	1	10	773	33	4,4
10.04.	6391	19,8	24	2	16	770	30	3,6
11.04.	6411	20,2	4	2	19	767	24	1,5
12.04.	6429	18,4	59	24	18	769	49	4,7
13.04.	6448	19,3	68	10	18	764	37	5,8
14.04.	6463	15,2	96	4	10	767	26	7,1
15.04.	6479	15,6	85	19	13	761	48	6,9
16.04.	6497	17,8	35	8	21	758	34	6,5
17.04.	6509	11,7	92	50	14	758	52	3
18.04.	6521	12,3	73	54	9	760	43	6,2
19.04.	6532	10,2	79	53	8	763	45	6,4
20.04.	6539	6,6	88	89	7	757	68	7,3
21.04.	6549	9,1	97	66	6	763	58	5,8
22.04.	6558	8,7	95	6	8	767	47	3,1
23.04.	6578	20,5	63	2	16	763	41	4,3
24.04.	6599	20,8	37	4	20	761	32	3,1
25.04.	6615	16,5	26	3	21	761	30	2,3
26.04.	6634	18,3	57	8	21	758	33	2,7
27.04.	6647	13,2	46	8	21	758	37	2,8
28.04.	6667	19,9	60	15	25,1	729	29	5
29.04.	6672	5,2	99	94	7	762	67	6,5
30.04.	6691	19,1	67	3	14	764	44	1,8

The experimental study was conducted, the obtained results were statistically processed. The functional dependence of generating electricity on meteorological parameters was identified, namely, cloudiness, probability of precipitation, air temperature, atmospheric pressure, humidity and wind speed [4, 10].

According to the results, the amount of obtained energy substantially depends on the indexes of cloudiness and humidity. However, there is no direct relationship with the indices of atmospheric pressure and wind speed. Somewhat less dependence is observed on air temperature and probability of precipitation. Based on the gained data, the curves of dependence of power generation with the help of light-absorbing panel on weather conditions were built, the functional dependence of power generation on the cloudiness in the area conditions of Ivano-Frankivsk were obtained (figure 3) and also the functional dependence of electricity generation by photovoltaic panels on relative air humidity for the area conditions of Ivano-Frankivsk was got (figure 4).

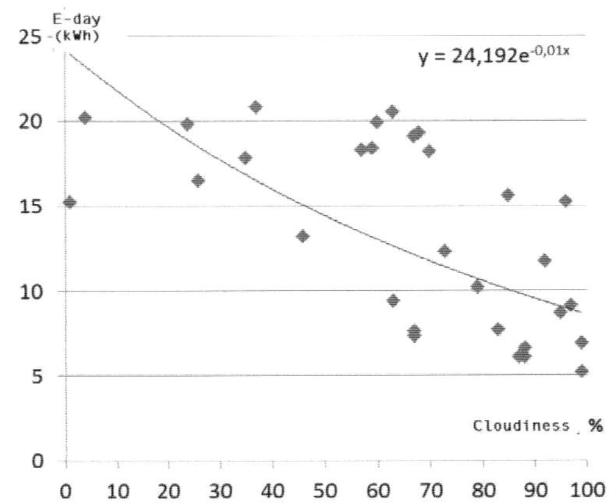

Figure 3. The functional dependency of electricity generation by the photovoltaic panels on the level of cloudiness in Ivano-Frankivsk

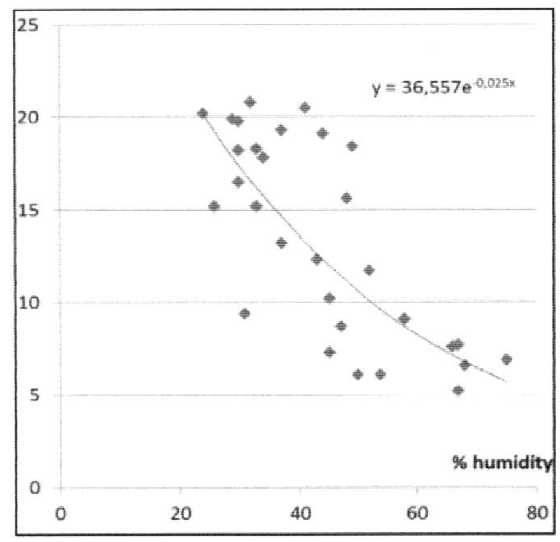

Figure 4. The functional dependency of electricity generation by the photovoltaic panels from the level of humidity in Ivano-Frankivsk

4. Methodology of assessment of SPP impact upon the environment

The assessment of the magnitude and significance of the impact on the environmental components is usually carried out in three phases:

Stage 1: Determination of first-line effects (screening);

Stage 2: Development of a complex of mitigation measures;

Stage 3: Evaluation of residual effects [2, 3].

The suggested method concerns the determination of the significance of the residual effects of renewable energy, in particular solar power plants, on the environment and is carried out after screening and mitigation measures. Due to the fact that the impact of many factors influencing the natural and socio-economic environment cannot be quantified, a rating method of impact assessment is proposed, which allows comparing different types of impacts using the matrix method. The methodology is based on the three specified parameters of influence: spatial, temporal and intensity of influence. The assessment of the impact is carried out on individual components of the environment, taking into account the normal operation of the facility and the probable emergency situations.

4.1. Methodology of environmental impact assessment

Assessment of impacts upon the environment is carried out according to individual components: climate and microclimate, air environment, geological environment, aquatic environment, soils, flora and fauna, protected objects [5].

The procedure for determining the significance of impacts on the environment. The significance of the impact is estimated by such parameters as spatial scale, time scale and intensity. The comparison of the significance values for each parameter is estimated according to the developed scales suggested in Tables 2-4.

For the natural environment, no effect is taken into account, as any activity will have an impact on the natural environment. Zero effect will only be in the absence of planned activities.

Table 2. Scale for estimating the spatial scale (area) of impact on the natural environment

Gradation	ПТК	Spatial boundaries of impact (km2)	Grade
Spot	Facies, tracts	Impact area up to 1 km^2	1
Local	Groups of tracts, terrain	Impact area up to 10 km^2	2
Local	Landscape	Impact area from 10 up to до 100 km^2	3
Regional	Landscape Districts, Provinces	Impact area more than 100 km^2	4

IOP Conf. Series: Materials Science and Engineering **749** (2020) 012033 doi:10.1088/1757-899X/749/1/012033

Table 3. Scale of estimation of time scale (duration) of impact on the natural environment

Gradation	**Time scale (duration) of impact**	**Grade**
Short-term	less than 3 months	1
Medium-term	from one season (more than 3 months) up to one year	2
Long-term	over a long period (more than 1 year but less than 3 years) usually covers the time frame of construction	3
Long-lasting (permanent)	more than 3 years	4

Table 4. The scale of the intensity of the impact on the natural environment

Gradation	Description of impact intensity	Grade
Minor	Changes in the natural environment do not exceed the existing limits of natural variability	1
Weak	Changes in the natural environment go beyond the limits of natural variability, the natural environment is completely self-healing.	2
Moderate	Changes in the natural environment, which exceed the limits of natural variability, violate the individual components of the natural environment. The natural environment retains the ability to self-healing	3
Considerable	Changes in the natural environment result in significant damage to the components of the environment and / or ecosystems. Separate components of the natural environment lose their ability to self-healing	4

The significance of the impacts is an integrated (integral) estimate. Determination of significance of influence is carried out in several stages.

Stage 1. To determine the significance of the impact on the individual components of the environment, it is necessary to use Tables 2- 4 with the criteria of impact. Significance of impact (type of activity) on the i-th component of the environment is determined as follows:

$$Q_i^{nat} = Q_i^{tnat} \cdot Q_i^{snat} \cdot Q_i^{jnat} \tag{1}$$

where: Q_i^{tnat} - A grade of time impact on the i-th component of the natural environment;

Q_i^{snat} - A grade of spatial impact on the i-th component of the natural environment;

Q_i^{jnat} - A grade of impact intensity on the i-th component of the natural environment.

Stage 2. The category of significance is determined by the interval of values depending on the grade, obtained in the calculation, as shown in table 5.

Table 5. Categories of significance of impacts on the natural environment

Category of significance, grade			**Categories of significance**	
Spatial scale	Time scale	Impact intensity	Grades	Significance
Spot 1	Short-term 1	Minor 1	1-8	Minor impact
Local 2	Medium-term 2	Weak 2		
Territorial 3	Long-term 3	Moderate 3	9-27	Moderate impact
Regional 4	Long-lasting (permanent) 4	Significant 4	28-64	Significant impact

The significance categories are the same for different components of the natural environment and can be compared with the identified component of the environment that is most affected.

Categories of significance are determined for all components of the natural environment. If the significance of the influence determined for a particular component of the natural environment (atmospheric air, animal world, etc.) is the only one, it is directly used to assess the resulting significance of the effect. However, in practice, one component of the natural environment can be influenced by different sources (activities), then the integrated (average) assessment of the significance of the impact on the i-th component of the environment will look like:

$$Q_{integr}^{nat} = \frac{\sum_{k=1}^{n} Q_{ki}^{nat}}{n} \tag{2}$$

where: Q_{ki}^{nat} - integrated grade of k-th impact (type of activity) on the i-th component of the natural environment; n - Number of influences (activities) on the i-th component of the environment.

4.2. Methodology for assessing impacts on the socio-economic environment

The assessment of the impact on the socio-economic environment as well as on the natural environment is carried out by separate components. Table 6 presents the composition of the socio-environmental components that can be considered in the process of assessment of the impact on the socio-economic environment from SPPs.

Table 6. Components of the socio-economic environment

Components of the social environment	Components of the economic environment
Safety and health of the population	Employment
Recreational resources	Land use
Historical and cultural resources	Saving organic fuel
Communal Infrastructure	…
Transport	
Waste	

The procedure for determining the significance of impacts on the socio-economic environment. Significance of influence is estimated by parameters of spatial, temporal scale and intensity scale. To assess the whole set of consequences of the planned activity on the social and economic conditions, a 4-level gradation with 1 to 4 points is accepted, with a negative and positive sign, both negative and positive factors of influence are ranked. Unlike the natural environment, there can be zero impact if negative effects are compensated by the same level of positive effects.

Each grading of influence on the components of the socio-economic environment is determined by the relevant criteria in Tables 7 -9.

Table 7. Scale of estimation of the spatial scale (area) of influence on the socio - economic environment

Gradation	Criterion	Grade
Zero effect	no impact	0
Spot impact	the impact is felt on the site of the object	1
Local influence	the impact is felt on the territory of adjacent settlements	2
Territorial influence	the impact is felt on the territory of one or more administrative districts	3
Regional influence	the impact is felt on the territory of the region	4

IOP Conf. Series: Materials Science and Engineering **749** (2020) 012033 doi:10.1088/1757-899X/749/1/012033

Table 8. The scale of the intensity of the impact on the socio - economic environment

Gradation	Criterion	Grade
Zero impact	no impact	0
Short-term impact	less than 3 months	1
Medium-term impact	from one season (more than 3 months) up to one year	2
Long-term impact	over a long period (more than 1 year but less than 3 years)	3
Long-lasting (permanent) impact	more than 3 years	4

Table 9. The magnitude of the intensity of impact on the socio-economic environment

Gradation	**Criterion**	**Grade**
Zero effect	no impact	0
Insignificant influence	positive and negative deviations in the socioeconomic environment correspond to existing fluctuations of the variability of this indicator before the implementation of the project	1
Weak influence	positive and negative deviations in the socio-economic environment outweigh the existing trends in changing living conditions in settlements	2
Moderate influence	positive and negative deviations in the socio-economic environment exceed the existing conditions of the middle-level level	3
Significant influence	positive and negative deviations in the socio-economic environment outweigh the existing conditions of the medium-regional or inter-regional level	4

Integral assessment of the impact on the socio-economic environment consists of three stages.

Stage 1. To determine the significance of the impact on the individual components of the socio-economic environment, it is necessary to use the tables of the criteria for the effects of Tables 7 - 9.

Significance of influence (type of activity) on the i-th component of the socio-economic environment is determined as follows:

$$Q_i^{se} = Q_i^{tse} \cdot Q_i^{sse} \cdot Q_i^{jse} \quad (3)$$

where: Q_i^{tse} - A grade of time influence on the i-th component of the socio-economic environment;

Q_i^{sse} - A grade of spatial influence on the i-th component of the socio-economic environment;

Q_i^{jse} - A grade of intensity of influence on the i-th component of the socio-economic environment.

Valuation points of significance are determined separately for negative and separately for positive spatial, temporal influences and intensity influences.

Stage 2. In the second stage, for each component under consideration, the integral grade is determined by summing up the complex assessment points of the significance of negative or positive influences.

Stage 3. The category of significance is determined by the interval of values depending on the score obtained during the calculation, as shown in table 10.

IOP Conf. Series: Materials Science and Engineering **749** (2020) 012033 doi:10.1088/1757-899X/749/1/012033

Categories of significance are the same for different components of the socio-economic environment and can be compared with the identified component of the socio-economic environment that is most affected.

The score obtained allows you to determine the level of influence (Insignificant, Moderate and Significant) on a specific component of the socio-economic environment.

Table 10. Categories of significance of impacts on the socio-economic environment

Final grade	The final impact
from +1 to +8	Slight positive impact
from +9 to +27	Moderate positive impact
from +28 to +64	Significant positive impact
0	Impact is absent
from - 1 to - 8	Minor negative impact
from –9 to –27	Moderate negative impact
from –28 to –64	Significant negative impact

Similarly, with an assessment of the impact on the natural environment, if the significance of the effect determined for a specific component of the socio-economic environment is the only one, it is directly used to assess the resulting significance of the impact. When acting on one component of the natural environment of various sources (types of activities), an integrated (average) assessment of the significance of the impact on the i-th component of the socio-economic environment will look like:

$$Q_{integr}^{se} = \frac{\sum_{k=1}^{n} Q_{ki}^{se}}{n} \tag{4}$$

where: Q_{ki}^{se} - Integrated point of k-th impact (type of activity) on the i-th component of the socio-economic environment; n - Number of influences (activities) on the i-th component of the socio-economic environment.

It is expedient to analyse the effects on certain components of both natural and socio-economic environments from different sources with the help of estimating matrices based on the matrix of Leopold. Integral (average) estimates for a particular component of the environment are used to determine the significance of the impact.

5. Conclusions

The results of experiments showed the dependence of the generation of energy by photovoltaic panels on meteorological elements of weather conditions, in particular cloudiness and humidity. At the same time, it has no direct dependence on atmospheric pressure and wind speed. The constructed maps of quantitative indicators of solar energy potential for located facilities providing renewable energy in the Carpathian region of Ukraine indicate the feasibility of introducing opportunities for using solar energy in the Carpathian region.

The method of determining the significance of the residual impacts of solar power plants on the environment and the socio-economic environment is proposed. It is performed after screening and mitigation measures. Due to the fact that the influence of many factors influencing the natural and socio-economic environment cannot be quantified, a ball method of impact assessment is proposed, which allows comparing different types of impacts using the matrix method. The methodology is based on the three specified parameters of influence: spatial, temporal and intensity of influence. The impact assessment is carried out on selected components of the environment, taking into account the

IOP Conf. Series: Materials Science and Engineering **749** (2020) 012033 doi:10.1088/1757-899X/749/1/012033

normal operation of the facility and probable emergencies: climate and microclimate, air environment, geological environment, water environment, soils, flora and fauna, protected objects.

References

[1] James A 2015 *Fact Sheet: 6 Things You Should Know About The Value Of Renewable Energy* [Електронний ресурс] Режим доступу http://thinkprogress.org

[2] Adamenko Y O, Arkhypova L M and Mandryk O M 2016 *Territorial standard quality hidroekosystem protected areas* Hydrobiological Journal, 52 (6) P 51–59 English version published by Begell House

[3] Arkhypova L M and Pernerovska S V 2015 *Forecasting water bodies hydrological parameters using singular spectrum analysis* Scientific bulletin of National Mining University, Scientific and technical journal number 2 (146) P 45-50

[4] *Database of meteorological data, weather conditions* [electronic resource] Access mode: gismeteo.ua

[5] Korchemlyuk M and Arkhypova L 2016 *Environmental audit of Ukrainian basin ecosystem of the Prut river* Scientific bulletin of National Mining University, 5 (155) P 98-106

[6] Twidell J and Weir T 2015 *Renewable Energy Resources* Taylor & Francis Ltd. ROUTLEDGE: London, United Kingdom 816 pages

[7] Edited by Boyle G 2012 *Renewable Energy: Power for a Sustainable Future* Oxford, United Kingdom, 584 pages

[8] Mandryk O M, Arkhypova L M, Pobigun O V and Maniuk O R 2016 *Renewable energy sources for sustainable tourism in the Carpathian region* IOP Conference Series: Materials Science and Engineering Volume 144, http://iopscience.iop.org

[9] *Surface meteorology and Solar Energy* https://eosweb.larc.nasa.gov

[10] *Weather Statistics in the Ivano-Frankivsk Oblast* [electronic resource] Access mode: www.rp5.ua

IOP Conf. Series: Materials Science and Engineering **749** (2020) 012034 doi:10.1088/1757-899X/749/1/012034

Modelling experiments with small-scale thermal system used for pig fattener heating

V Erdelyi[1a], J Toth[2], L Janosi[1] and I Farkas[2]

[1] Department of Mechatronics, Szent István University, Gödöllő, Hungary
[2] Department of Physics and Process Control, Szent István University, Gödöllő, Hungary

E-mail: [a] erdelyi.viktor.ferenc@gek.szie.hu

Abstract. This study presents the ongoing research conducted in the field of thermal systems which are commonly used in agriculture and also in other fields to utilize the solar thermal energy. This paper describes a small-scale physical system that is controlled by hardware-in-the-loop (HIL) method. The solar thermal input is realized by a heating element, which can provide a 60 W power in total. The input energy is controllable via Pulse Width Modulation (PWM) to mimic the stochastic changes in weather. The heat storage unit is a 1 litre in volume isolated tank, which contains also the heat exchanger unit. The working fluid in this system is water which is forced by custom made peristaltic pumps both on the "collector" side and on the output. In this way the mass flow rate of the pump can be controlled on the heat exchanger and on the user side, as well. The user side is modelled by a water-air heat exchanger forcing cooled air by a varying speed fan. The temperatures are measured on several points of the system which are used to control the flow rates. Experiments were performed with various controlling algorithms in order to find an optimal solution to provide heat for a pig fattener. The recent status of the test results is shown in the case of On-Off and PID control methods in contrary.

1. Introduction

In this section, the literature overview of the topic will be presented with discussing the studied topic actuality and the issues of the use of solar energy resources.

1.1. Actuality of the topic to be studied

Nowadays - because of the overpopulation of world - it is too pressing to permit a longer delay [1] mitigating food crisis. More than 1.9 billion people exists on the Globe of 6.7 billion who are suffering of malnutrition. The number of people existing near to critical starvation exceeds 600 million [2].

As a scenario the population of the World could reach 9.6 billion inhabitants by the time 2050. This increase requires to enlarge the food production by 70 % [3] and still we will be able to keep only the present condition.

Besides, we are in a shortage of food in the World we have less and less area where we could produce that [4].

So, the biggest challenge is the growing demand from population growth [5] and climate change [6].

The better utilization of arable land and energy seems to be a good solution.

In our current society, it is customary, and even expected, to get cheap, high-quality meat every day, so it is important to produce it quickly and at low cost with large commercial conditions.

The use of renewable, especially solar energies in livestock farming should lower the production costs, which could result in a more affordable food source.

Pig meat is one of the most selling animal protein sources in Europe. Because pigs are highly sensitive to environmental parameters such as air quality and above all to temperature [7], it is important to know what production parameters and technologies are needed for the most efficient production. Therefore, a small-scale physical system was implemented that is controlled by hardware-in-the-loop (HIL) method.

This method gives the possibility to compare various controlling algorithms to find an optimal solution to keep the thermal parameters of a pig barn optimal.

1.2. Solar energy issues

The development of the solar systems is highly speeded up in the recent years. For that reason, a lot of experiments were carried out and a lot of prototype systems were built. Generally saying, it is expensive to build a physical system to do experiments. With the aid of the computer-aided modelling methods these costs can be reasonably reduced, in spite of that such algorithms were created for a specific task and they not provide too much flexibility.

The model-based design is a possible way to resolve this issue [8] This designing method uses mathematical models and the simulation of them to reduce the cost and to speed up the test phase [9]. This approach is useful to analyse and develop an existing solution, as well [10].

The extension of the model-based design is the hardware-in-the-loop (HIL) simulation, where real-time data is used for the model, this is achieved by using embedded systems and sensors. The "hardware" part can be either FPGA devices or more common embedded systems, such as Arduino or Raspberry Pi [11]. There are two main approaches to make a HIL simulation, one where the simulation runs on a host machine and the embedded system acts like an input. In this way, a complex simulation can be built, without the hardware limitations of the embedded system. The other method is where the whole simulation runs on the embedded system, this way the system becomes portable, but it requires more advanced hardware [12].

The presented simulations were performed using a custom-made Simulink library called SimSolar software, which provides a block-oriented modelling environment [13].

2. Materials and methods

In the following section the used devices and models will be presented.

2.1. Physical layout

First, a small-scale physical model of a solar collector unit was created, which has to be controlled accurately. For this reason, an electric heater unit was used to be able to precisely dose the energy which is representing the solar thermal energy input. The model was made so, that there will be a room for further improvements, which are either pre-planned or needed based on the data of the tests. In this experiment only the temperatures on the input of the collector, on the output of the collector and the temperature of the heat storage tank were measured.

The model utilizes 8 DS18B20 digital temperature-sensors, three in the heat storage unit, one in the collector unit, on both collector in and out measurement points, one on the user output, which is the water-air heat exchanger and one ambient temperature measurement point.

Figure 2 shows the block diagram of the experimental device assembly while figure 3 shows the actual experimental system. Two FDM (Fused Deposition Modelling) printed peristaltic pumps were used, to vary the flow rate in both collector, and user loop.

Since peristaltic pumps are operating on the principle of positive displacement, therefore they can be used to precisely control the flow rates with the use of stepper motors as a driving solution. To drive the stepper motors, A4988 stepper driver-boards were applied.

IOP Conf. Series: Materials Science and Engineering **749** (2020) 012034 doi:10.1088/1757-899X/749/1/012034

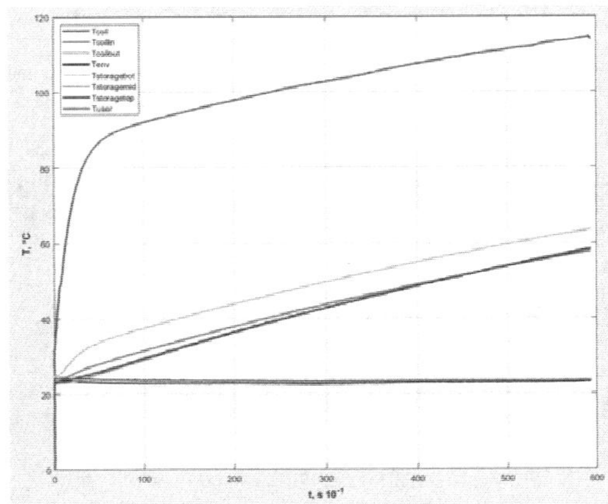

Figure 1. Values of the temperature sensors

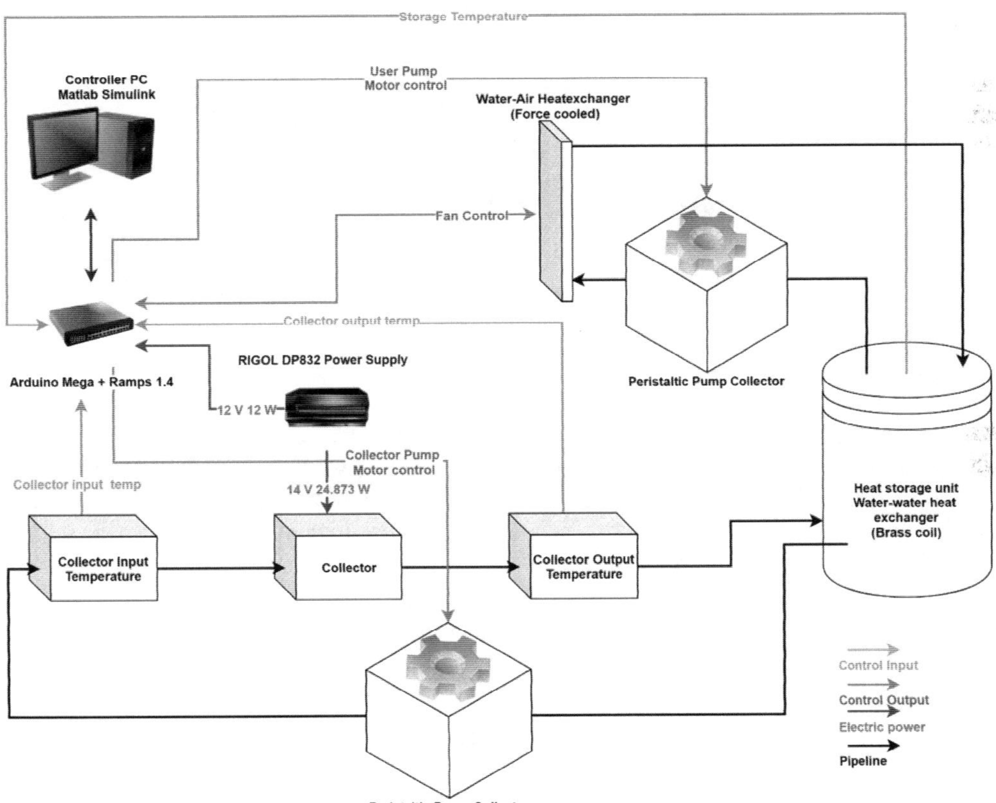

Figure 2. Block diagram of the small-scale model

As a heat storage unit, a polypropylene plastic bucket insulated by roof insulation material is used, with a built-in a brass coil, which is used to exchange heat between the storage and collector loop.

First, we measured a heating cycle. We used 50W in average to heat the collector circuit, and we drove the pump of the collector circuit at 300 1/min., which translates to approx. 0.08 l/min flowrate. In this experiment we used all the sensors. Figure 1 shows the results of this test.

The user pump wasn't applied in this test.

IOP Conf. Series: Materials Science and Engineering **749** (2020) 012034 doi:10.1088/1757-899X/749/1/012034

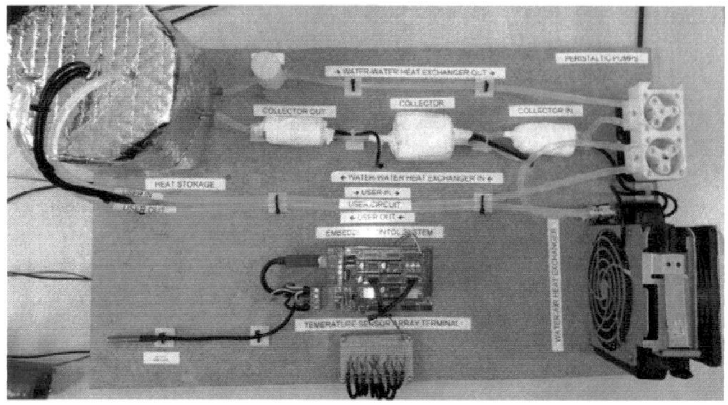

Figure 3. Assembled experimental system

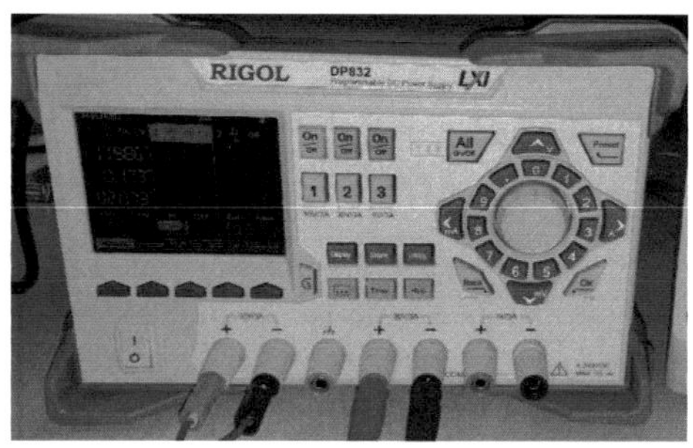

Figure 4. Power supply

The power is provided by a RIGOL DP832 three channel linear lab bench power supply (Figure 4). This supply made possible to constantly monitor the used energy to power the heating element. Since it is a variable supply, the output power can easily be varied

2.2 Arduino firmware

The program running on the Arduino (firmware) controls the stepper motors of the pumps, collects the temperature readings of the sensors and handles the communication with the Simulink system as it is shown in figure 5. The data of the sensors is read by a 10-bit analogue-digital converter and sent to Arduino board via one-wire communication. The Arduino then sends the data array to the Simulink model, where the raw-data-temperature conversion happens.

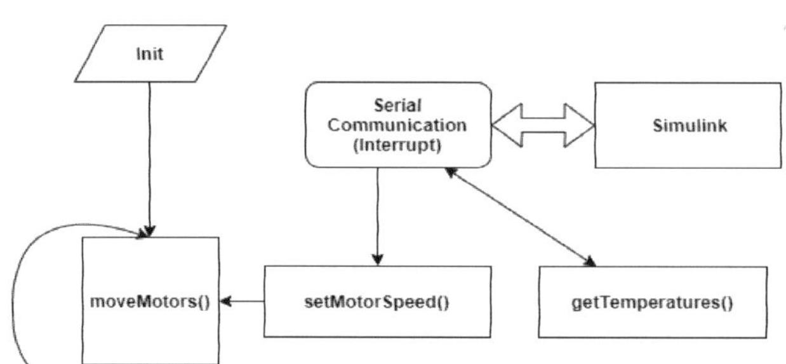

Figure 5. Flow chart of the firmware

The communication is a standard full duplex serial communication (RS232) with a baud rate of 9600 running form an interrupt routine, this way the control of the stepper motors is mostly uninterrupted. This solution was necessary in order to provide the reliable speed of the motors.

IOP Conf. Series: Materials Science and Engineering **749** (2020) 012034 doi:10.1088/1757-899X/749/1/012034

2.3 On-Off controller process

The HIL solution was first used in an On-Off controlled process which can be seen in figure 6.

The model contains the necessary components of the serial communication (configuration, read, write), the logic of the communication, the temperature converting routines and the On-Off controller unit itself.

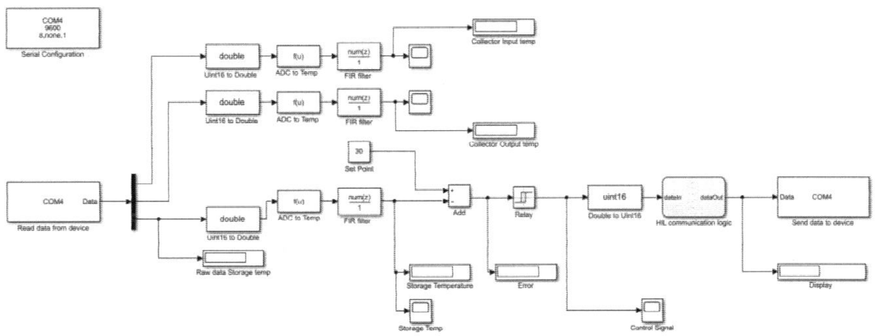

Figure 6. Simulink model of the On-Off controlled system

The data of the signals of the communication build from byte strings packed from 2 bytes, which is the standard 'uint16' format (16 bits), but the blocks of the mathematical computations uses double precision floating point types, which are 'doubles', so the necessary conversions had to be made.

The readings of the temperatures contain some noise, so a simple Finite Impulse Response filter was used in order to compensate this effect, this solution can be interpreted as a moving average.

The control logic of the HIL simulation contains 4 steps, as it is shown in figure7:

1. Initialization, 2 - *get the temperature of the storage,* 3-*set the speed of the motor,* 4-*go to step 2*

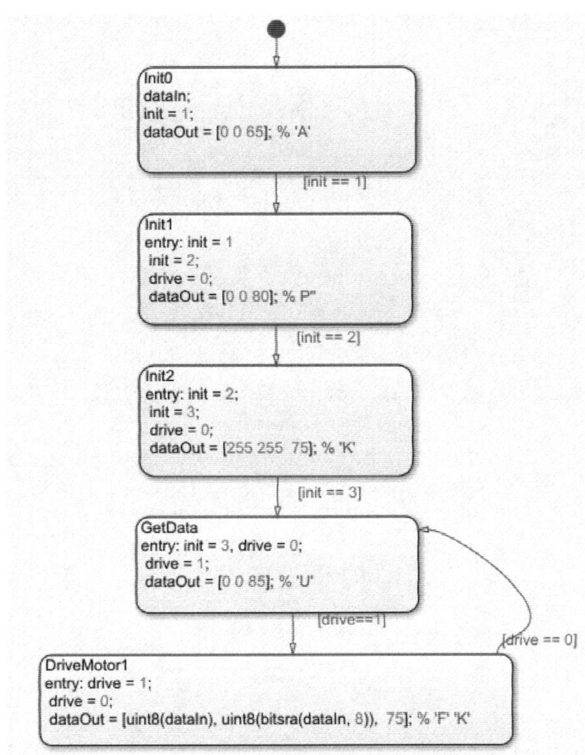

Figure 7. Communication logic

IOP Conf. Series: Materials Science and Engineering **749** (2020) 012034 doi:10.1088/1757-899X/749/1/012034

To achieve the stable operation a special toolbox was used: Stateflow. The Simulink environment has no built-in solution for a so called, 'state machine', but this type of communication requires one. The state machine is a simple system, where the operation of the machine depends only on the internal state of the system, like a Turing machine. The Stateflow toolbox enables this approach.

The chosen controller was an On-Off controller, which has a binary controlling algorithm, because there are only two possible outputs, usually a low (0) and a high (1) level signal, the actual value of the 'high' and the 'low' signals depends on the application. The hysteresis prevents the controller for the rapid oscillations. With the introduction of the error function, which can be written as:

$$\varepsilon(\tau) = r(\tau) - y(\tau). \tag{1}$$

The operation of the controller can be described as follows:

$$u(\tau) = \begin{cases} 1, & s = 1 \\ 0, & s = 0 \end{cases}. \tag{2}$$

where s is the internal state of the controller, it depends on the previous state as follows:

$$s_\tau = \begin{cases} 1, & if\ s = 0\ \&\ \varepsilon(\tau) < -h/2 \\ 0, & if\ s = 1\ \&\ \varepsilon(\tau) > h/2 \\ s_{\tau-1}, & otherwise \end{cases} \tag{3}$$

The controlled parameter in this case is the flowrate of the pump on the user side. The flowrate can be controlled by the rotational speed of the motor as the following.

$$Q = V_g n. \tag{4}$$

2.4 *PID controller*

The PID (Proportional–Integral–Derivative) controller was the second trial to be implemented. The PID is a more sophisticated, continuous controlling algorithm. The governing equation of the PID controller:

$$u(\tau) = P\,\varepsilon(\tau) + I\int_0^\tau \varepsilon(t)\,dt + D\,\frac{d}{d\tau}\,\varepsilon(\tau) \tag{5}$$

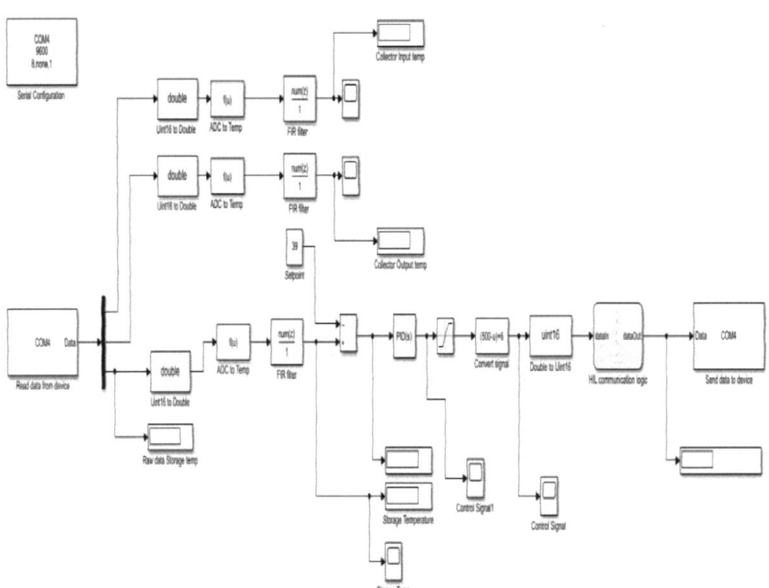

Figure 8. Simulink model of the PID controlled system

IOP Conf. Series: Materials Science and Engineering **749** (2020) 012034 doi:10.1088/1757-899X/749/1/012034

The base of the HIL system is the same, only the controller has changed as seen in figure 8. This is one of the main advantages of the block-oriented modelling design, that the change of the whole models means only the replacement of some of its blocks

3. Results and discussion

The figures 8-11 show the results of the simulation trials. The temperature of the storage (blue), the temperature of the collector input (red), the temperature of the collector output (orange) and the setpoint can be seen in figure 9 in case of the On-Off controller. Figure 10 shows the output of the On-Off controller.

The temperatures, with the same colouring, of the PID model are in figure 11 and the output of the PID controller is in figure 12.

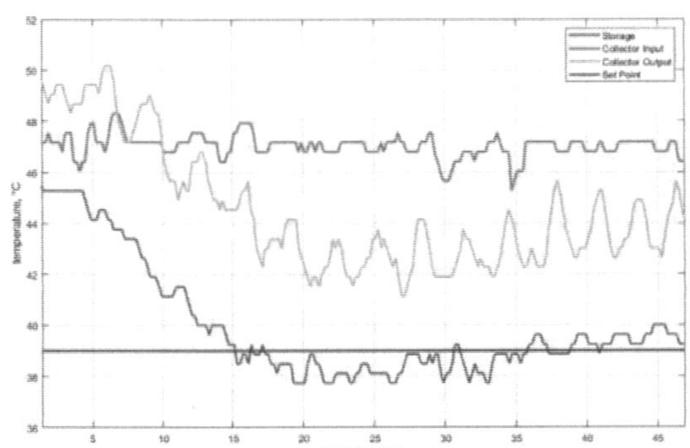

Figure 9. Temperatures in the function of time (On-Off)

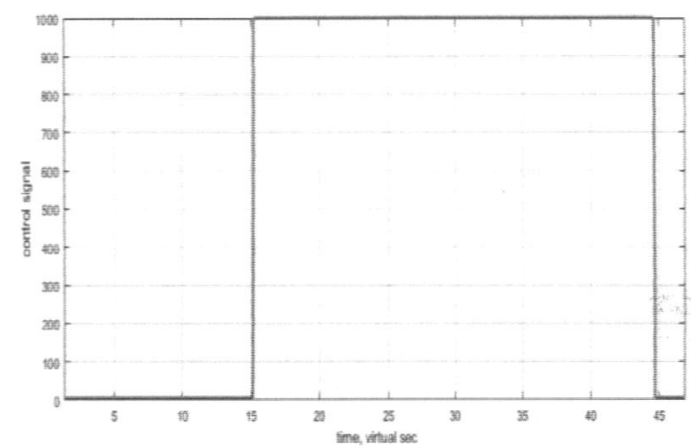

Figure 10. Control signal (On-Off)

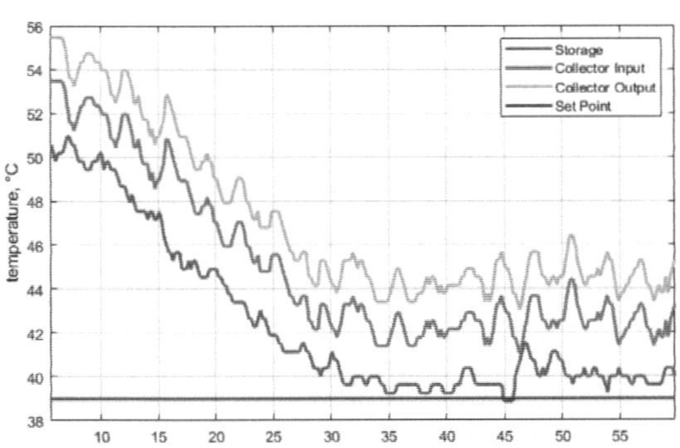

Figure 11. Temperatures in the function of time (PID)

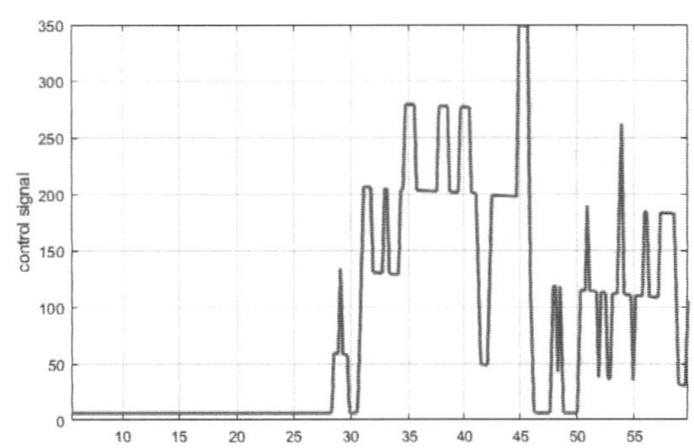

Figure 12. Control signal (PID)

4. Conclusion

In conclusion, it can be stated, that these applications have a very large time constant, so they are hard to control. In our cases the experiments showed, that PID controller can be very unstable when controlling the temperature of the heat storage unit. This is due the inaccurate measurement of the temperatures. So, the first improvement will be to use a sophisticated temperature measurement device to acquire data from the system. The PID method, in addition, is very sensitive to the working point. It means that when the P, I and D parameters of the controller are tuned, and so the system gets into another setpoint or different starting conditions, the control can drastically overshot. The On-Off controller, in this case, was a better solution then the PID. In conducting further experiments, more algorithms will be tested. It is assumed, that due to the large time constant of the system the model-based control would be the optimal solution.

Nomenclature

D	coefficient of the derivative member
h	hysteresis
I	coefficient of the integral member
n	rotational speed of the motor, 1/s
P	coefficient of the proportional member
Q	flowrate, m^3/s
r	reference signal
s	state of the controller
u	control signal
V_g	geometric volume, m^3
y	output of the system
ε	error
τ	time, s

References

[1] Glaeser B 1989 *Umweltpolitik zwischen Reparatur und Vorbeugung. Eine Einführung am Beispiel Bundesrepublik im internationalen Kontext* Westdeutscher Verlag GmbH Oplade pp.14-15

[2] Szakály S 2004 *Táplálkozási dilemmák és az élelmiszerek fejlesztésének világstratégiai irányai Élelmiszer* Táplálkozás és Marketing, évf. 1 sz.. 1-2

[3] Northoff E 2016 *2050 A third more mouths to feed* Food and Agriculture Organization of the United Nations

[4] EEA Report. 2006 *Urban sprawl in Europe, The ignored challenge* European Environment Agency Report No 10 ISSN 1725-9177

[5] Boserup E 2017 *The conditions of agricultural growth: The economics of agrarian change under population pressure* Routledge pp.4-7

[6] Zhang P, Zhang J and Chen M 2017 *Economic impacts of climate change on agriculture: The importance of additional climatic variables other than temperature and precipitation* Journal of Environmental Economics and Management 83 pp.8-31.

[7] Cross A J, Keel B N, Brown-Brandl T M, Cassady J P and Rohrer G A 2018 *Genome-wide association of changes in swine feeding behaviour due to heat stress* Genetics Selection Evolution 50 p.1

[8] Paterno F 2012 *Model-based design and evaluation of interactive applications* ISBN: 978-1-447-10445-2

[9] Erdélyi V and Jánosi L 2017 *Digital twin and shadow in smart pork fetteners*, Proceedings of the 5 th International Scientific Conference on Advances in Mechanical Engineering Debrecen Hungary ISBN 978-963-473-304-1 pp.142-146

[10] Bilčík M, Božiková M, Petrović A, Malínek M, Cviklovič V, Olejár M and Ardonová V 2018 *Analysis of selected photovoltaic panels operating parameters as a function of partial shading and intensity of reflected radiation* Acta Technologica Agriculturae Vol. 21 pp.14-17.

[11] Velasquez D R, Collazos VT and Mines J M 2017 *A low-cost hardware-in-the-loop real time simulation of control systems* Proceedings of the 2017 IEEE 24th International Congress on Electronics Electrical Engineering and Computing INTERCON 2017 ISBN 978-150906362-8

[12] Ilyas A, Ayyub M, Khan MR, Jain A and Husain M 2018 *Realisation of incremental conductance the MPPT algorithm for a solar photovoltaic system* International Journal of Ambient Energy Volume 39 Issue 8 ISSN 0143-0750 pp.873-884

[13] Tóth J and Farkas I 2017 *Developing a Simulink library for solar energy applications* R&D in Mechanical Engineering Letters Gödöllő Hungary Vol. 16 HU ISSN 2060-3789 pp.89-95

IOP Conf. Series: Materials Science and Engineering **749** (2020) 012035 doi:10.1088/1757-899X/749/1/012035

Analysis of the extrusion process parameters of high strength aluminium alloys used in the aerospace industry

V Hotea and J Juhasz[a]

Technical University of Cluj-Napoca, Faculty of Engineering, Department of Mineral Resource, Material and Environment Engineering, Baia Mare, Romania

E-mail: [a] jozsef.juhasz@cunbm.utcluj.ro

Abstract. The paper approach some aspects of the most important parameters that influence the process of extrusion and the quality of material exiting from the die such as extrusion ratio, working temperature and speed of deformation. An important advantage gained by using extrusion of high strength aluminium alloys is that different microstructures may be obtained by varying process parameters, such that optimization of processing conditions leads to improved thermo-mechanical properties. Among the multiple process parameters in aluminium extrusion, the extrudate temperature is the most important because it determines the quality of the product and indicates the possibility of increasing extrusion speed.

1. Introduction

Aluminium alloys are the raw material for aircraft structural parts for over 80 years due to their well-known performance, well-established manufacturing design methods and reliable inspection techniques. The Aluminium Association maintains a widely recognized system for each category of aluminium alloys, as described in ANSI H35.1, Alloys and Temperatures for Aluminium [1]. This standard date back to 1957 (American Standard H35.1-1957) when only the names of deformable alloy systems that were developed by the Aluminium Association were considered. This standard was re-enacted in 1962 by adding names of temperamental systems as an effect. Good high-temperature strength is attained by the addition of copper (up to 4%) and/or nickel, manganese, or iron up to one percent each. Good chemical resistance is shown by alloys with additions of magnesium, manganese, or a combination of magnesium and silicon. Machinability is greatly improved by the addition of lead and bismuth up to 0.6% each. A fine-grained structure is obtained especially through the addition of titanium and boron (up to 0.1%) [2]. Also, the homogenization treatment given to the alloy further improves the distribution of alloying element in the aluminium alloy's composition.

From the 7xxx series, there are Al-Zn-Mg-Cu alloy systems containing zinc, magnesium and copper, as well as additives such as chromium, manganese, or zirconium, and impurities such as iron and silicon. These alloys have a short hot-melt temperature due to a high deformation resistance and an early melting tendency at high processing temperatures [3, 4].

The extrudability, which can be measured by the maximum extrusion speed, is one of the most important factors influencing the cost and efficiency of the extrusion parameters of the process. The temperature and velocity as well as the state of stresses in the deformation zone, mainly in the die region, a significant role in improving the extrudability of a particular alloy. Various evaluations of the extrudability of high-strength aluminium alloys were studied by Zasadzinski and Misiolek who presented the results of extrusion investigations of alloys that are difficult to deform by applying controllable container temperatures [5].

In a recent study, the extrudability of 7xxx series alloys was measured as the maximum extrusion speed before the hot break or the crack speed in the extrudate section. Their resins show that the extrudability of these alloys is strongly dependent on some alloying constituents such as Mg and only slightly dependent on the zinc content. Adding percentages of copper and zirconium also reduces extrusion speed [6].

Despite the fact that control systems and infrared thermometers are commonly used in the aluminium extrusion industry, the Temperature Process Control (TPC) System is the first to effectively merge these two technologies to deliver true automatic temperature-based control in a system that is easy to use [7].

2. Fundaments and principles of aluminium alloys extrusion

Extrusion is a plastic deformation process in which a block of metal (billet) is forced to flow by compression through the die opening of a smaller cross-sectional area than that of the original billet as shown in figure 1. Extrusion is an indirect-compression process. Indirect-compressive forces are developed by the reaction of the workpiece (billet) with the container and die; these forces reach high values. The reaction of the billet with the container and die results in high compressive stresses that are effective in reducing the cracking of the billet material during primary breakdown from the billet [8].

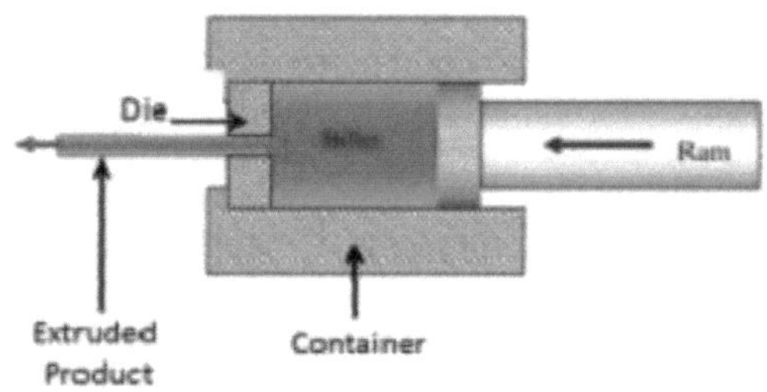

Figure 1. Principle of direct extrusion [8]

The most important and common method used in aluminium extrusion is the direct process. Figure 1 show the principle of direct extrusion where the billet is placed in the container and pushed through the die by the ram pressure. These processes can be summarized as follow [9]:

- First billet or ingot (metal work piece of standard size) is produced. This billet is heated in hot extrusion or remains at room temperature and placed into a extrusion press (Extrusion press is like a piston cylinder device in which metal is placed in cylinder and pushed by a piston;
- Now a compressive force is applied to this part by a plunger fitted into the press which pushes the billet towards die;
- The die is small opening of required cross section. This high compressive force allows the work metal to flow through die and convert into desire shape;
- The extruded part removes from press and is heat treated for better mechanical properties.

Direct extrusion finds application in the manufacture of solid rods, bars, hollow tubes, and hollow and solid sections according to the design and shape of the die.

3. Study and analysis parameters of extrusion process

The most important parameters that influence the process of extrusion and the quality of material exiting from the die are as follows [10]:

- Extrusion ratio (E.R.);
- Working temperature (Tb);

- Speed of deformation (Ve);
- Alloy flow stress.

The correlation between these variables is shown in figure 2.

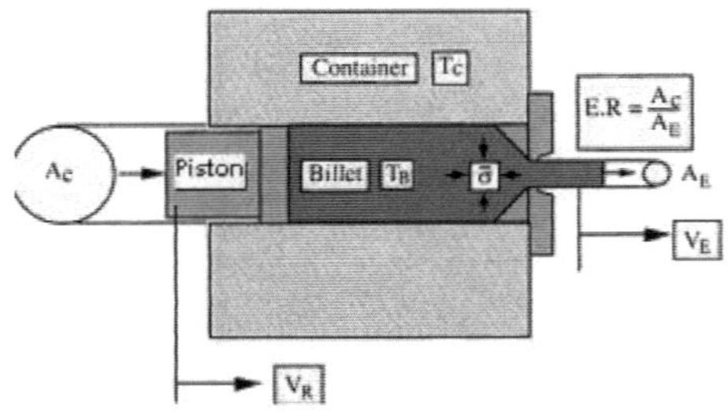

Figure 2. The main variables that influence the force required to produce extrusion and extruded quality [10]

3.1. Extrusion speed

The response of a metal to the extrusion processes can be influenced by the deformation velocity. Increasing the speed of the ram produces an increase in the extrusion pressure. The temperature developed at extrusion increases as the ram speed increases. This increase is due to the fact that the degree of stress is directly proportional to the speed of the piston and the magnitude of the heat generated is proportional to the stress rate. When the ram speed is lower, the more time will be available for the flow-generated heat energy. Heat conduction is more pronounced on aluminium due to its higher conductivity.

The start-up parameters are set from the Extrusion Speed Process Controller to Start, and its correction, as shown in figure 3.

Figure 3. Control panel of extrusion speed and temperatures at start-up [Source: UAC Europe SRL]

The most used alloys for small- and large-shaped extruded sections in the aircraft industry is 2024 and 7075, which offers a high strength to-weight ratio.

The chemical composition, heat treatment and their extrudability are shown in table 1.

Table 1. High strength aluminium alloys and extrudability

Alloy	**Heat treat**	**Major Alloying elements**				**Extrudability (a)**
		Mn	**Mg**	**Cu**	**Zn**	
2024	Natural aging	0.3-0.9	1.2-1.8	3.8-4.9	0.25	15
7075	Artificial aging	0.30	2.1-2.9	1.2-2	5.1-6.1	10

IOP Conf. Series: Materials Science and Engineering **749** (2020) 012035 doi:10.1088/1757-899X/749/1/012035

The starting parameters for the alloys in the 2xxx and 7xxx series are shown in tables 2 and 3.

Table 2. The starting parameters for alloys AA 2xxx series

Extrusion Ratio (%)	Temp. billet (^{0}C)	Temp. container (^{0}C)	Extrusion speed (mm/min.)
<60	460	468	1000
40-60	430	450	1000
40-60	400	430	1000
20-40	380	430	1000
<20	380	410	1000

Table 3. The starting parameters for alloys AA 7xxx series

Extrusion Ratio (%)	Temp. billet (^{0}C)	Temp. container (^{0}C)	Extrusion speed (mm/min.)
<100	443	443	1000
70-100	420	440	1000
40-70	390	400	1000
<40	360	370	1000
<40	300	320	1000

As can be seen from Tables 2 and 3, the starting extrusion speed for the 2xxx and 7xxx series alloys is the same (1000 mm / min) as it's safe to avoid crack defects. The typical extrusion exit speeds achievable for different types of high strength aluminium alloys are shown in Table 4 [10], and are a direct indication of the extrudability of each alloy [11].

As can be seen from table 4, that the extrusion exit speed for several brands of high-strength aluminium alloys varies and is dependent on the cross-section shape and the extrusion ratio [12]

Table 4. Typical extrusion speed for high strength aluminium alloys [10]

Alloy	Heat treat or not	Temp. (^{0}C)	Exit speed (mm/min.)
2014-2024	Treat	420-450	1.5-3.5
7001	Not treat	370-415	0.5-1.5
7075, 7079	Not treat	300-460	0.8-2.0
7049, 7150, 7178	Not treat	300-440	0.8-1.8

The extrusion outlet of the extrusion press is illustrated in figure 4.

Figure 4. Output extrudate from the extrusion press (Source: UAC Europe SRL)

3.2. Working Temperature

Temperature is one of the most important extrusion parameters. The flow demand is reduced if the temperature is increased and the deformation is therefore easier to achieve, but at the same time the maximum extrusion speed is reduced because the localized temperature can lead to the incipient melting temperature of the billet.

Despite the fact that control systems and infrared thermometers are commonly used in the aluminium extrusion industry, the Temperature Process Control (TPC) System is the first to effectively merge these two technologies to deliver true automatic temperature-based control in a system that is easy to use. Williamson Corporation and SAI Automation have partnered to develop this innovative system which integrates Best Practices Standards with simple and effective closed loop controls to effectively optimize press performance and enable a process for continuous improvement. Unique TPC features include [7]:

- Real-time temperature-based closed loop control;
- Continuous optimization of profile temperatures at the platen exit for isothermal control;
- Continuous optimization of billet feed temperatures;
- Automatic compensation for varying operating conditions;
- Recipe and order management modules that organize and apply Best Practices Standards for each die and order;
- Advanced Utilities and Reports to automatically capture critical process data;
- Flexible design to integrate with existing control systems;
- Advanced options to optimize tapered billet heating, nitrogen die cooling, and quench rates;
- Installation and training in less than one week with little or no downtime.

3.3. Extrusion ratio

The extrusion ratio (ER) for a multi-hole die is defined by the relationship:

$$ER = \frac{A_C}{n(A_E)} \qquad (1)$$

in which: A_C - the area of container, n - the number of symmetrical holes in the die, A_E - the area of extrudate shape.

The extrusion ratio of a extruded form is a clear indication of the amount of mechanical work that will take place as the extrudate form.

Actual demand is a function of the extrusion ratio and, finally, the extrusion pressure required for extrusion is a function upon request. If the extrusion ratio of the profile is low, the amount of plastic deformation is also low. As a result, the amount of work done during extrusion will be less. In extruded aluminium with a low extrusion ratio, the structure will be similar to the cast aluminium casting (thicker grains). This structure will be mechanically weak and, as a result, extruded dies with an extrusion ratio of less than 10 to 1, which cannot be guaranteed to satisfy the mechanical and physical properties of the material specifications.

4. Conclusions

An important advantage gained by using extrusion of high strength aluminium alloys is that different microstructures may be obtained by varying process parameters, such that optimization of processing conditions leads to improved thermo-mechanical properties.

The most important process variables affecting both the quality of the extruded and the productivity of the extrusion process is working temperature and extrusion speed. Among the multiple process parameters in aluminium extrusion, the extrudate temperature is the most important because it determines the quality of the product and indicates the possibility of increasing extrusion speed.

IOP Conf. Series: Materials Science and Engineering **749** (2020) 012035 doi:10.1088/1757-899X/749/1/012035

The extrudability, which can be measured by the maximum extrusion speed, is one of the most important factors influencing the efficiency of process extrusion parameters. The temperature and velocity as well as the state of stresses in the deformation zone, mainly in the mould region, play a role significant in improving the extrudability of a certain type of high strength aluminium alloy.

References

[1] Aluminum Association Secretariat 1997 *ANSI H35.1 Alloy and Temper Designation Systems for Aluminum* Aluminum Association Washington DC
[2] http://docplayer.net/21620621-Chapter-2-production and processing of aluminum.html
[3] Strawbridge D J, Hume-Rothery W, Little A 1948 *J. Inst. Met.* Vol 191 (No. 74) 1947-1948
[4] Hatch J.Ed. 1984 *Aluminum: Properties and Physical Metallurgy* ASM International pp.424
[5] Zasadzinski J and Misiolek W 1988 *Estimating Optimal Speed/Temperatures to Maximize Hot Extrusion Exit Speed* Proc Fourth International Aluminum Extrusion Technology Seminar Vol 1 Aluminum Association and Aluminum Extruders Council
[6] Asboell K, Jensrud O, Reiso O and Jacobsen D 1996 *Extrudability and Mechanical Properties of Some 7xxx-Series Alloys* Proc. Sixth International Aluminum Extrusion Technology Seminar Vol 2 Aluminum Association and Aluminum Extruders Council
[7] https://www.williamsonir.com/wp-content/uploads/2018/
[8] http://www.mech4study.com/2017/04/extrusion process
[9] https://www.williamsonir.com/wp-content/uploads/2018/04/Aluminum_Extrusion-Application
[10] Saha P K 2000 *Aluminum Extrusion Technology* http://www.asm-intl.org ISBN 0-87170-644-X pp.153
[11] Wilm A 1911 *Metall* Vol 225 (No. 8)
[12] Dix E H 1945 *Trans. AIME* Vol 130 (No. 35)

IOP Conf. Series: Materials Science and Engineering **749** (2020) 012036 doi:10.1088/1757-899X/749/1/012036

Precessional gear-box research regarding vibration activity behaviour

I Malcoci[a], N Trifan and R Ciobanu

Technical University of Moldova, Department Fundamentals of Machines Design, Chisinau, Republic of Moldova

E-mail: [a] iulian.malcoci@bpm.utm.md

Abstract. An ideal dynamic system in general and precessional transmissions in particular should not generate any vibration, because vibration is equal with loss of energy and bigger sound level emission. In our previously research we were focused on different vibro-acoustical aspects regarding kinematical and power precessional planetary gear-box when precessional reducer is minimal and maximal filled with cooling and lubrication liquid at different operational speed (power precessional transmission K-H-V Type) and vibro-acoustical behaviour for kinematical precessional transmission 2K-H Type. In this paper I will analyse vibration behaviour for power precessional gearbox 2K-H Type regarding vibrations that occurs at bearings and main shaft misalignment. Vibration research and analysis will be made using GUNT PT500 Machinery Diagnostic System. Data acquisition will be made using acceleration sensor type IMI603C01 for FFT analysis in GUNT PT500.04 software. Obtained results will be compared with German standard VDI-2058 "Limit value for vibration severity".

1. Introduction

The purpose of modern-day machines in general and gearbox in particular is to carry out needs-based on maintenance, good working conditions and to minimize the repair and other servicing downtimes and dysfunctionalities. This increases the overall equipment effectiveness and optimizes the cost. The mechanical condition of a machine or a machine component can be accurately diagnosed from the nature and extend of vibration they generate. The aim is to detect damage as it occurs, allowing scheduled repairs or maintenance to be carried out in time [1].

Vibration in planetary precessional gear box occurs at bearings, gear wheels, misaligned shafts, imbalance rotating parts, couplings. If damage occurs, not only the dynamic processes change, but also the forces that act on system components [4].

2. Vibration generation sources in the planetary precessional gear

Precessional planetary transmission is a mechanical system subjected to mechanical oscillations generating vibrations and noise and can be divided into the following groups: dimensional deviations of sliding contact elements, dimensional deviations at machine parts that make rolling contact, static and dynamic imbalance and assembly errors vibration [1].

IOP Conf. Series: Materials Science and Engineering **749** (2020) 012036 doi:10.1088/1757-899X/749/1/012036

2.1. Dimensional deviation of sliding contact elements

Precessional gear units that engage with the engaging frequency $\omega_z = z \cdot \omega_1$ where z teeth number and ω_1 the angular velocity of the main shaft, are a source of vibration and noise and a transmission quality indicator. The roles of the satellite, which form a kinematic twist-axle roller, are also a source of vibration:

- dimensional and shape deviations of the rollers and axles;
- changing the position of the rollers in the solicited and unsolicited area of the satellite crowns as well as the stiffness of the contacts;
- uneven motion of tapered rollers as a result of different stresses depending on the position.

2.2. The Dimensional deviation at machine parts that make rolling contact

Precessional planetary transmission typically includes 3 pairs of bearings: the main shaft bearing, the crankshaft, which support the precession and rotation movement of the satellite shaft, and the driven shaft. The separate level of vibration and bearing noise influences the general level of pre-planetary planetary transmission and its competitiveness. The bearings act through different effects (direct or indirect) at the vibration and noise level of other elements (shafts, satellite block, housing) [4].

The roller bearing contact may have the following causes:

- changing the position of the rolling elements (rollers or balls) in the loaded or unloaded area of the bearings, in relation to the dimensional deviations and the number of rolling elements. Also, periodically modifying the stiffness of the contacts and the system by favouring the occurrence of parametric type vibrations;
- non-uniform motion of the rolling body due to different and complex loads (especially in the satellite block bearings) depending on the position; this uneven motion and strain also leads to friction and collisions of the wheel bodies or cages;
- making roll-over contacts on surfaces with dimensional, shape and position deviations: diameters, frontal or radial tread patterns, eccentricity, ovality, corrugation, roughness;
- the movement of rolling body over impurities placed on contact surfaces or localized defects or damage (Pitting tipping, abrasive wear, fingerprints, etc.)

2.3. Static and dynamic imbalance

When the centre of mass of a moving or rotating element does not coincide with the centre of rotation (or precession), an imbalance is created, which may be in a plane called static or multi-plane imbalance, known as dynamic imbalance. Generally, in any machine, the shafts have a small imbalance that will generate a sinusoidal vibration with a frequency equal to that of the rotation motion.

The central deviation of the masses of the satellite block from the precession centre "O" with an "a" value generates phasic vibrations (Figure 1). Solution: static balancing of the satellite block at the design stage.

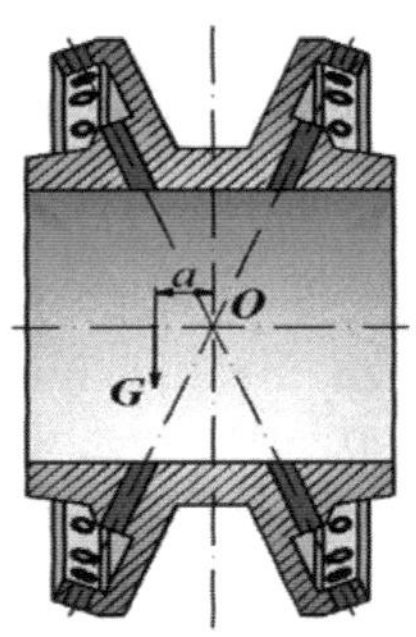

Figure 1. Satellite block imbalance

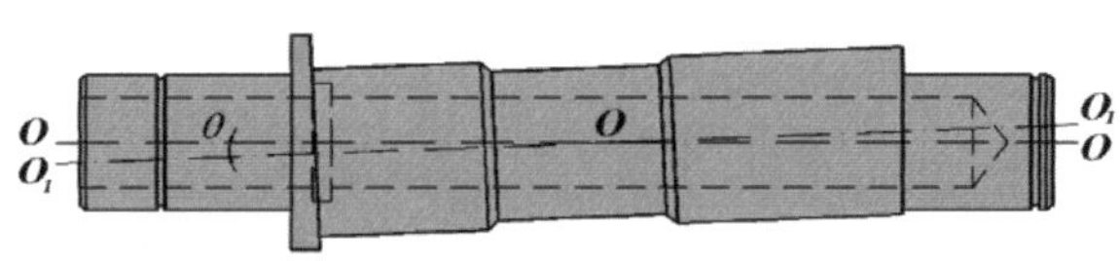

Figure 2. The crank shaft imbalance

IOP Conf. Series: Materials Science and Engineering **749** (2020) 012036 doi:10.1088/1757-899X/749/1/012036

In precessional transmissions, the imbalance is generated by the satellite and the crank shaft. The functional slope of the O_1-O_1 axis with the angle θ to the O-O axis of the precessional planetary transmission crank (Figure 2) also leads to the generation of vibrations. Solution: balancing the crank shaft. One of the solutions is shown in figure 2, when a cavity is made in the crank-shaft, the axis of which is inclined at an angle γ greater than the angle θ, and hole diameter is determined with formula $d = D\sqrt{tg\theta / tg\gamma}$

2.4. Assembly vibration errors

The components of the "satellite block and crank shaft" node can be executed perfectly and balanced but assembling may cause different assembly errors (excessive play or too high grip). One of the most common assembly faults is the non-coincidence of the satellite precession centre (point O in figure 3) with the intersection point of the O_1-O_1 axis of the crank (inclined part) and O-O of the drive shaft. An effective solution to compensate for this error is to install the satellite and crank shaft, on radial bearings with cylindrical roller, which in turn is an additional source of vibration.

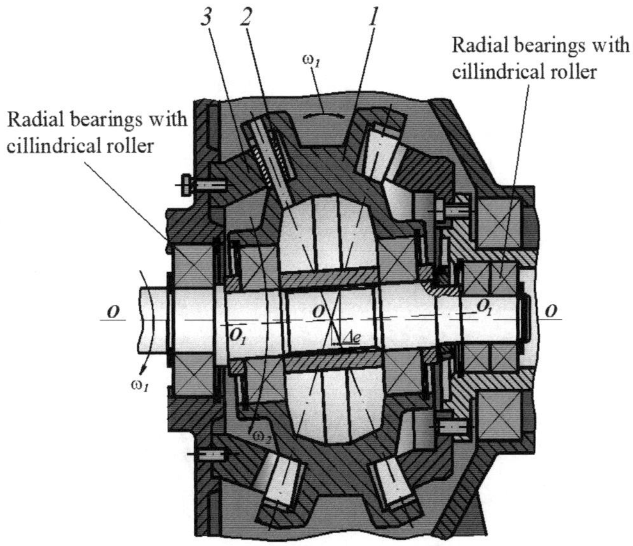

Figure 3. Satellite block and crank shaft node

There are various constructive solutions, next in figure 3 it is shown one of them [4]. The auto-aligning process of satellite block 1 in the precessional planetary transmission can be presented in the following way. The constant component of the primary errors of the precessional planetary transmission elements, which does not depend on the reciprocal position of the parts in the transmission, generates some deviation Δe of the satellite block 1 from the precession planetary transmission axis. Regarding this new position of the satellite block 1, the precession movement with the frequency ω_1 (Figure 3) and the frequency rotation ω_2, after a trajectory determined by the harmonic first of the actual error in the satellite block gear (radial and frontal impact of the crowns): $\omega_2 = \omega_1 / z_2$ – where z_2 represents the number of rollers of the satellite crown, in contact with the fixed central toothed wheel 3.

The installation of the satellite block 1 on radial bearings, with cylindrical roller give the possibility of self-alignment, which will allow to obtain compensation of the manufacturing and installation errors of the engaging elements, which perform in reduction of the vibration and noise level.

3. Precessional transmission vibration frequencies identification

3.1. Planetary precessional transmission type 2K-H

Precessional planetary transmissions, like any gear with toothed wheels, are characterized by various excitatory forces that generate various oscillations. We will then examine two structural types of 2K-H and K-H-V planetary planetary transmissions, which have different kinematic structures and different vibration generators [4].

In order to detect vibration sources in the precessional planetary transmission, it is necessary to determine the oscillations of the dynamic elements, which would allow the processing and analysis of vibrograms. The crank shaft rotates with the input frequency ω_1, so the crankshaft assembly and assembly errors will be pronounced around these frequencies:

$$\omega_1 = \frac{\pi n_1}{30}(s^{-1}). \tag{1}$$

The satellite block will also perform a spherical-space motion around the precession center (the O-O crankshaft intersection point (Figure 4) with the O_1-O_1 axis of the crankshaft inclined axis and the conical shaft axes of the crowns) with the frequency ω_1 and rotating motion around its geometrical axis with angular velocity:

$$\omega_2 = \frac{\omega_1}{i_1} = \omega_1 \cdot \frac{z_2}{z_1 - z_2}, \tag{2}$$

where z_1 is the teeth number of the central wheel 1; z_2 - the rollers number of the satellite crown block.

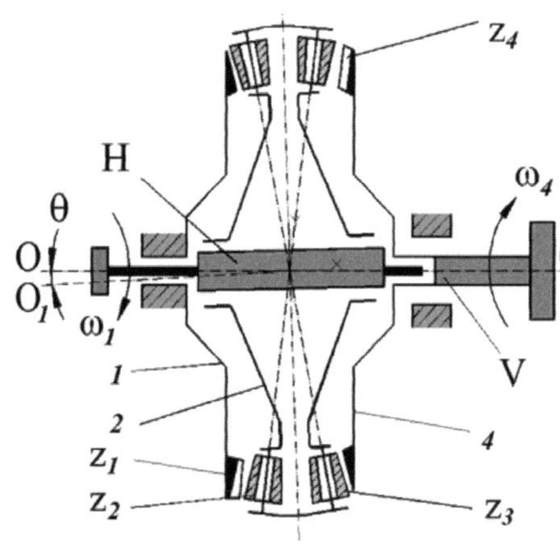

Figure 4. 2K-H kinematic scheme

The bearing rings of the "satellite-crankshaft" block will rotate with the frequencies:
- the inner ring, rigidly installed on the crankshaft, will rotate at the crankshaft angular speed ω_1;
- the outer ring, rigidly installed in the satellite block, will rotate at the angular speed ω_2*.

Case 1: If the crank shaft and the satellite block rotate in the same direction $z_1=z_2+1$ and $\omega_2^* = \omega_1 - \omega_2$.

Case 2: If the crank shaft and the satellite block rotate in different directions $z_1=z_2-1$and $\omega_2^* = \omega_1 + \omega_2$.

In this case z_1 is the teeth number of fixed central wheel 1, z_2 - the number of crowns 2 of the satellite block, which contact the fixed wheel 1.

The driven toothed wheel 4 will rotate at the angular speed:

$$\omega_4 = \frac{\omega_1}{i_2} = \omega_1 \frac{z_2 z_4 - z_1 z_3}{z_2 z_4}, \qquad (3)$$

where z_3 is the number of rolls of the second crown of the satellite block 2, which engages the driven toothed wheel 4, z_4 – the teeth number of the central gear wheel 4.

3.2. Planetary precessional transmission type K-H-V

From structural point of view, the K-H-V precessional transmissions (figure 5) include an additional vibration generator node, namely the connecting mechanism V, which can be made in the form of synchronous or fingers couplings [4].

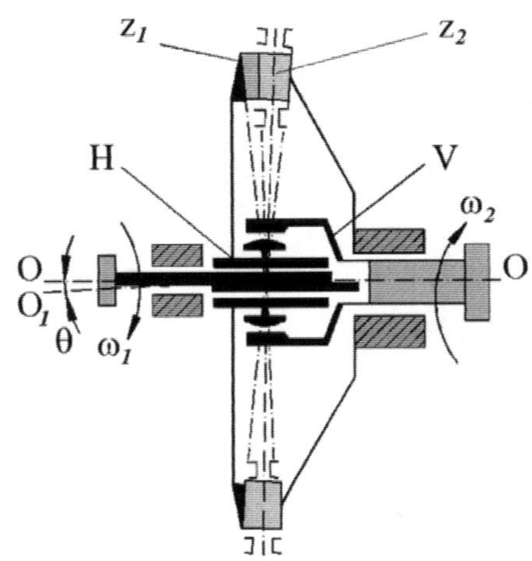

Figure 5. K-H-V kinematic scheme

The crankshaft rotation frequency is analogous to the 2K-H scheme, with only one precession gear (one toothed wheel and one satellite roller). The rotation frequency of the driven shaft is determined with the relation:

$$\omega_2 = \frac{\omega_1}{i} = \pm\omega_1 \frac{z_2}{z_1 - z_2}, \qquad (4)$$

where z_1 is the teeth number of central wheel; z_2 - the rollers number of satellites crowns.

Case 1: $z_1 < z_2$, $\omega_2 = -\omega_1 \frac{z_2}{z_1 - z_2}$.

Case 1: $z_1 > z_2$, $\omega_2 = \omega_1 \frac{z_2}{z_1 - z_2}$.

In vibro-acoustic diagnostic practice there are various methods to minimize the vibration and noise level for technical and energetic machines. The mechanical transmissions used in various machines, installations are sources of vibration and high frequency noise. The most efficient but also the most expensive method for obtaining silent transmissions is the method of processing the parts with very high precision or the static and dynamic balancing of the mobile elements. For precessional planetary transmissions we developed and research several methods for decrease vibration and sound level [1, 2, 4].

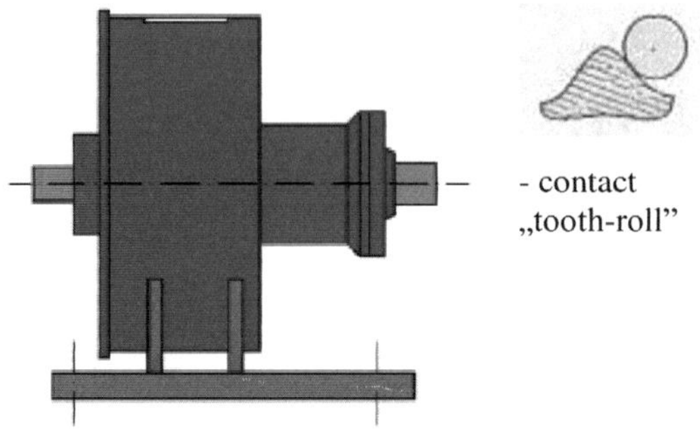

a) Power K-H-V precessional transmission

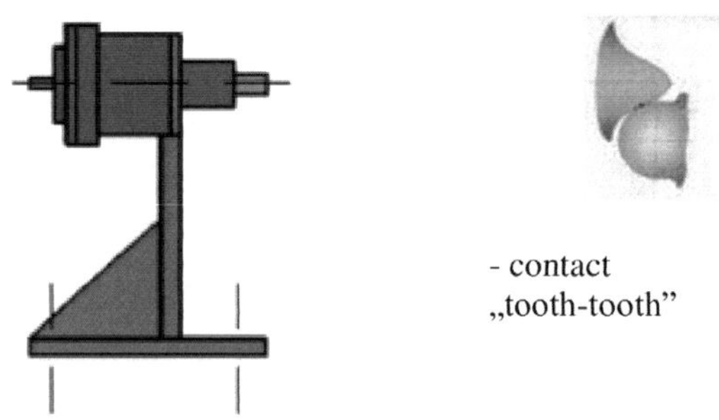

b) Kinematical 2K-H precessional transmission

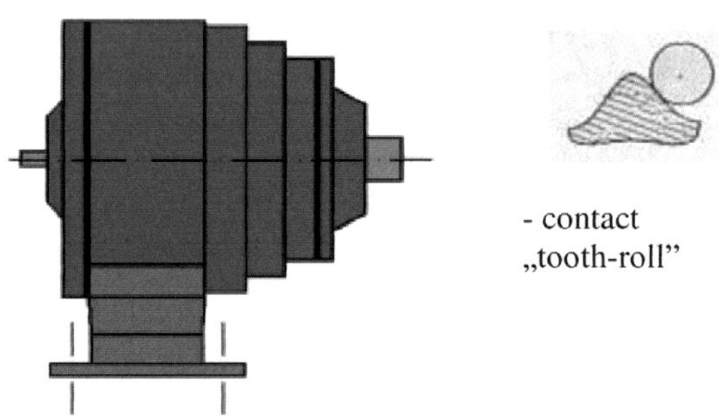

c) Power 2K-H precessional transmission

Figure 6. Researched precessional gearbox

4. Performed and future research

For the experimental part of the research, three types of precessional planetary reducers were selected, one of which was a 2K-H kinematic reducer in two satellite constructions made of metal powders and plastic mass [2,3] and two 2K-H and KHV power reducers [1]. The noise level was measured with the Brüel & Kjæer analyzer Type 2250 and the severity of the vibrations was measured using the PT500 and PT500.04 software of the German company GUNT [5].

Acknowledgements

Performed vibro-acoustical analyses and research on power K-H-V precessional transmission (Figure 6a) when precessional reducer is minimal and maximal filled with cooling and lubrication liquid at different operational speed [4] and on kinematical 2K-H precessional transmission (Figure 6b) when was researched two main cases. First case with satellite wheel made from plastic materials and second case with satellite made from powders materials. For whole cases kinematical and power planetary precessional reducers have demonstrated a good and silent operation at different loading regimes and speed [2,3], in accordance with the standards [6], with specification that precesional reducer with satellite wheel from plastic material is more silent.

Further vibro-acoustical research will be carried out on power 2K-H precessional transmission (figure 6c) where the transmission ratio $i=-10,5$, nominal torque $T=20Nm$ and operation speed $n=800min^{-1}$.

References

[1] Dulgheru V and Malcoci Iu 2018 *Dynamic model elaboration for the planetary precessional gear reducer against vibration* doi.org/10.1051/matecconf/201817806009

[2] Bodnariuc I and Malcoci Iu 2016 *Vibroacoustical diagnosis of planetary precessional kinematikal transmission* DOI:10.1088/1757-899X/147/1/012003

[3] Malcoci Iu and Dicusara I 2016 *Planetary Precession Gear with Thin-Layered Ruber Metal* Nonconventional Technologies Review Nr. 2 June ISSN 2359-8646 pp.35-38

[4] Malcoci Iu 2015 *Vibro acoustic research of planetary precessional transmissions* PhD Thesis UTM CZU: 621:534.6(043.2) pp.135

[5] Abraham D, Boxhamer J, Mittash P 2011 *Instruction manual GUNT PT500* Gerätebeu GmbH Barsbütel Germany

[6] *** 1964 *Standards of evaluation for the mechanical vibrations of machines* VDI 2058, Verlang Düsseldorf (FRG)

IOP Conf. Series: Materials Science and Engineering **749** (2020) 012037 doi:10.1088/1757-899X/749/1/012037

Homogenization heat treatment effects over the structure of 2724 cast aluminium alloy semi-products

A Pop[a] **and G Iepure**

Technical University of Cluj-Napoca, Faculty of Engineering, Department of Mineral Resource, Material and Environment Engineering, Baia Mare, Romania

E-mail: [a] aurica.pop@cunbm.utcluj.ro

Abstract. This project showcases the effects of the homogenization treatments applied to 2724 cast aluminium alloy semi-products, as well as the results regarding the resistance and plasticity improvement of these semi-products. These results have proven the validity of the initial processing hypothesis, which states that certain modifications of the crystalline structure can lead to the improvement of the characteristics mentioned above. A highly resistant 2724 aluminium alloy sample, representing the Al-Cu-Mg system, was used for this experiment. At a structural level, the alloy is made of aluminium based solid α solution grains and a series of binary, ternary or complex intermetallic phases that can either be soluble or insoluble. This type of alloy is mostly used in the machine building or aeronautics industry. The results are according to the microstructural observations, the homogenization heat treatment effect leading to the removal of any chemical heterogeneity (dendritic segregation) and a balanced structure made of polyhedral crystals and a homogeneous dispersion throughout the whole mass of the alloy.

1. Introduction

The homogenization heat treatment has the purpose of removing any chemical inhomogeneity (dendritic segregation) and reaching a balanced polyhedral crystals structure. The chemical inhomogeneity can be found in single-phase alloys as well as by-phasic alloys, thus making it necessary for the inhomogeneity to be eliminated (diminished), so it becomes necessary to remove the primitive limits of the cast grains as well as modifying the fibrous structure of the semi-products, seeing as how these are the main factors which determine the characteristics variation.

Highly resistant aluminium alloy extruded products used in the aeronautical industry are being subjected to various tests. Under pressure, these products must show a good correlation between characteristics such as mechanical resistance, plasticity, tenacity, fatigue resistance and corrosion resistance.

The purpose of heat treatments is to modify the initial structure and characteristics. The homogenization heat treatment applied to cast aluminium semi-products was the object of many scientific researches.

The present paper focuses on the aluminium alloys heat treating process. The material subjected to this experiment was a 7150 Zr 0.13% alloy. The results for the single stage conventional homogenization (figure 1 a, c) were compared to the new two-stage homogenization (figure 1 b, d), the first being pre-treated at 250ºC.

IOP Conf. Series: Materials Science and Engineering **749** (2020) 012037 doi:10.1088/1757-899X/749/1/012037

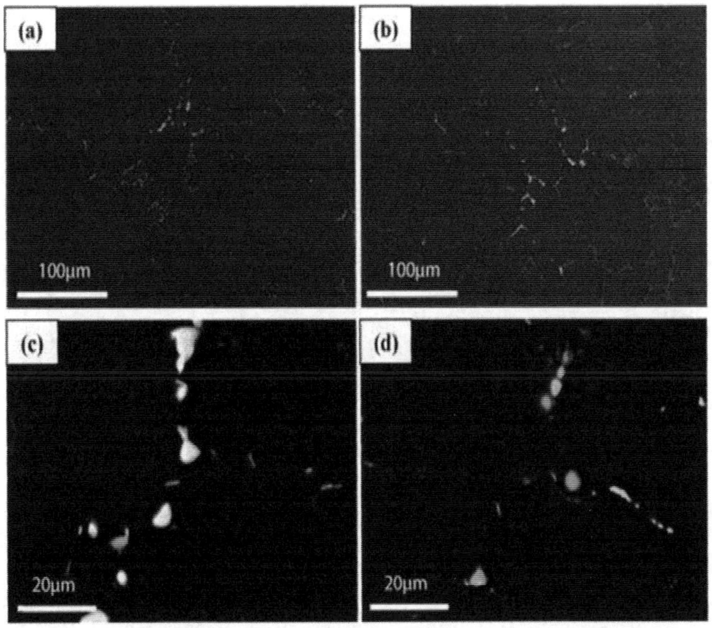

Figure 1. SEM pictures of the 7150 alloy's homogenized microstructure; (a), (c) single stage homogenized sample; (b), (d) two-stage homogenized sample. [1]

The heat resistant Al-Cu-Mg-Fe-Ni aluminium alloy, based on the 2024 alloy, has a complex chemical structure that was reached by dropping the amount of Si impurities and adding Fe and Ni. The 2D70 aluminium alloy is considered to be the most important structural material used in aerospace applications and it has the same room temperature resistance as the 2024 alloy and it can support even higher temperatures than the 2024 alloy. In order to reach an adequate homogenization heat treatment it is necessary to study the cast microstructural intermetallic compounds as the heat treatment goes on. The conventional homogenization heat treatment was done at a temperature of 490°C for 16 hours and the double homogenization heat treatment was done at 490°C for 16 hours, followed by 530°C for 16 hours. [2]

The industrially modified AA5083 aluminium alloy was analysed in the paper [3] and then subjected to a multiple stages homogenization heat treatment. The heat treatments took place at a 430-555° temperature interval. The industrially modified AA5083 aluminium alloy's microstructure was investigated using the optical techniques SEM, TEM and microanalysis. The effect that the microsegregations had when applied to the structure of the precipitates was analysed. The constituent components of the Al16 (Fe,Mn) and Mg2Si particles were formed in the inter-dendritic regions, while the n-Al11(Mn,Cr)4 dispersions formed in the dendritic cores. Results of the initial cast alloy's electrical resistance as well as the electrical resistance of the heat treatment affected areas are shown in the paper.

The alloys studied in the paper (6061, 2219, 2014) [4] took form by being melted in an induction furnace and cast into cast iron ingot shaped matrixes. These ingots eventually homogenized at a temperature of about 540°C.

The AA 5052 cast aluminium alloy was homogenized at a temperature of 470°C for 15 hours and air cooled. Five different heat treatments were applied to the laminated alloy at temperatures of 220, 250, 300, 350 and 380°C for 4 hours each. Microstructures and the laminated aspect were observed along the middle sections, parallel to the rolling direction, of the treated samples [5].

IOP Conf. Series: Materials Science and Engineering **749** (2020) 012037 doi:10.1088/1757-899X/749/1/012037

The 7B04 cast alloy's [6] crystalline grains have an average size of 115μm and α (Al) primary and secondary phases' eutectic structure network situated between granular and interdendritic grains (figure 2).

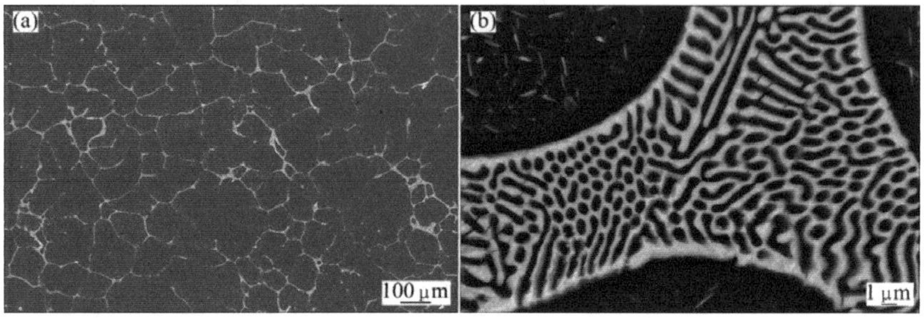

Figure 2. Microstructural view of the 7B04 cast alloy: (a) dendritic structure; (b) secondary phases [6]

The impact of the electrical current on the toughness and on the morphology of the secondary phases of the 7B04 aluminium alloy, homogenized at 380°-465°C for 2 hours, were closely observed by measuring the electric conductivity, performing hardness tests, X-ray diffraction, SEM and EDS.

The homogenized sample's area fraction is smaller in secondary phases than the sample that was homogenized under normal conditions at the same temperature. The results show that by raising the temperature from 380°C to 465°C, the electrical conductivity of the sample that was homogenized under normal conditions dropped from 34.9% IACS to 28,7% IACS, hardness grew from 96 HV to 146 HV and the zonal fraction of the secondary phase dropped from 4.5% to 1.89%. [6]

2. Materials and experimental method

The main alloying elements for aluminium are Cu, Mg, and Zn to which Mn, Ni, Cr and Fe are added. Alloying's purpose is to improve the mechanical properties of aluminium. The most widely used alloys are the Al-Si, Al-Mg, Al-Cu-Mg, Al-Mg-Mn, Al-Mg-Si and Al-Zn-Mg-Cu systems.

The material used for the research is a 2724 highly resistant aluminium alloy of the Al-Cu-Mg system. At a structural level, the alloy is made of α solid solution grains and a series of binary, tertiary or complex intermetallic phases, soluble or insoluble.

This alloy is mostly used in the aeronautical and machine constructing industries, having good mechanical characteristics, high fatigue resistance and good resistance to corrosion.

The microstructural analysis was done by microscopic observation of the cast semi-products samples which were polished and attacked with an attack reagent.

Hardness was determined by an F scaled Rockwell durometer.

3. Experimental Results

Experimental research focuses on the impact that the homogenization heat treatments have on the structure and the characteristics of the cast aluminium semi-products, especially the 2724 ones.

Figure 3 shows a 2724 aluminium alloy sample, collected from a 150mm diameter cast semi-product, will be initially used to emphasise a crystalline structure.

Figure 4 showcases the microstructural level of the 2724 alloy sample before the heat treatment. The grains are big, elongated according to the hereditary cast grains which are subdivided in aggregate subgrades.

The homogenization heat treatment was made in a hot air furnace, the semi-products being placed at an equal distance from one another in order to be homogenized uniformly, the temperature was permanently monitored using 5 K-thermocouples

The thermocouples were mounted at the ends of the semi-products, leaving a distance equal to half of the semi-products' diameter (if the mounting was made at the end where the casting began) or a maximum equal distance (if the mounting was made at the end where the casting ended).

Figure 3. Research used 2724 alloy sample

Figure 4. The microstructural level of the 2724 alloy sample before the heat treatment

The homogenization treatment was applied to 3 batches of samples at the temperatures of 470°C (1st batch of samples), 480°C (2nd batch of samples) and 490°C (3rd batch of samples), 7 hours heating, 16 hours maintaining, adding up to a total of 23 hours of treatment.

The 2724 alloy sample (Fig. No.3) was tested for hardness using a CV100 lab Rockwell hardness tester. An acting force of 588N was used on a 1/16 inch diameter ball for 5 seconds. In table no. 1 we can see the values of the Rockwell hardness as well as the equivalent for the Brinell hardness, and to make sure that the results are real, the hardness tests were done 4 more times on the same product ending with an overall average. The results are shown in table 1.

Table 1. Rockwell and Brinell hardness values for the 5 tests as well as their average

Sample identification	Hardness	I	II	III	IV	V	Average Hardness, HRF	Equivalent HB(500kgf)
Sample 1	HRF	73	75	75.3	74	75.6	74.58	71
Sample 2	HRF	75.2	77	77.5	77	76	76.5	72
Sample 3	HRF	77,1	81,8	79	78,8	77,5	78,8	73

Figure 5. The 2724 alloy sample at a microstructural level after the homogenization treatment

The hardness tests showed that the 3rd batch of samples, heated at 490°C (homogenization temperature of 490°C and varied between 485 and 499°C), had the best results. Samples were collected, from the 3rd batch of the treated samples, for elongation tests and microscopical analysis (figure 6).

The elongation test results are shown in table 2, determining the elongation at break and the elasticity module.

Figure 5 showcases the homogenized sample alloy. This is a balanced structure made of polyhedral crystals. The result of the heat treatment is the removal of any chemical heterogeneity (dendritic segregation).

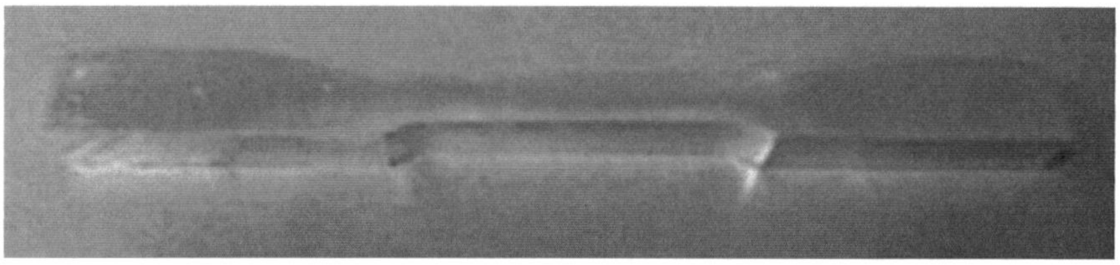

Figure 6. Elongation at break sample

In order to emphasize the way polycrystalline metals act when under mechanical stress, elongation at break is used as a reference point.

Table 2. Elongation at break results

Sample identification	Rm, MPa	A, %	E, MPa
Sample 3	250,77	12	16427

4. Conclusions

When picking aluminium alloys with the purpose of making products for the aeronautical industry, one must keep in mind that these alloys need to be highly resistant to mechanical stress, corrosion resistance, high fatigue resistance as well as a highly resistant to heat.

The experimental results, regarding the improvement of the resistance and plasticity characteristics as a result of the homogenization heat treatments, for the analysed cast alloy semi-products showed that certain modifications of the crystalline structure can lead to the improvement of these characteristics.

By applying homogenization heat treatments to the semi-products, most dendritic segregations were removed and the optimal temperature for the heat treatment to succeed was set around 490□C.

Best hardness and elongation at break values as well as a satisfying plasticity were reached throughout the homogenization heat treatments applied to the aluminium alloy samples.

References

[1] Starke E A Jr, Khalfalla Y and Benyounis K Y 2016 *Aluminium Alloys: Thermal Treatments* Reference Module in Materials Science and Materials Engineering

[2] Guo-jun Wang, Bai-qing Xiong, Yong-an Zhang, Zhi-hui Li and Pei-yue Li 2009 *Microstructural characterization of as-cast and homogenized 2D70 aluminium alloy* International Journal of Minerals Metallurgy and Materials vol. 16 August 2009 pp.427-431

[3] Radetic T, Popovici M and Romhanji E 2012 *Microstructure evolution of a modified AA5083 aluminium alloy during a multistage homogenization treatment* Materials Characterization vol. 65 March 2012 pp.16-27

[4] Narender K, K. Gopal Kishan Rao and Krishna N G 2013 *Thermo physical properties of wrought aluminium alloys 6061, 2219 and 2014 by gamma ray attenuation method,*

Thermochimica Acta vol. 569 10 October 2013 pp.90-96

[5] Bo WANG, Xian-hua CHEN, Fu-sheng PAN, Ian-Jun MAO and Yong FANG 2015 *Effects of cold rolling and heat treatment on microstructure and mechanical properties of AA 5052 aluminium alloy* Transactions of Nonferrous Metals Society of China vol. 25 August 2015 pp.2481-2489

[6] Li-Zi HE, Pin-feng JIA, Lin ZHANG and Jian-zhong CUI 2016 *Evolution of secondary phases and properties of 7B04 aluminium alloy during DC homogenization* Transactions of Nonferrous Metals Society of China vol. 26 February 2016 pp.319-327

IOP Conf. Series: Materials Science and Engineering **749** (2020) 012038 doi:10.1088/1757-899X/749/1/012038

Consideration of the actual stage of nanomaterials used in dentistry

C M Pop[1a], A Pop Vadean[2] and I Tascu-Stavre[3]

[1] Technical University of Cluj-Napoca, Faculty of Engineering, Department of Engineering and Technology Management, Baia Mare, Romania
[2] Technical University of Cluj-Napoca, Department of Mechatronics and Machine Dynamics, Cluj-Napoca, Romania
[3] Technical University of Cluj-Napoca, Faculty of Science, Baia Mare, Romania

E-mail: [a] corynutza_30@yahoo.com

Abstract. The present paper is meant to be a short presentation of the actual stage of nanomaterials used in dentistry. It shows some theoretical consideration regarding characteristics of some nanomaterials used in dentistry. In recent years, increasing demands for dental reconstruction, have led to the development of new materials that have multiple applications in dental medicine and which must meet aesthetic requirements, biocompatibility requirements as well as hardness or durability. Surpassing the extraction period, dentistry has entered the "restorative era" in which it is desirable to introduce new materials that will lead to the most natural restorations with the physical and chemical closest properties to those of the dental tissues. The discovery of such materials has become a definite necessity in order to achieve consistent progress in this respect. Interest in using nanomaterials, composites with nanomaterials in restorative dentistry, as well as reconstruction, is constantly increasing. It is therefore important that the mechanical properties and not only, of these materials be carefully studied and especially improved

1. Introduction

Nanotechnology has a strong impact on all areas (medicine, engineering, communications, the environment, cosmetics), being a multidisciplinary field based on the belief that new scientific discoveries will make it possible to manipulate and control each atom in different structures. Restorative dental medicine undergoes a period of rapid change as the pattern of the appearance and evolution of carious malosis changes, as well as the discovery of new materials and techniques.

The domain is interdisciplinary and complex, involving the simultaneous use of knowledge in several areas and scientific disciplines such as clinical / medical research, biology, chemistry, physics, mathematics, engineering, and robotics. Nanomedicine encompasses several interlocking and mutually reinforcing areas: nanomaterials and nanodispensers, molecular nanotechnology, nanoscale imaging, innovative drug delivery systems, nanotoxicolog.

Research into nanomedicine allows a deeper understanding of human body functioning at the molecular level, meeting patients' requirements in diagnosing and treating diseases and restoring certain functions of affected organs or tissues. In order to improve the characteristics of some nanomaterials used in dental medicine, in this paper I tried a thorough investigation of both components, nanocomposite materials and dental biomechanics.

IOP Conf. Series: Materials Science and Engineering **749** (2020) 012038 doi:10.1088/1757-899X/749/1/012038

2. The current state

In recent years, increasing demands for dental reconstruction have led to the development of new materials that have multiple applications in dental medicine and which have to meet several aesthetic requirements of biocompatibility as well as of hardness or resistance over time. Interest in the use of nanomaterials, composites with nanomaterials in restorative dental restoration and reconstruction is on the rise. That is why it is important that the mechanical properties and not only of these materials be carefully studied and especially improved. By diversifying dental restoration and reconstruction materials with an increasingly complex chemical composition can increase the risk of adverse reactions. Therefore, the dentist needs to know the structure and properties of the materials he uses and adapt them to each case. This progress in the field of composite nanomaterial technology has significantly changed the options of physicians. Composite materials combine into a single product, some components, which usually do not naturally associate. At small dimensions the materials have properties different from those on a macroscopic scale [1]. For example, nanoscale materials have a higher mechanical strength than macroscopic samples of the same material.

The ability of a material to deform without it. Breaking depends on the chemical bonds that keep the atoms together. In principle, ties chemical properties are the same, both in the nanoscale and in the macroscopic scale. But the evidence macroscopic include many defects, cavities, missing atoms, etc. which limits properties mechanical. The density of these defects can be much smaller on a nanometer scale, and like resulting nanomaterials become more resistant.

Different types of composite nanomaterials used in dental medicine differ in viscosity. Flow (fluid) composites are applied by syringes directly into small cavities, and condensable composites are designed for medium or large cavity reconstitution. The composite materials differ from each other, especially by the size of the filler particles.

The analysis of new methods and conceptual models on the synthesis of nanocomposites, however, should start from the identification of the disadvantages of the existing methods.

Nanotechnology, on a scale of less than 100 nm, can be said to be the artistic engineering that allows the design, properties and performance to be achieved on finished materials.

Remarkable studies have been made in this area over the past few years, going from theory to clinical studies. At present, there is a range of applications of these nanomaterials in different fields, which is to be improved both quantitatively and qualitatively.

Table 1. Some current and future applications of nanomaterials

Developing	In the initial phase	Existing products
Nanocrystalline drugs for light absorption	Molecular labeling using quantum dots of CdSe	Antibacterial storage elements based on Ag
Insulin inhalable	"Carriers" for low solubility drugs	Gold for labeling and detection
Nanospheres for inhalable and injectable drugs	Instruments for implants with hydroxyapatite	Magnetic contrast imaging agents based on $Fe2O3$
Computational biochemical computational nanometric modeling at hydrophobic-hydrophilic interface between tooth and composite material	Embedding nano-fibers dispersed through resin for reinforcement	Solar screens with ZnO and $TiO2$
Detecting viruses using quantum dots		
Magnetic particles for repairing the human body with prostheses or artificial parts		
Bone growth promoters		

IOP Conf. Series: Materials Science and Engineering **749** (2020) 012038 doi:10.1088/1757-899X/749/1/012038

Table 2. Applications of nanomaterials in dental medicine and products on the market

Field	Existing material
Restorative dental medicine	Ketac^{TM} , Ketac N100 Filtek Supreme XT Premise TM Ceram XT
Regenerative dental medicine	Ostim Nano-bone
Periodontology	Arestin
Dental prevention	NanoCareGold
Orthodontics	KetacTM Filtek Supreme Plus Universal
Prosthetics	Nanotech elite H-D plus
Implantology	Nanotite^{TM}
Endodontics	AH plus ^{TM} Guttaflow

The latest nanocomposites used in restorative dentistry due to their special properties (abrasion resistance, hardness, tear resistance, good adhesion to dental structures, etc.) are nanocomposites which containing a resin type matrix and nanoparticles or metal-ceramic for dental implants.

The interest in the field of these materials also concerns the possibilities to manipulate their structure, in the desire to modify and improve the mechanical, chemical or electrical properties. Following studies, it has been observed that changes in properties can be observed when the particle size reaches a level called critical size.

Table 3. Changes in material properties depending on the particle size

The studied property	Particle size (nm) where the change of properties is observed
Catalytic activity	<5
Transforming magnetic materials from hard to fine	<20
Changing the refractive index	<50
Production of consolidation and quenching	<100
Change in hardness and plasticity	<100

Although current nanotechnology-based products are modest, the international scientific community believes that this multidisciplinary field can influence people's lives through applications in most fields, including dental medicine, with the necessary potential. Determination of mechanical properties, physical and biological properties of these nanocomposite materials is a necessary step to assess the ability of these composites to cope with clinical needs.

Table 4. The specialized equipment required for the characterization and production of nanomaterials

Naming	Characteristics
JEOL 5600 LV Electronic Scanning Microscope, equipped with Oxford Instrument X-ray Spectrometer	Enlarge: x300000
Planetary mill Fritsch, Pulverisette 4	- possibility of changing the speeds of the containers and the disc; - possibility of grinding in a controlled atmosphere; - afferent soft.
Planetary mill Fritsch, Pulverisette 6	- programmable disk and container speeds; - lower energy range (especially for mechanical milling).
Heat treatment furnace	-temperature working, up to 1700 ° C; - working ability in controlled atmosphere.
INEL Equinox 3000 Difractometer	Angular range can be recorded instantly
Plasma sintering Equipment	Control of temperature and electric current pulse Electric current 1500-2000 A
Hydraulic press	-100tf
Metallographic microscope	Maximum 1000x zoom, dedicated photo / video camera, Soft Olympus Analysis
Analytical balance Precision 200	Sensitivity 0.0001g
Particle analyzer (LASER Analysette 22 Nanotec)	Determination of granulometric distribution between 10nm and 2000μm

2.1. *Manufacture of nanostructures*

Taking models from nature, researchers have developed two possible strategies for obtaining of nanostructured materials [2]. The top-down strategy (from sea to small) consists of reducing the size of larger particles by grinding and sifting. Bottom-up strategy [3] (from low to high) describes the development of nanoparticles starting from atoms and molecules, passing through the phenomenon of controlled-gel sol-gel or pyrolytic flame. Both strategies show the inconvenience of particle agglutination [4]. Nanoparticles have a larger area than the volume and hence a larger surface energy. Untreated they tend to agitate rapidly in 0.5 μm (500 nm) microparticles in diameter, losing its property. Nanoparticle surfaces need to be inactivated chemically in order to benefit from their special properties.

Nanotechnology offers more possibilities for obtaining nanostructures. Among the most used techniques we list: i) Self-assembling; ii) Lithography at nanometric scale; iii) Transfer of a print; iv) Inkjet printing; v)Compliance by demolition(release); vi) Hybrid techniques; vii) Microscope recording with local probe. In the following, we will detail some of these method.

2.1.1. Self-assembling. Is an inexpensive method of building materials and assemblies nanostructured. It is essential to mention that supramolecular structures are self-assembling without external intervention. These materials are formed by interactions interatomic and intermolecular, other than metallic, ionic or covalent bonding forces, traditional. This opens a new horizon for the development of nanostructures based on assembling molecular constituents. Thus, after molecular chemistry, founded on bonds which are established between atoms to form molecules, is another field, chemistry supramolecular - based on the association of two or more chemical species [5].

2.1.2. Transferring a pattern. Is a technique that starts from an object made by conventional or UV lithography, and a mold is produced resulting in a mold. One is applied elastomeric material (often polydimethyl and PDMS lax) which, after crosslinking, forms which can then serve for the replication of several (nano) structures where it started.

2.2. Classification of composites

- Organic matrix composites (OCM);
- Metal matrix composites (MMC) made of alloys, magnesium, titanium, nickel, copper, etc.
- Ceramic matrix composites (CMC) made of silicon carbides, aluminum oxides, ceramic materials. [6]

The second classification criterion refers to reinforcement: these include particle reinforcements, filaments, continuous fiber composites, and composites containing fabrics.

Bone tissue in the human body is also a composite. It is made up of a hard but crumbly material called hydroxyapatite (calcium phosphate) and a soft and flexible material called collagen (a protein). Collagen cannot give the bone system the resistance it needs, but in combination with hydroxyapatite this becomes possible.

Most composite materials consist of two phases: a phase called a matrix that is continuous and surrounds the second phase, called the dispersed phase. The properties of the composites depend on the properties of the component phases, the quantities in which they are present, as well as the shape and size of the particles and their orientation.

The bond between the fibres and the matrix is created during the composite material manufacturing phase. It has a fundamental influence on the mechanical properties of the material

The properties of these materials are given by:

- particle size;
- the volumetric fraction of nanoparticles;
- granulometric distribution of particles.

2.3. Properties of composite materials reinforced with nanoparticles

The use of these materials is constantly increasing, due to the advantages they have, namely:

- Costs are lower compared to fibres;
- It is possible to obtain an isotropic material in the composite matrix;
- Matrix embedding technologies are simpler.

The size of these particles is very important because, for nanoscale particles, the relationship with the matrix manifests at the atomic level, which prevents the deformation of the matrix in the vicinity of each particle, and the matrix transfers to the particles much of the voltage to which the nanocomposite is subjected. [7]

The 10 to 100 nanometer-sized particles, evenly dispersed, give the composite material high mechanical strength and hardness, but make plastic deformation limited due to the fact that it prevents the movement of the dislocations.

2.3.1. The mechanical properties of composite materials reinforced with nanoparticles. The properties of these composites depend on the characteristics of the matrix, the size of the reinforcement particles, the nature of the particles, the way they are dispersed. Among the most important features of particle-reinforced composites, we mention the following:

- Traction tear strength - this can be caused by matrix breakage, breakage at matrix and particle interface, or tearing of reinforcement particles at the action of an external force;
- Compressive strength - this characteristic depends entirely on the nature and properties of the matrix, because under the action of compressive force, the composite begins to deform plastic.

2.3.2. Nature and properties of reinforcing particles. The main groups of reinforcement particles are:

- Metallic oxide particles (Al_2O_3, SiO_2, TiO_2, MgO, etc.)
- Metallic carbide particles (SiC, TiC etc.)

Table 5. Properties of some particles used to obtain composite materials

Material	Density g / cm2	Expansion coefficient 10-6 / ° C	Traction resistance MPa	Young's module GPa
Al2O3	3.98	7.92	221	379
C	2.18	-1.44	-	690
SiC	3.21	5.40	-	324
MgO	3.58	11.61	41	317
SiO2	2.66	1.08	-	73
TiC	4.93	7.60	55	269

According to research in this field, the most common are systems where the particles are with reduced size (<100 μm) in proportions of 10% -30%.

2.4. New aspects followed in manufacturing dental nanocompositions

Most of the efforts made in recent years in research have focused on adhesive systems, composites, ceramics and dental aesthetics (teeth whitening techniques). The future dental medicine composites is becoming more and more promising. Silanic agents long used in composite manufacturing are optimized and much better controlled to obtain an excellent adhesion interface [8]. Loading with composite filler, which directly influences the properties of the material (wear resistance, for example) is increasingly based on nanomaterials (nanoparticles). Nanoparticles cause the appearance very small distances between the composite components, increasing wear resistance a resin matrices [9]. Also, some nanofillers induce chromatic changes and radioactivity in systems. The current interest is to move from photopolymerization techniques to those of homogeneous polymerization. Despite the positive interest sparked among practitioners photopolymerization methods, they do not provide a certain depth of polymerization or a degree of conversion, especially in hard-to-reach cavities, so are being tested polymerization methods [10].

Generally, a larger amount of filler leads to increased hardness, elasticity, fracture and wear resistance. We will stop on the last aspect. The force itself does not explain the relationship between the filler and the wear resistance. Wear in the case of composites used in dentistry involves several different mechanisms [11]. In oral environment wear on occlusal surfaces is caused by abrasive particles with dimensions ≈ 0.1 μm found in food and which it is assumed to be silicon-based. The composite matrix is subject to wear, but not to the fillet which is tougher.

Thus, the protection of the matrix is possible by designing particles of load that is in close contact to protect the particle impact matrix abrasives from food. These are called microprotection [12] and have long been studied for composites with microparticles. It is obvious that this has improved for microhybrid composites and newer ones for nanohybrids.

In recent years, the increasing demands for aesthetic reconstructions have led to the development of new classes of composites for direct reconstitution, with excellent physical, mechanical, aesthetic and durable properties. The latest material is composites containing a conventional resin matrix in which were embedded nanoparticles or nanocomponents (clusters).

Clinicians need to understand exactly the nanometric concept, which is equivalent to 103 μm, in other words 1μm = 1000 nm. In dental medicine the nanofiller is a particle inorganic in the approximate size of 40 nm = 0.04 μm. This size, however, is not is innovative in terms of dental

IOP Conf. Series: Materials Science and Engineering **749** (2020) 012038 doi:10.1088/1757-899X/749/1/012038

composites, as microfilaments have been touched this size of the filler since the 70's. What is absolutely new is the possibility of nano fillers to improve inorganic phase loading.

An increase in filler content implies an improvement in mechanical properties [13]. The objective of nanotechnology is to develop a dental restorative material that can be used in any area (anterior or posterior) of the dental arcade. This material must combine the qualities of a microfil composite with that of a hybrid composite: through finishing and polishing results in a smooth and weather-resistant surface and mechanical properties to provide them with resistance to stress [14].

3. Types of dental nanocomposites

3.1. Nanocomposites of the polymer-ceramic type

Polymeric composites are those materials having in their composition at least two components of different chemical structure, thermodynamically compatible, mutually dispersed or associated, at least one of which is a polymeric phase. [15]

They have the following advantages:

- they are lighter due to the low density of the matrix;
- shows the resistance / density ratio higher than classical materials;
- they have good resistance to fatigue;
- they have resistance to the corrosive action of environmental factors;
- they have low energy consumption with regard to the technology of forming the parts;
- a high performance / price ratio.

Matrix and reinforcement usually have opposite properties. Thus, from a soft matrix and a hard fitting, or from a hard matrix and a soft reinforcement, a composite can be obtained several times more resistant to the weaker component. The distribution of the reinforcement in the matrix may be different; the particles can be evenly or unequally distributed, in linear strings or agglomerations on certain planes, resulting in variations in mechanical properties. Polymeric composites with good mechanical and tribological characteristics have to meet certain requirements, such as low wear and friction coefficients, good mechanical characteristics at high temperatures [16].

From the studies carried out, it was shown that the introduction of polymer addition materials meant to reduce the negative effects induced by the temperature increase of the parts in contact, the interaction of the surfaces, the applied loads and the working environment. Particles and fibers, especially short, are used to add polymers, but solid lubricants, such as PTFE and graphite, can also be used to reduce the friction coefficient.

3.2. Nanocomposites of metal-ceramic type

Ceramic nanomaterials are generally refractory polycrystalline compounds, such as: silicates, metal oxides, carbides, refractory hydrides, sulphides, selenides, with atomic and ionic bonds, whose complex crystalline structure is obtained by sintering. General for their hardness and wear resistance, in applications such as hip and tooth joint surfaces, but also for bone restoration initiator properties as bone-binding surfaces in implants.

Ceramic materials have recently received particular attention as potential candidates for manufacturing bone replacement / tooth replacement implants. They are already widely used in dentistry due to their aesthetic appearance, high resistance to compression and lack of reaction to human body fluids, although they are currently not widely used in nano-level composites.

Bio ceramic from calcium phosphates is the closest bio ceramic to the composition of the natural bone. The stabilization of the phases in the system, especially in the C3P - apatite zone, is of great interest for bio ceramic materials.

Metallic materials can be sterilized by common techniques, while ceramic materials only by dry heat. It can be seen that the biocompatibility and mechanical properties of the materials are inverse, the materials with high mechanical strength being biotolerated and the bioactive ones having properties reduced mechanical properties.

Table 6. Physical and mechanical properties of ceramic materials used in orthopedic and implantological prosthetics

Ceramic material	Porosity %	Density g/cm3	Young's Modulus GPa	Resistance to Compression MPa	Resistance to Tearing MPa	Resistance to Flexure MPa
Aluminium oxide	≈0	3,93-3,95	380-400	4000-5000	350	400-560
Hydroxyapatite	0,1-3,0	3,05-3,15	7-13	350-450	38-48	100-120
Vitreous carbon	-	1,4-1,6	-	-	-	70-200
Zirconium dioxide	≈0	4,9-5,56	150-190	1750	-	150-700

3.3. Nanocomposites based on hydroxyapatite
Hydroxyapatite ceramics are widely used for reparative operations in the human body due to the fact that it has excellent biocompatibility. This material has numerous applications in implants such as hip bonds, bone substitutions, dental implants. However, low mechanical strengths and low tensile strength are restrictive conditions when used. In order to eliminate these inconveniences, in the hydroxyapatite ceramics, reinforcement materials are introduced, among which the zirconium dioxide most commonly used for this purpose [17].

3.3.1. Ceramo-hydroxyapatite. The starting material is pentachlorohydroxy-triphosphate, (Ca5 (OH) (PO4) 3); Synthesis process results in ceramic hydroxyapatite.

Physical and Mechanical Properties [18]

• mechanical resistance - is very high, given by the ceramic structure. Dense ceramics have greater resistance than porous ceramics.
• ph - neutral - 7.
• electric charge - neutral
• chemical - the structure corresponding to the naturally occurring bone mineral hydroxyapatite.

3.3.2. Hydroxyapatite powders doped with silver. It is well-known that in surgery the risk of infection of post-operative wounds is an important issue. To prevent such problems, broad spectrum antibiotics are now being used against which microbial strains have developed resistance. An alternative to the use of antibiotics could be the use of silver, an element known from ancient times to possess antibacterial properties and disinfectants. Thus, there are some studies on silver-doped hydroxyapatite that could reduce the risk of post-operative infections (for example: after installing a dental prosthesis or an implant).

Following the doping of hydroxyapatite with silver, a single mineralogical phase is formed, namely hydroxyapatite. Silver also replaces calcium in the hydroxyapatite structure, and there are studies that have shown that with the increase in silver concentration, the average size of crystallites decreases.

Vitreous ceramics have ceramic constituents based on oxyapatite, fluoropatite, hydroxyapatite etc., which activate tissue-implant reactions and lead to stimulation of bone regeneration in the implanted area. Bioactive composites can also be used as superficial bioactive glass layers deposited on metallic surfaces.

A new type of composite material is used in implantology; it consists of a metal substrate (Ti-Al-V) which is coated with bioactive hydroxyapatite. This type of composite facilitates the infiltration of bone tissue by providing a biological binding of the tissue to the ceramic implant. As a result, a very good anchoring of the inorganic bone implant is achieved. The bending strength of the HA / Ti composite is about 150 MPa, so with high mechanical properties, which provides sufficient resistance to the inorganic implant, which is included in the hard bone tissue and thus ensures the biological and functional stability of the implant.

4. Anatomical and biomechanical aspects of teeth

First of all, we must say that the teeth are the toughest structures of the body, namely the dental enamel. Therefore, the characteristics of the materials for restoration or reconstruction are difficult to obtain because they must have parameters at least similar to those of the natural tooth, both mechanically, as well as biocompatibility and dental aesthetics.

Dental treatments of any kind are supposed to restore the morphology of maxillary dental system functions lost or altered following various specific conditions. Such an objective is impossible to achieve in its ideal sense by virtue of the limited possibilities of the human being.

The teeth are formed in two parts:

1. Dental crown - This is the visible part of the tooth cavity. Hence the name of the prosthetic work "dental crown" because this part of the tooth is trying to replace it.
2. Dental Radiation - It is the part located in the maxillary bone that assures the implantation of the tooth. A tooth may have one or more roots. Also, their shape and size differ from tooth to tooth.

Denture plays a vital role in digesting food, helping to fragment it into smaller pieces by biting and mastication. Dental enamel consists of 98% hydroxyapatite. Hexagonal apatite crystals are tightly wrapped and aligned perpendicular to the tooth surface. The excellent hardness of the tooth enamel provides excellent resistance to abrasion and wear; the high content of ceramic material makes it very tough but with a low tear strength. A tooth that would only consist of enamel would be prone to rupture, so enamel is the garment of a softer substrate but more resistant to tearing, the dentin. Dentine has a chemical composition similar to cortical bone, having also similarities in structure.

The root of the tooth is covered with cement, a fibrillated bone substance with a porous structure. The pulp occupies the central cavity and contains circularly aligned collagen fibres, nerve cells and blood vessels, the periodontal membrane fixes the root of the tooth into the alveolar bone.

Application of engineering principles at the level of living systems (biomedical engineering) opened a new era in the diagnosis and therapy of various diseases. Biomechanics studies the response of living organisms to different demands.

The various types of fixed prosthetic restorations that solve a wide range of edentations (supun)subject maxillary dental system to various force, which translate through a series of displacements in different directions or plans of the space (translation, rotation) and / or deformations (flexion, torsion), the way of response of the involved tissues being different.

5. Conclusions

Nanotechnology has the power to revolutionize the scientific world, allowing scientists to manipulate matter at atomic or molecular level using physical, chemical, biological or engineering techniques. It allowed researchers to understand the relationship between macroscopic properties and the structure of materials.

Dental nanocomposites have to meet a number of requirements: to have the color, resilience and hardness similar to natural teeth, to be easy to apply and to allow predictable results every time. They also have to be biocompatible and adhere to enamel and dentin. Composites with nanofilling, state-of-the-art nanoparticle materials have outstanding properties in terms of aesthetics and gloss over time as well as mechanical strength and predictability.

The development of completely new and innovative technologies, nanomaterials and nanosystems have led to a new generation of products that will improve the quality of life and the environment in the coming years. The progress of science at the nano level through the smaller dimensioning of the particles that are part of the restorative composite materials is preceded by the so-called "dento-mimetic" reconstructions, which will follow the reproduction of the dental enamel in detail. Embedding in the matrix structure of composites nanoparticles with antibacterial and remineralization based on fluoride or amorphous calcium phosphate represents the latest trends in nanotechnology with applications in dental medicine currently in the experimental phase.

As a coating of a metal, hydroxyapatite offers the mechanical integrity of the metallic device and the biocompatibility of the mineral composition of the bone. Hydroxyapatite coatings can eliminate the need to use cements to fix the device to the existing bone and minimize the appearance of corrosion products resulting from direct metal-body fluid contact. Because worn fragments of cements and corrosion products are the main sources of irritation and inflammation of tissues associated with implants, the deposition of hydroxyapatite could intensify the healing process and reduce the convalescence period.

It can be said that nanotechnology applied in dental medicine can be categorized as an innovative project, thus having a strong potential to revolutionize the diagnosis and treatment, as well as regeneration of tissues.

References

[1] Chiroiu V, Ştiucă P, Munteanu L and Donescu Ş 2005 *Introducere în nanomecanică* Bucureşti: Ed. Academiei Române
[2] Wautelet M et. al. 2003 *Les Nanotechnologies* Paris Ed. Dunod
[3] Dagani R 2000 *Building from the bottom up* Chem Eng News pp.27-32
[4] Heindl D and Erdrich A 1998 *Filler based on silicon dioxide, method for its preparation, and its use* US 5,852,096 Heraeus Kulzer GmbH
[5] Freitas R A Jr. 1999 *Nanomedicine, Volume I: Basic Capabilities* Georgetown Landes Bioscience http://www.nanomedicine.com/
[6] McKeen L W 2009 *Fatigue and Tribological Properties of Plastics and Elastomers* 2nd Edition William Andrew
[7] Nica I, Cimpoeşu N, Rusu V, Andronache M and Ştefănescu C 2012 *Structural properties of nanofilled and microfilled restorative composites* Materiale Plastice pp.176-180
[8] Antonucci J M, Dickens S, Fowler B O, XU H K X and Mc Donough W G 2003 *Chemistry of silanes: interfaces in dental polymers and composites* Trans Acad Dent Mater pp.81-109
[9] Gedik R, Hurmuzlu F, Coskun A, Bektas O O and Ozdemir A K 2005 *Surface roughness of new microhybrid resin-based composites* J Am Dent Assoc pp.1106-1112
[10] Watts D C and Cash A J 1991 *Determination of polymerization shrinkage kinetics in visible-lightcured materials: methods development* Dent Mat. 60. Cook WD pp.281-287
[11] Pflug K 1999 *Dental materials having a nanoscale filler* WO 99/17716 Dentsply Corp
[12] Bayne S C, Tayler D F and Heyemann H O 1992 *Protection hypothesis for composite wear* Dent Mater pp.305-309
[13] Taylor D F, Kalachandra S, Sankarapandian M and McGrath J E 1998 *Relationship between filler and matrix resin characteristics and the properties of uncured pastes* Biomaterials pp.197-204
[14] Mitra S B, Wu D and Holmes B N 2003 *An application of nanotechnology in advanced dental materials* J Am Dent Assoc 34 pp.1382-1390
[15] Bowen R L 1956 *Use of epoxy resins in restorative materials* J Dent Res pp.360-369
[16] Friedrich K 1993 *Advances in composite tribology* Elsevier
[17] Virendra B D 2004 *Contemporary Dental Materials* New Delhi, London Oxford University Press
[18] Nica I, Aluculesei A, Baciu C, Rusu V and Agop M 2011 *Characterization of some dental composites,* Metalurgia International XVI pp.38-42

IOP Conf. Series: Materials Science and Engineering **749** (2020) 012039 doi:10.1088/1757-899X/749/1/012039

Aspects concerning the geometric elements of the rolling deformation area

E Pop[a] and J Juhasz

Technical University of Cluj-Napoca, Faculty of Engineering, Department of Mineral Resource, Material and Environment Engineering, Baia Mare, Romania

E-mail: [a] popelena1@yahoo.com

Abstract. The paper presents an experimental and practical study of aluminium strips lamination, calculating the conformations that define plastic deformation by lamination. A sample from aluminium alloy in the shape of a band was rolled and we calculated the logarithmic deformations in lamination. We started the lamination process from a semi-finished product, with a surface for the cross-section section (S_0) and it was carried out in three drawings up to a final section (S_3). The deformation assessment and measurement tool, the logarithmic degree of deformation gives the extent of the plastic deformation at each drawing, and the calculated total deformation degree is useful to quantify the deformation preformed for the entire lamination process over the three drawings. If analysed in this way, the plastic deformation process, the material undergoes a higher reduction at the last drawing and an ecruisation of the band at its margin also. The preliminary usage of the coefficients defining lamination has a particular importance for practical objectives, in the carrying out of another lamination process.

1. Introduction

In the process of lamination, plastic deformation is achieved by continuously drawing the metallic material between the cylinders, by changing the dimensions of the material (decreasing the thickness, increasing the length and increasing the width), through the continuous modification of the lamination area.

The presentation and verification of calculation formulas for technological operations in the lamination of metallic materials implies the establishment of kinematic relations (evolution of geometric shape and deformation degrees) between the undistorted part and the deformed part, namely the prediction of the metal flow in the successive conduct of rolling operations. With the help of certain experiments of deformation through lamination performed in the laboratory, a large variety of improved properties through lamination are highlighted [1]. There is an increase in the resistance of the laminated materials and their structure is composed of fine grains. For a good progress of the lamination process, in order to obtain the desired shape and properties, it is necessary to understand and control the metal flowing mechanism in the deformation zone. The flowing direction of the metal, the size of the deformation and the temperatures, largely determine the properties of the products processed by lamination. Being influenced by the process variables, the metal flow in lamination determines the mechanic properties associated with the deformation in the deformation zone and to the undesired formation or generation of flaws (no-filling, cracks or creases, at the surface or within the adjacent layers). In order to investigate the progress of the metal layers in lamination, bands covered with modelling clay were laminated [2].

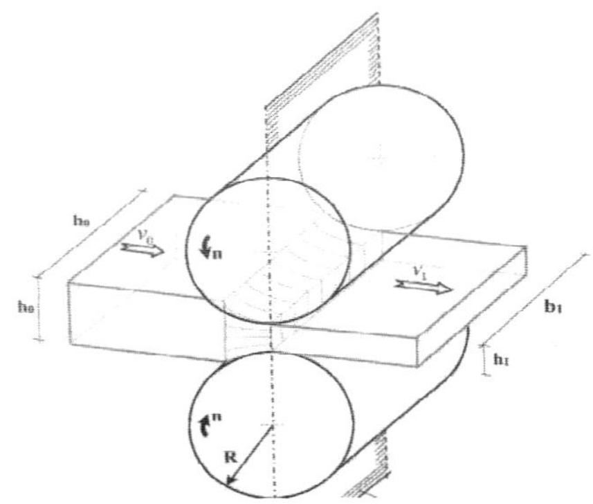

Figure 1. The idealised structure of the deformation zone [6].

During the lamination process, friction between the cylinders and the deformed material in the deformation zone is inevitable and it determines the quality of the products' surface, the value of the lamination force, the energy consumed within the process etc. [3]

By analysing the relative distribution of the deformations in the deformation zone, answers are searched for to the immediate distribution of the deformation in the lamination process in an equilibrium state.

Various researchers determined the reason for the formation of flaws by analysing the plastic deformation in the plates' lamination zone [4]. To solve the problems related to the quality of the margins or the flaws of the margins they have built a thermo-mechanical model with finite elements (FEM) to simulate the warm lamination process of plates. Depending on the relation between the Lode parameter and the types of plastic deformation and the distribution of the values obtained in the post-manufacturing for the Lode parameters, they determined the types of flaws in lamination and they analysed the state of tensions at the exit from the deformation zone in lamination.

The lamination process is one of the most popular processes in the industry for processing materials where almost 80% of the metallic equipment have been exposed to lamination, at least once during the manufacturing process. As a result, many scientists have tried to adjust the quality and quantity of products by optimising this process and identifying the parameters which influence them in obtaining the highest quality products. In the paper written by the researcher [5] the lamination process was analysed through different analytical and numerical methods, such as the snow plate method, the gliding field method, the higher limit method, the border element method and the finite element method. In comparison with other analysis methods of the lamination process, the finite element is the most practical and the most precise [5].

2. Expression of deformation in the rolling of bands

Changing the size of the material, depends on the state of tension [6] and the deformation, induced in the material, by the deformation zone. The evolution of the deformation is quantified on the idealized structure of the deformation zone (figure 1) associated with the shape of the working cylinders and the laminate. The cross-section section of the work cylinder is considered to be a perfect circle. The line that unholds the centre of the cylinders is considered to be perpendicular to the direction of lamination and remains so during lamination [6]. The distance between the cylinder centres is considered to remain unchanged, ignoring the elastic deformation of the cajan. The rotation speed of the working cylinders is constant and after the mechanical load on them increases and the slowing occurs under high torsion loads. The laminate material is rectilinear, its sides make straight angles, the thickness and width are uniform, as well as the roughness of the surface is the uniform.

During lamination, the metallic material is not subjected to the simultaneous plastic deformation on its entire length, but only on a certain, relatively small portion, which is in the space between the cylinders. The lamination process is characterized by a shrinking of the height of the laminate before entering between the cylinders (h_0) at a height value (h_1) after the output between the cylinders. This transformation is associated with increasing the width (from b_0 the b_1) and the length of the laminate (from l_0 the l_1). This change of form has been evaluated [6], [8] by size, with the significance of some coefficients that define the plastic deformation suffered by the material:
reduction coefficient $\eta = h_0/h_1$, the coefficient of width $\beta = b_1 /b_0$, the coefficient of lengthen $\lambda = l_1/l_0$.

In this case, I intended to laminate a band sample, aluminum alloy, 2000 series Al-Cu-Mg, on the lab mill with duo rolling and I will follow the geometric elements of the lamination zone and the entire lamination process and thus conclusions can be drawn on the lamination process.

During lamination, the metal volume prior to lamination V_0 equals the volume of metal after lamination V_1 it is known that [7] $\eta = \beta \cdot \lambda$. This means that the volume of metal deformed on the height of the laminate is equal to the deformed in the transverse direction (width) and longitudinal (lengthwise). Technologically, these coders give the extent of the plastic deformation capacity of the metallic material between the working cylinders, establish the connection with the size of the deformed metal door, making possible the constructive development of laminated products.

Table 1. Calculation of the coefficients which define the plastic deformation on lamination of the sample.

Parameters	Undeformed, initial sample	First drawing	Second drawing	Third drawing
Length, l [cm]	6.00	7.10	8.20	11.50
Width, b [cm]	1.25	1.32	1.52	2.45
Thickness, h [cm]	0.75	0.60	0.45	0.20
Surfaces of the cross-section section, S [cm^2]	0.94	0.79	0.69	0.49
Reduction Coefficients η_1, η_2, η_3		1.25	1.33	2.25
Widening Coefficients β_1, β_2, β_3		1.06	1.15	1.60
Lenghtening Coefficients λ_1 λ_2, λ_3		1.18	1.15	1.40
$\eta = \beta \cdot \lambda$		1.25	1.33	2.25
Partial coefficient for lenghtening, λ_1, λ_2, λ_3		1.18	1.15	1.40
$\lambda_t = \frac{S_0}{S_3}$			1.92	
$\lambda_t = \lambda_1 \cdot \lambda_2 \cdot \lambda_3$			1.92	

In many lamination processes, as with the lamination of aluminium alloy strips, the evolution of the cross-sectional form of the laminate is more suggestive than the mere modification of each of the three dimensions, therefore the coefficient of lengthness is used not as a ratio of the length of the laminate, but of the surfaces of the cross sectional, before lamination S_0 and after lamination S_1. Knowing that:

$$S_0 = h_0 \bullet b_0, \tag{1}$$

$$S_1 = h_1 \bullet b_1 \tag{2}$$

We calculated in the table 1 the partial coefficient of lengthing λ_1, λ_2 , λ_3 Which give the measure of the plastic deformation at each passage:

$$\lambda_1 = \frac{S_0}{S_1} ; \lambda_2 = \frac{S_1}{S_2} ; \lambda_3 = \frac{S_2}{S_3} \tag{3}$$

And the total coefficient λ_t It is useful to quantify the deformation made on the whole process of lamination on the three drawings:

$$\lambda_t = \frac{S_0}{S_3} = \frac{S_0}{S_1} \frac{S_1}{S_2} \frac{S_2}{S_3} = \lambda_1 \cdot \lambda_2 \cdot \lambda_3 \tag{4}$$

IOP Conf. Series: Materials Science and Engineering **749** (2020) 012039 doi:10.1088/1757-899X/749/1/012039

In our case, the length coefficients are equal to the partial lengths of lengthing. In view of the evolution of the size of the laminate, the numerical values of the three coefficients are always higher than one, so they are super unitarian. In the practice of lamination, the maximum values of these coefficients, relative to a single drawing, do not exceed the value (1.8 ÷ 2.0), limitation imposed by the plastic deformation capacity of the metallic material, the construction of the rolling line and the power of the engine drive. To these the following can be added: the presence of defects, an uneven metallographic structure, or even an uneven distortion of the various portions of the laminate [6].

3. Geometric elements of the deformation zone on lamination

The deformation zone is the part where lamination is carried out and is characterized by the following geometric elements:

-The height and width of the cross-section section, at the entrance and exit of the metal between the working cylinders (h_0, b_0, h_1, b_1), the grip angle (α) and the contact length at the metal cylinder interface (l_c) according to figure 2.

-The difference between the front and exit laminate height of the deformation zone, called absolute reduction in the case of aluminium alloy tape lamination [6] is:

$$\Delta h=h_0-h_1 \tag{5}$$

-The difference between the width of the laminate on the output and the entrance to the deformation zone, named absolute widening

$$\Delta b=b_1-b_0 \tag{6}$$

-AB circle arc, after which contact is made to the metal cylinder interface, called the contact arc

-The projection of the horizontal contact arc (on the rolling direction) c is the length of the geometric warp zone:

$$l_c=\sqrt{R\,\Delta h} \tag{7}$$

-The angle at the centre of the circle that is associated with the cylinder-metal contact, called α contact angle, is determined from the relationship (8) or (9) for low angle values $\alpha = (10\text{-}15)^0$

$$\cos\alpha=1-\frac{\Delta h}{D} \tag{8}$$

$$\alpha=\sqrt{\frac{\Delta h}{R}} \tag{9}$$

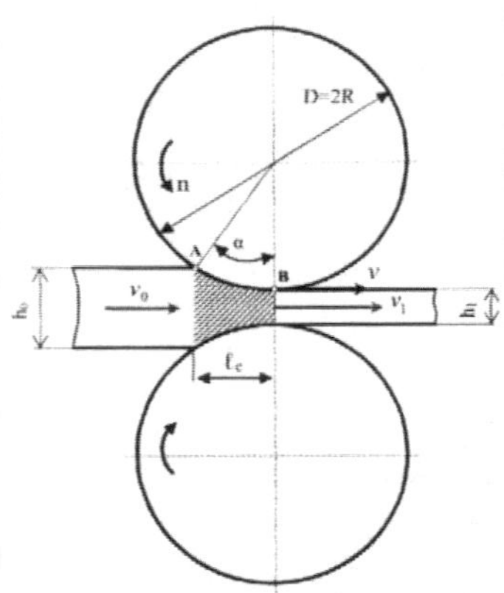

Figure 2. The geometric elements of the deformation zone on lamination [6]

Table 2. The calculation of the logarithmic deformations on the lamination of an alluminium alloy sample.

Parameters	Undeformed, initial sample	First drawing	Second drawing	Third drawing
Length, l [cm]	6.00	7.10	8.20	11.50
Width, b [cm]	1.25	1.32	1.52	2.45
Thickness, h [cm]	0.75	0.60	0.45	0.20
Logarithmic deformation for height $\delta_h=\ln\frac{h_0}{h_i}$		0.22	0.29	0.81
Logarithmic deformation for width $\delta_b=\ln\frac{b_i}{b_0}$		0.05	0.14	0.47
Logarithmic deformation for lenght $\delta_l=\ln\frac{l_i}{l_0}$		0.17	0.14	0.34
$\delta_t = \ln\frac{S_0}{S_i}$		0.17	0.14	0.34
$\delta_t = \delta_1 + \delta_2 + \delta_3$			0.65	
$\delta_t = \ln\frac{S_0}{S_3}$			0.65	

For larger plastic deformations (ε > 15%) Laminate material has undergone appreciable dimensional changes and is more loopless to characterize the process of lamination by logarithmic deformations according to table 2. When the lamination process begins from a semi-manufactured with the surface of the cross section (S_0) and runs in three drawings, up to a final section (S_3) an evaluation tool and measure of deformation is the degree of logarithmic distortion, effective and total.

If we analyse the process of plastic deformation in this way, we can see that at the last drawing the material has undergone a higher reduction but also a belt gauge at its edge.

In the stable process of lamination, normal and friction forces act on the entire length of the contact arc at the cylinder-metal interface. The metal between the cylinders shall be deformed in the direction of minimal resistance, but the friction forces have opposite meanings at the entrance and exit of the deformation zone (figure 3). The vertical section where contact friction forces change meaning is called a neutral section, represented by the angle at the centre of the cylinder, called neutral angle (γ_n). In the absence of the action of the posterior and previous tension in the laminate, the neutral angle can be calculated with the expression [6], [8]:

$$\sin \mu n = \frac{1}{2} \sin \alpha_i - \frac{1-\cos \alpha_i}{2\mu_i} \tag{10}$$

where: μ_i is the coefficient of friction at the cylinder-metal interface.

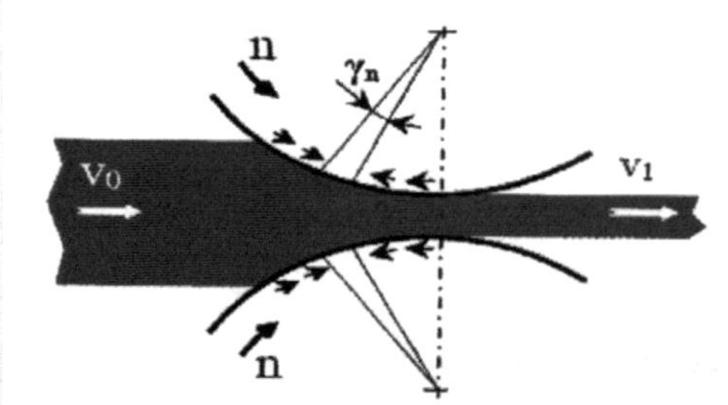

Figure 3. The action of the friction forces at the metal cylinder's interface.

The deformation zone, in the longitudinal direction, is divided, relative to the position of the neutral section, in a delay zone (in the vicinity of the entrance plane, the friction forces are oriented in the direction of lamination) and another advance (in the vicinity of the plan output of the metal between the cylinders, the friction forces have the reverse direction of the rolling direction). The research and development of lamination processing involves a thorough understanding of manufacturing processes, the influence of variables, such as friction conditions, material properties and the geometry of the product to be processed.

4. Conclusions

The requirements for high quality products, as well as for reducing production and exploitation costs, led to a certain deformation on a passage of the material, taking into account the calculation of the coefficients which define the plastic deformation. We have calculated the coefficients that define the plastic deformation through lamination, and it can be seen that at higher deformation coefficients, at the third passage, the laminate material was unevenly deformed on the edges and defects occurred in the form of cracks. By introducing algorithms for the automatic calculation of deformation coefficients, the shape of passage, numerical modelling, physical and numerical simulation in the creation of optimal geometry, a strong representational framework of synthesis and analysis is provided. The calculation of logarithmic deformations when laminating an aluminium alloy sample makes it possible to present a lamination program to create an adequate flow of the deformed material in the drawing sequence. Following the calculations of the coefficients defining the plastic deformation presented above, it can be said that they represent, the most important contributions in the characterization of the transitions. In the evolution of the lamination process, as is observed, the outer shape of the processed product changes and suffers an increase in the longitudinal dimension (lengthening), a reduction in height and increase in width, which mainly depends on the reporting of the laminated product to the initial profile. This improves both the process state and the consistency of the products, minimizes costs and makes controlling the deformation easier. Starting from simple laboratory experiments, calculating the coefficients that define lamination, understanding the relations between the plastic deformation mechanism and chemical, mechanical, thermal, metallographic phenomena, modelling of the lamination process can be obtained.

References

[1] Matruprasad R, Surjya P K, Shiv B and Singh 2015 *Cross Rolling: A Metal Forming Process*
[2] Fox M N and Sutcliffe M P F, 2008 *Modelling material transfer in metal rolling using plasticine* Available online at www.sciencedirect.com Wear 264 pp.18-25
[3] Pernis R and Kasala J 2011 *Deformation distribution in rolling zone* Acta Metallurgica Slovaca, Vol. 17 No. 3 ISSN 1335-1532 pp.150-157
[4] Bian Hongxia, Zhu Liang and Hou Guoqing 2013 *Advanced Materials Research Online* Trans Tech Publications Switzerland Vol. 668 ISSN 1662-8985 pp.437-441
[5] Shahani A R, Setayeshi S, Nodamaie S A, Asadi M A and Rezaie S 2009 *Prediction of influence parameters on the hot rolling process using finite element method and neural network* Journal of Materials Processing Technology 209 ISSN 0924-0136 pp.1920-1935
[6] Nistor L 2016 *Simularea proceselor de laminare a metalelor* Editura U.T.PRES Cluj Napoca ISBN 978-606-737-169-7
[7] Eriksson C and Lundberg S E 2004 *Working range of roll pass sequences in wire rod rolling with interstand tensions* Ironmaking and Steelmaking vol.31 ISSN 1743-2812 pp.181-192
[8] Wusatovski Z 1972 *Bazele laminarii* Editura Tehnica Bucuresti
[9] Byona S M, Na D H and Lee Y 2009 *Effect of roll gap adjustment on exit cross sectional shape in groove rolling. Experimental and FE analysis* Journal of Materials Processing Technology 209 ISSN 0924-0136 pp.4465-4470
[10] Kaneko S et al. 2003 *The spread rolling of IF steel strip.* Journal of Materials Processing Technology 141 ISSN 0924-0136 pp.365-369

IOP Conf. Series: Materials Science and Engineering **749** (2020) 012040 doi:10.1088/1757-899X/749/1/012040

Research on the behaviour of the metallic material in the draw-plate while drawing the wire

E Pop

Technical University of Cluj-Napoca, Faculty of Engineering, Department of Mineral Resource, Material and Environment Engineering, Baia Mare, Romania

E-mail: popelena1@yahoo.com

Abstract. This research introduces the theoretical and experimental study of the plastic deformation through drawing of the copper wire, the presentation of the microstructures resulting after each drawing, certain possible flaws which occur within this plastic deformation process and the conclusions on the grains` flip-over and elongation. During the drawing process, the metallic material undergoes some important transformations in the draw-plate. The equiaxed grains are compressed in transversal direction and elongated in the drawing direction. The wire drawn is hardened due to the elongation of the grains and due to the development of a substructure of dislocations in the deformation grains. The theoretical presentation of the topics in discussion is then followed by a calculation of the reduction and the Δ parameter which causes less unused work if it is low value, limits the cold hardening of the wire and allows the possible execution of a higher total reduction between two annealing processes. A smaller angle for the drawing favours the coating of the lubricant and lengthens the contact area. Thus, a better cooling is obtained for the wire - draw-plate interface and the lubricant film is more stable.

1. Introduction

The materials play an important role in the technical progress which is favourably influenced by the modification of the attributes of certain materials through plastic deformation or by the appearance of new materials.

A unified, fundamental and quantitative description of the structure and the behaviour of the materials undergoing various loadings is accomplished with the help of the theory of plastic deformation. This aims to produce, manufacture and rationally use the materials by acknowledging the correlation laws between the chemical composition, the atomic or molecular arrangement, the microstructure and the physico-mechanical properties of the materials as well as the possibilities to modify these properties chemically, thermically or through plastic deformation [1].

The solid bodies can be deformed by outer forces applied on them. Other factors, such as the hydrostatic pressure, humidity, temperature, magnetic fields, etc. may act upon the body together with these exterior forces. Their action modifies the entire process of plastic deformation as well as the mechanic and physical properties of the body. The plasticity theory describes the stress-deformation relation in a plastically deformed body [2] when the outer factors acting on it are known. However, the study of a deformation process by applying scientific knowledge and the scientific support given by modern engineering and the use of advanced presentation tools may change the perception of the whole process. With the help of these laboratory experiments of drawing a great variety of properties can be highlighted. Luksza J, Majta J, Burdek and Ruminski M studied the behaviour of the wires in plastic deformation during both single- and multi-pass-drawing [3].

An inquiry in the process of wire drawing in a steady state and the study of the effects of different variables of the process over the drawing and deformation parameters lead to more and more performant deformation processes with superior products. If we compare the drawing deformation for several materials (copper, aluminium and steel) [4], general characteristics true for the drawing process and specific to each material can be determined.

Many process variables, besides heat from deformation process, affect the quality of drawn products. U.S. Dixit and P.M. Dixit studied the effect of process variables, such as reduction ratio, semi-angle of the die, and coefficient of friction of the interface, on the quality of the drawn product [4].

In order to identify the effect of inclusions, Norasethasopon and Yoshida studied the effect of inclusions, relative latency and reduction, maximum hydrostatic stress and maximum die pressure during the copper wire drawing process [5].

The thermal effect on the drawn products has been studied by Changsun M. and Naksoo K. based on the final product dimensional predictions, based on plastic deformation theories. Thus, the practical results were twice as close to the calculated dimensions if the thermal effect was taken into account, if this effect was not taken into account [6].

2. Flaws occurring at wire drawing

We have plastically deformed through drawing, on the table by means of draw-plates with different drawing angles which are presented in table 3.1 the wire undergoing this plastic deformation operation was prepared as a number of samples, with the initial dimension d0 presented in the table 3.1 Following the first drawing, the samples had the d1 dimension.

During the drawing process, the metallic material undergoes important transformations in the draw-plate. The equiaxed grains are compressed in transversal direction and elongated in the drawing direction [7]. The wire drawn is hardened due to the elongation of the grains and due to the development of a substructure of dislocations in the deformation grains [8].

For the wire - draw-plate interface, such deformations occur not only because there is friction and lubrication in an area with temperatures rising rapidly and also due to the formation of a new lateral surface of the wire. At the wire drawing, for one passing with a 30% reduction of the transversal section surface, there is a 20% growth of the lateral surface of the wire [9]. A new lubricant film needs to be applied if the wire drawing speed is high.

When considering that the passing of the material through the wire drawing is performed so that each volume element with flat surfaces maintains these flat surfaces even after the drawing, we may state that a simplified transport occurs as it would happen if a wire would be strained through simple traction, without draw plate [10]. During such an ideal process, the deformation is considered to be homogenous or even.

In reality, the plastic deformation of the wire during the drawing through the draw-plate is not homogenous and even. When the material comes into contact with the wall of the draw-plate first, its flowing must be changed to a direction which is parallel to the wall of the draw-plate. This causes shearing within the wire. When the drawing of the wire is finished, the flow of the material must be changed again to a direction parallel to the drawing direction. This causes shearing which is opposed to the one resulting from the initial contact with the wall of the draw-plate. These two shearing results are balanced out and cannot be externally measured.

The shearing results for the deformed material fibres are higher in external regions and the deformation is uneven [11].

The grains nearing the draw-plate contact area are more deformed, more sheared than those in the centre, and the grains in the central axial area are just slightly more deformed and hardened. This can lead to the occurrence of some flaws specific to the drawn wire: cone, cup (figure 2.1 and figure 2.2) or dog-leg type fractures etc.

IOP Conf. Series: Materials Science and Engineering **749** (2020) 012040 doi:10.1088/1757-899X/749/1/012040

Knowing the crystalline structure helps us to control and predict metallographic behavior relative to metal performance in different processing processes. If we understand the structure of metals we can evaluate and modify their properties. All this is necessary to make the most suitable selection for specific applications under conditions determined by demand, temperature and environmental influences [12].

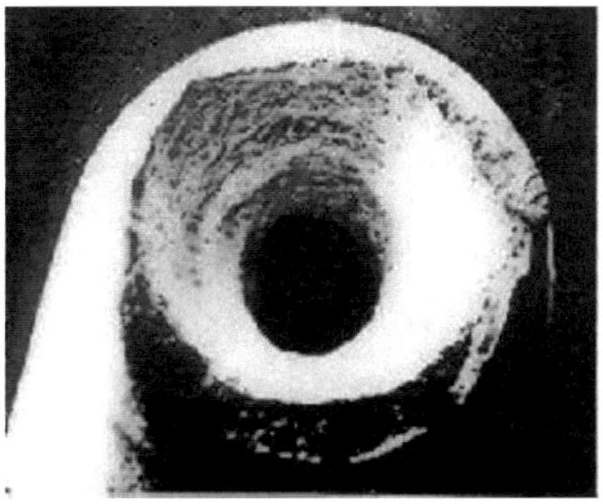

Figure 2.1. Drawn wire, cup flaw, 30 X [13]

Figure 2.2. Drawn wire, cone flaw, 50 X [13]

The microscopic analysis of the copper wire in its initial state indicates a structure with equiaxed grains as presented in figure 2.3, a. After the first drawing, the grains in the outside wire have an elongated structure as it appears in figure 2.3, b. In the final state when the wire has hardened, the grains have a slightly elongated structure as presented in figure 2.3, c.

a. b. c.

Figure 2.3. Microstructure of the copper wire in initial state and after the two drawings through the draw-plate

During the cold plastic deformation of the drawn copper wire, a crushing of the crystalline grains occurred with remarkable influences over the properties of the metallic materials. Simultaneously with the crushing of the crystalline grains and the accumulation of internal residual tensions, as we have already shown, there is a reorientation and elongation of the crystalline grains in the direction of the deformation. All these determine the so-called cold-hardening state and the phenomenon itself leads to an off-balance stare. As Scott D. A 'says, there are many questions that cannot be answered simply by examining the microstructure' [14].

The higher the degree of cold hardening, the more affected the attributes are. The increase in the yield limit and the fracture strength are accompanied by a greater diminishing of the deformation capacity (tensile elongation) caused by the cold-hardening. These may disturb the deformation, wire-drawing process to such extent that, at a certain point, it must be stopped. In order to continue the plastic deformation processing, it is necessary to first operate a full annealing to restore the cold hardened structure and to remove internal tensions.

Processing, properties and microstructure are interdependent. The microstructure depends on the composition and processing process. Microstructure is responsible for mechanical properties [1].

IOP Conf. Series: Materials Science and Engineering **749** (2020) 012040 doi:10.1088/1757-899X/749/1/012040

The size of the grains significantly influences the mechanical properties of the metals. The large size of the grain shows us strength, toughness and low ductility. A fine grain wire has better mechanical properties than a large grain [15].

3. Expressing the reduction and the Δ parameter for the wire drawing

The calculated value of the reduction is used in numerical form, but in technology constructions calculations, in this case the wire drawing process with different drawing angles being characterised, it is possible to express it in percentages, as presented in the table 3.1. The deformation of the wire material during the wire drawing process has been calculated by using the formula with expresses the surface reduction of the transversal section. (figure 3.1) [1].

$$r = \frac{S_0 - S_1}{S_0} = 1 - \frac{S_1}{S_0} \tag{3.1}$$

$$r = 1 - \frac{d_1^2}{d_0^2} \tag{3.2}$$

A deeper analysis on the influence of the geometry of the active zone over the behaviour of the deformed material was possible in 1955, after Wistreich [16] laid the foundations of the understanding of the plastic deformation process. With all the apparent simplicity, numerous elements in the drawing process were not clear enough and have become the subject for detailed research [17]. The type of the material flow and the tension state are being studied for various materials. In our case, the material is cold plastically deformed electrolytic copper. We have calculated the Δ parameter with the dimensions indicated in figure 3.1 and the formula 3.4.

$$\Delta = \frac{d_0 + d_1}{d_0 - d_1} \cdot \sin \alpha \tag{3.3}$$

or expressed depending on the reduction (r):

$$\Delta = \frac{1}{r}(1 + \sqrt{1 - r})^2 \sin \alpha \tag{3.4}$$

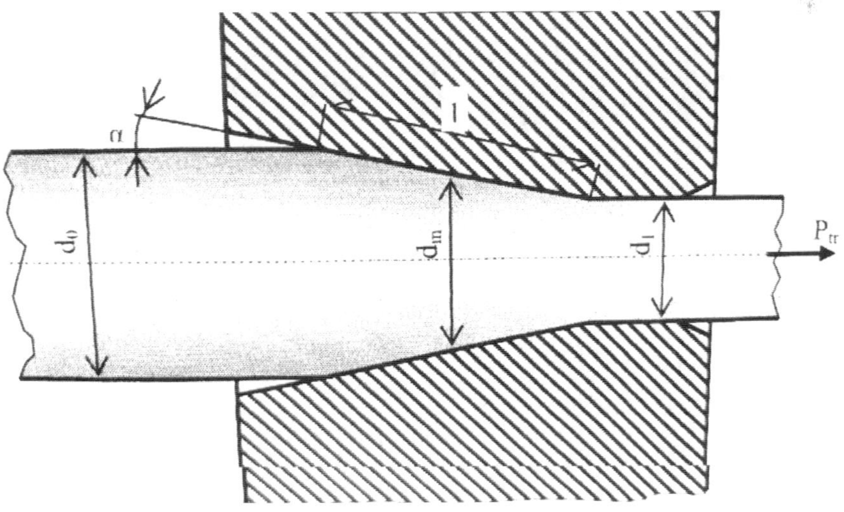

Figure 3.1. Position of the wire in the draw-plate, the geometry of the active zone [1]

Table 3.1. Values for the reduction r and the Δ parameter for different angles

d0 [mm]	d_1 [mm]	r [%]	α_1 [^{0}]	Δ_1	α_2 [^{0}]	Δ_2
2,8	2,6	14	5	2,35	7	3,29
2,8	2,4	27	5	1,13	7	1,58
2,8	2,1	44	5	0,61	7	0,85

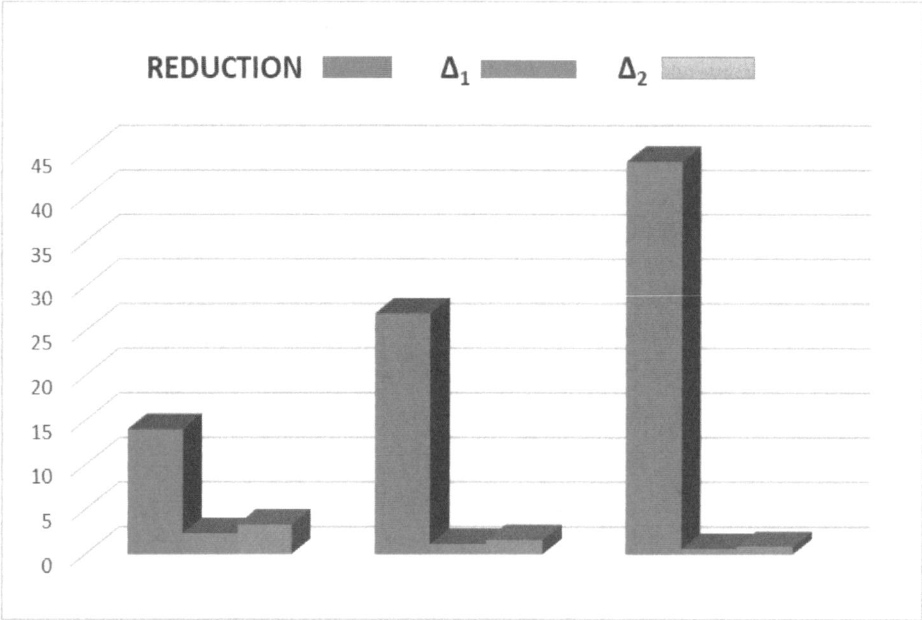

Figure 3.2. Values for the reduction r and the Δ parameter at wire drawing for $\alpha_1 = 50°$ and $\alpha_2 = 70°$

4. Conclusions

Wire drawing with a low value of Δ parameter causes less unused work, limits the cold hardening of the wire and allows the possible execution of a higher total reduction between two annealing processes.

Lower values of this parameter can be obtained for a construction of the draw-plate with small angles (α) and high reductions (r).

At wire drawing, in order to improve lubrication, it is necessary to decrease the value of the angle (α). A smaller angle favours the lubricant's application and lengthens the contact area. In this way, a better cooling for the wire - draw-plate interface is produced and the lubricant film is more stable.

For high speed wire drawing, the recommended angle is α = (5-7)0, especially for the last drawings, when the wire begins to harden. Thus, a higher length of the wire drawing zone in the draw plate is obtained and a more intense lubrication. All these lead to improvements in the wire's attributes and to an increase in the productivity for the plastic deformation through wire-drawing.

References

[1] Nistor L 2008 *Treating of metallic materials* Editura U.T. PRES Cluj Napoca
[2] Cazimirovici E 1981 *Theory of Plastic Deformation* Editura D.P. Bucuresti
[3] Luksza J, Majta J, Burdek M and Ruminski M 1998 *Modelling and measurements of mechanical behaviour in multipass drawing process* Journal of Materials Processing Technology (80-81) pp.398-405

[4] Dixit U S and Dixit P M 1995 *An analysis of the steady-state wire drawing of strain-hardening materials* Journal of Materials Processing Technology Volume 47 Issues 3-4, pp.201-229

[5] Norasethasopon S and Yoshida K 2006 *Influences of inclusion shape and size in drawing of copper shaped-wire* Journal of Materials Processing Technology (172) pp.400-406

[6] Moon C and Kim N 2012 *Analysis of wire-drawing process with friction and thermal conditions obtained by inverse engineering* Journal of Mechanical Science and Technology 26 (9) pp.2903-2911

[7] Juhasz J 201 *Study on non-distructive ultrasound control* Scientific Bulletin of North University Center of Baia Mare Series D, Mining, Mineral Processing Non-ferrous Metallurgy Geology and Environmental Engineering Volume XXXI No. 2

[8] Hotea V 2018 *Characterization and susceptibility to intergranular corrosion of aa2024 aluminum alloy* Scientific Bulletin of North University Center of Baia Mare Series D, Mining, Mineral Processing Non-ferrous Metallurgy Geology and Environmental Engineering Volume XXVIII No. 2 pp.7-11

[9] Juhasz J 2014 *Advanced technological gases purification system*, Scientific Bulletin of North University Center of Baia Mare Series D, Mining, Mineral Processing, Non-ferrous Metallurgy, Geology and Environmental Engineering Volume XXVIII No. 1

[10] Hotea V 2016 *Metalurgical processes gas emisions and the greenhouse effect* Scientific Bulletin of North University Center of Baia Mare Series D, Mining Mineral Processing Non-ferrous Metallurgy Geology and Environmental Engineering Volume XXX No. 2

[11] Pop E 2008 *Processes of plastic deformation processing, obtaining of geosynthetic materials* Risoprint Cluj-Napoca

[12] Moon C and Kim N 2012 *Analysis of wire-drawing process with friction and thermal conditions by inverse engineering* Journal of Mechanical Science and Technology 26(9) pp.2903-2911

[13] *** Breaking Encyclopedia

[14] Scott D A 1991 *Metallography and microstructure of ancient and historic metals* Printing: Tien Wah Press Ltd. Printed in Singapore

[15] Felder E, Levrau, C, Mantel, M, Truong N G and Dinh 2012 *Identification of the work of plastic deformation and the friction shear stress in wire drawing* Wear Volumes 286-287 pp.27-34

[16] Wistreich J G 1955 *Investigation of the mechanics of wire drawing* Proceedings of the of Mechanical Engineers, Volume 169 issue 1 Issue published June 1 pp.654-678

[17] Betzalel A 1963 *Analysis of wire drawing and extrusion through conical dies of small cone angle* Journal of Engineering for Industry 85(1) pp.89-95

AUTHOR INDEX

Institute of Physics
Dirac House, Temple Back
Bristol BS1 6BE UK

ISSN: 1757-8981
ISBN 978-1-7138-2658-3

9 781713 826583

CURRAN ASSOCIATES INC.
proceedings.com